3rd Global Congress on Construction, Material and Structural Engineering (GCoMSE 2021)

IOP Conference Series: Materials Science and Engineering
Volume 1200

Online
23-24 August 2021

Editors:

Nor Haslinda Abas
Mohd Azuan Zakaria
Norfaniza Mokhtar
Goh Wan Inn
Seyed Jamalaldin Seyed Hakim
Alvin John Lim Meng Siang

ISBN: 978-1-7138-4659-8
ISSN: 1757-8981

Printed from e-media with permission by:

Curran Associates, Inc.
57 Morehouse Lane
Red Hook, NY 12571

Some format issues inherent in the e-media version may also appear in this print version.

Institute of Physics
Dirac House, Temple Back
Bristol BS1 6BE UK

Phone: 44 1 17 929 7481
Fax: 44 1 17 920 0979

techtracking@iop.org

Additional copies of this publication are available from:

Curran Associates, Inc.
57 Morehouse Lane
Red Hook, NY 12571 USA
Phone: 845-758-0400
Fax: 845-758-2633
Email: curran@proceedings.com
Web: www.proceedings.com

TABLE OF CONTENTS

PREFACE

It is our great pleasure to introduce the proceedings of the 3rd Global Congress on Construction, Material and Structural engineering (GCoMSE 2021) that has been organized by Jamilus Research Centre, Faculty of Civil Engineering and Built Environment, Universiti Tun Hussein Onn Malaysia (UTHM), on the 23-24 August 2021. The GCoMSE series has been organized since 2017, and 2021 is its third edition, scheduled to take place biannually.

This congress aims to provide a platform for scientists, researchers, academicians, and industrial professionals to share and exchange their knowledge, expertise, and experiences as well as current research outcome. The conference theme of "Transforming Construction, Fostering Sustainable Structure and Material" really emphasizes on the efforts in improving the sustainability of civil engineering and the built environment. These fields are among the essential elements in creating a sustainable world in order to promote balanced economic growth and enhancing the global quality of life holistically.

Due to the Covid-19 pandemic, manuscripts in various disciplines such as Structural Engineering, Material Engineering, Transportation, Infrastructure, Construction, and many more were presented virtually to keep everyone safe. Many memorable experiences were made among the participants especially for the first timer in a virtual conference. The enthusiasm of the presenters was outstanding, and the warm and friendly virtual atmosphere encouraged a good exchange between disciplines and between established and early career researchers.

On behalf of the organizing committee, sincere appreciation is expressed to all keynote speakers, chairpersons, scientific committee, and reviewers for the efforts in preparing the manuscripts, reviewing, and managing the sessions, respectively. Also thanks to all authors and participants for their heartful contributions to the GCoMSE 2021.

Sincerely,

Ir. Dr. SALLEHUDDIN SHAH BIN AYOP

Chairman of GCoMSE 2021

IOP Conf. Series: Materials Science and Engineering 1200 (2021) 011001 doi:10.1088/1757-899X/1200/1/011001

CONFERENCE COMMITTEE

Chairman	:	Dr. Sallehuddin Shah Ayop
Deputy Chairman	:	Dr. Norashidah Abd Rahman
Secretary 1	:	Dr. Sushilawati Ismail
Secretary 2	:	Dr. Tong Yean Ghing
Treasurer	:	Salmiah Selamat
Publication and Indexing	:	Dr. Nor Haslinda Abas
	:	Dr. Mohd Azuan Zakaria
	:	Dr. Norfaniza Mokhtar
	:	Dr. Goh Wan Inn
	:	Dr. Seyed Jamalaldin Seyed Hakim
	:	Dr. Alvin John Lim Meng Siang
Publicity and Promotion	:	Dr. Ali Tighnavard Balasbaneh
	:	Dr. Adel Ali Saeed Abduh Algheeti
	:	Dr. Nasradeen Ali Khalifa Milad
Website	:	Dr. Basil David Daniel
Programme	:	Dr. Zainorizuan Mohd Jaini
	:	Dr. Nor Hazurina Othman
	:	Dr. Masni A. Majid
Logistic	:	Dr. Mohd Hanif Ismail
	:	Dr. Shahrul Niza Mokhatar
	:	Dr. Muhammad Fikri Hasmori
	:	Dr. Nicholas Anting Guntor
	:	Dr. Mohd Azlan Mohd Yusoff
	:	Dr. Mohammad Soffi Md Noh
	:	Mohd Khairy Burhanudin
Certificates	:	Dr. Noorwirdawati Ali
	:	Dr. Sharifah Salwa Mohd Zuki

Editors

Dr. Nor Haslinda Abas

Dr. Mohd Azuan Zakaria

Dr. Norfaniza Mokhtar

Dr. Goh Wan Inn

Dr. Seyed Jamalaldin Seyed Hakim

Dr. Alvin John Lim Meng Siang

Scientific Committee

Assoc. Prof. Dr Azeanita Binti Suratkon (Universiti Tun Hussein Onn Malaysia, Malaysia)

Assoc. Prof. Dr. Riduan Bin Yunus (Universiti Tun Hussein Onn Malaysia, Malaysia)

Dr Asmah Alia Bohari (Universiti Teknologi Mara Sarawak, Malaysia)

Dr. Ade Faisal (Universitas Muhammadiyah Sumatera Utara, Indonesia)

Dr. Ahmad Zaki (Muhammadiyah University of Yogyakarta, Indonesia)

Dr. Ahmed Elamin (University of Greenwich, United Kingdom)

Dr. Ali Awaludin (Gadjah Mada University, Indonesia)

Dr. Alvin John Meng Siang (Universiti Tun Hussein Onn Malaysia, Malaysia)

Dr. Ashfaque Ahmed Jhatial (Mehran University of Engineering and Technology, Pakistan)

Dr. Chong Siaw Yah (Universiti Tunku Abdul Rahman, Malaysia, Malaysia)

Dr. Faisal Sheikh Khalid (Universiti Tun Hussein Onn Malaysia, Malaysia)

Dr. Fathoni Usman (Universiti Tenaga Nasional)

Dr. Goh Wan Inn (Universiti Tun Hussein Onn Malaysia, Malaysia)

Dr. Hairuddin Mohammad (Universiti Tun Hussein Onn Malaysia, Malaysia)

Dr. Hamidun Mohd Noh (Universiti Tun Hussein Onn Malaysia, Malaysia)

Dr. Hassan Ismail (Politeknik Merlimau, Malaysia)

Dr. Izudinshah Abd Wahab (Universiti Tun Hussein Onn Malaysia, Malaysia)

Dr. Josef Hadipramana (Muhammadiyah University of Sumatera Utara, Indonesia)

Dr. Khairunisa Muthusamy (Universiti Malaysia Pahang, Malaysia)

Dr. Lee Yee Ling (Universiti Tunku Abdul Rahman, Malaysia)

Dr. Lee Yee Yong (Universiti Malaysia Sarawak, Malaysia)

Dr. Lee Yoeng Huei (Curtin University, Malaysia)

Dr. Louisa Cheung Liu Yess (Ramboll Hong Kong Limited, Hong Kong)

Dr. Masni A. Majid (Universiti Tun Hussein Onn Malaysia, Malaysia)

Dr. Mazizah Ezdiani Mohamad (University College of Technology Sarawak, Malaysia)

Dr. Mia Wimala (Universitas Katolik Parahyangan, Indonesia)

Dr. Mohammad Mohammadhasani (Building, Housing and Road Research Center, Iran)

Dr. Mohd Azuan Zakaria ((Universiti Tun Hussein Onn Malaysia, Malaysia)

Dr. Mohd Hanif Ismail (Universiti Tun Hussein Onn Malaysia, Malaysia)

Dr. Mohd Hanifi Othman (Universiti Tun Hussein Onn Malaysia, Malaysia)

Dr. Mohd Norazam Yasin (Universiti Tun Hussein Onn Malaysia, Malaysia)

Dr. Muhammad Fikri Hasmori (Universiti Tun Hussein Onn Malaysia, Malaysia)

Dr. Mustafa Klufallah (University of Reading Malaysia, Malaysia)

Dr. Mustaqqim Abdul Rahim (Universiti Malaysia Perlis)

Dr. Noor Aina Misnon (Universiti Pertahanan Nasional Malaysia, Malaysia)

Dr. Noor Nabilah Sarbini (Universiti Teknologi Malaysia, Malaysia)

Dr. Noorasyikin Mohammad Noh (Universiti Tun Hussein Onn Malaysia, Malaysia)

Dr. Noorwirdawati Ali (Universiti Tun Hussein Onn Malaysia, Malaysia)

Dr. Nor Haslinda Abas (Universiti Tun Hussein Onn Malaysia, Malaysia)

Dr. Nor Hazurina Othman (Universiti Tun Hussein Onn Malaysia, Malaysia)

Dr. Norashidah Abd Rahman (Universiti Tun Hussein Onn Malaysia, Malaysia)

Dr. Norfaniza Mokhtar (Universiti Tun Hussein Onn Malaysia, Malaysia)

Dr. Norhafizah Salleh (Universiti Tun Hussein Onn Malaysia, Malaysia)

Dr. Norhaiza Nordin (Infrastructure University Kuala Lumpur, Malaysia)

Dr. Nur Farhayu Ariffin (Universiti Malaysia Pahang, Malaysia)

Dr. Putri Zulaiha Razi (Universiti Malaysia Pahang, Malaysia)

Dr. Radzi Ismail (Universiti Sains Malaysia, Malaysia)

Dr. Rendy Thamrin (Andalas University, Indonesia)

Dr. Rita Irmawaty (Hasanuddin University, Indonesia)

Dr. Saffuan Wan Ahmad (Universiti Malaysia Pahang, Malaysia)

Dr. Seyed Jamaldin Syed Hakim (Universiti Tun Hussein Onn Malaysia, Malaysia)

Dr. Sharifah Akmam Syed Zakaria (Universiti Sains Malaysia, Malaysia)

Dr. Sharifah Salwa Mohd Zuki (Universiti Tun Hussein Onn Malaysia, Malaysia)

Dr. Suaathi Kaliannan (Lincoln University College Malaysia, Malaysia)

Dr. Tong Yean Ghing (Universiti Tun Hussein Onn Malaysia, Malaysia)

Alamsyah (Politeknik Negeri Bengkalis, Indonesia)

Mohd Fairus Yusof (Universiti Tun Hussein Onn Malaysia, Malaysia)

Siti Khalijah Yaman (Universiti Tun Hussein Onn Malaysia, Malaysia)

Zev Aljauhari (Politeknik Bengkali, Indonesia)

IOP Conf. Series: Materials Science and Engineering 1200 (2021) 011002 doi:10.1088/1757-899X/1200/1/011002

Peer review declaration

All papers published in this volume of IOP Conference Series: Materials Science and Engineering have been peer reviewed through processes administered by the Editors. Reviews were conducted by expert referees to the professional and scientific standards expected of a proceedings journal published by IOP Publishing.

- **Type of peer review:-Double-blind**
- **Conference submission management system: http://confbay.com/**
- **Number of submissions received: 38**
- **Number of submissions sent for review: 37**
- **Number of submissions accepted: 36**
- **Acceptance Rate (Number of Submissions Accepted / Number of Submissions Received X 100): 94.74%**
- **Average number of reviews per paper: 3**
- **Total number of reviewers involved: 59**
- **Any additional info on review process:**

1. A preliminary review was conducted by the main editor and only approved papers were sent for peer-review.
2. Manuscripts are selected based on the reviewer's comments on the criteria given to them, as below:
   - Accept (publishable as it is)
   - Return to author for VERY MINOR revisions
   - Return to author for MINOR revisions
   - Return to author for Major revisions
   - Reject
3. Reviewers were asked to rate/provide feedbacks to the following key points related to scientific content, quality and presentation:
   - Abstract manifest the whole paper.
   - The theoretical/conceptual framework.
   - Methodology (sampling, design, instrumentation, statistical analysis, etc.).
   - Presentation of the paper in terms of writing style.
   - Presentation of the paper in terms of clarity/readability.
   - Things to be improved (Grammar, Paper format, Writing structure).
   - Comment to Author which includes technical, quality and presentation criteria of the manuscript.

4. We also run plagiarism checking using Turnitin software and remind the authors to ensure the similarity index should not be more than 25%.

- **Contact person for queries: Nor Haslinda Abas (nhaslin@uthm.edu.my)**

The Compressive Strength and Water Absorption of Railway's Concrete Sleepers Containing Palm Oil Fuel Ash (POFA) as a Cement Replacement Material

A S Nurfarhanna[1], A Suraya Hani[2*], O Mohamad Hairi[2], J Zalipah[3], AH Noor Azlina[3], S Norhafizah[3], A Anizahyati[4]

[1] Department of Transportation Engineering Technology, Faculty of Engineering Technology, University Tun Hussein Onn Malaysia, 84600, Pagoh, Johor, Malaysia
[2] Department of Civil Engineering Technology, Faculty of Engineering Technology, University Tun Hussein Onn Malaysia, 84600, Pagoh, Johor, Malaysia
[3]Faculty of Civil Engineering and Built Environment, Universiti Tun Hussein Onn Malaysia, Parit Raja 86400, Batu Pahat, Johor, Malaysia
[4]Faculty of Civil Engineering, UiTM Shah Alam, 40450 Shah Alam, Selangor, Malaysia
Corresponding author: suraya@uthm.edu.my

Abstract. Railway's concrete sleepers demand high consumption of cement which generates higher energy assumption and carbon emission. Meanwhile, in Malaysia, around 100 tonnes of palm oil fuel ash (POFA) were disposed of in the landfill, which endangering environmental health. However, this POFA have pozzolanic properties that can be employed as cementitious material. Therefore, this study aimed to produce a sustainable concrete sleeper by using POFA as a cement replacement material focusing on the compressive strength and water absorption performance. Concrete samples with a strength grade of 55MPa and w/c of 0.35 were prepared with three design mixes containing 0% (control), 20%(POFA20), and 40%(POFA40) of POFA. For the compressive strength test, a compression machine was used. Meanwhile, the water absorption was measured at atmospheric pressure. Both tests were conducted at 7 and 28 days of curing age. The results show that as the curing age increases, their water absorption and compressive strength improves, indicating a pozzolanic reaction. In terms of POFA content, the water absorption increases by 14% and 54% for POFA20 and POFA40, respectively. Meanwhile, the compressive strength reduced by 39% for POFA20 and 67% for POFA40. Since POFA20 meets the standards, it is however applicable in slab tracks.

1. Introduction

Demand for concrete sleepers is increasing with the increasing trend of demand for high-speed railway transportation. For 212 km long double-track high-speed railway projects, about 710,000 concrete sleepers are needed [1]. The world's railway lines are estimated to use around 2 billion sleepers, with a yearly demand of ten million new sleepers [2]. Meanwhile, by 2020, the production of cement is expected to reach 492,83 million tons, producing around 396,88 million tons of carbon emissions [3]. The cement content used in concrete sleepers ranges from 300 to 700 kg/m^3 [4][5][6]. This higher cement consumption leads to environmental impacts since approximately 650 to 900 kg of CO_2 will be produced per one ton of cement which contributed to 7% of the total global CO_2 emissions [7][8][9]. This concern has allowed [4][5][6][10][11][12] to use alternative resource materials, for instance, crumb rubber and mix plastic as a replacement aggregate. However, the replacement of cement or aggregate with crumb rubbers and plastic waste tends to reduce the strength of the concrete sleepers.

In Malaysia, up to 100 million tons of palm oil biomass was expected to increase in the year of 2020. Consequently, the environmental and health risks from the disposal of waste materials might also worsen [13][14]. As the waste materials have been dumped in the open air, it creates harmful environment [15][16]. The industry had faced the challenge from the 1990s where more than 1000 tons of POFA has been dumped into lagoons and landfills as shown in figure 1 [14].

[16] started studying the potential use of POFA as cementitious materials. However, they found that POFA has low pozzolanic properties, poor workability, and reduces the compressive strength of concrete. The low POFA pozzolanic reaction may be due to the high volume of unburned carbon due to the incompleteness of the burning process. Then, [17] grind the POFA to a finer particle at 10.1μm and use it as a cement replacement material in high strength concrete (HSC), the result shows that the compressive strength increased with an increasing percentage of POFA, producing HSC of strength around 79.8 to 85.9MPa, which was higher than control concrete. Comparing the raw POFA to treated POFA, raw POFA has lower silica dioxide, SiO_2 content, and higher loss on ignition, LOI than treated POFA. Besides, it also contains impurities such as fibers and coarse particles from incomplete combustion in the palm oil mill. Therefore, POFA is cannot be used directly in a raw form and required further treatment such as grinding and heating to enhance its pozzolanic properties [18][19][20][21][22].

Figure 1. A disposal site for POFA at Muar palm oil mill.

2. Materials

The materials that have been used were Ordinary Portland Cement (OPC), Palm Oil Fuel Ash (POFA), river sand, coarse aggregate, and tap water. POFA was used as a cement replacement material to improve the noise absorption in railway concrete sleepers.

2.1. Ordinary Portland Cement (OPC)

The OPC used for this study was CEM 1 type with a minimum strength class of 52.5 which in compliance with [23].

2.2. Palm Oil Fuel Ash (POFA)

POFA was collected from the palm oil mill located at Muar, Johore. The raw POFA was dried in the oven at 105 ± 5 °C for 24 ± 1 hr to remove the moisture content and reduce its carbon content. Figure 2 shows the POFA change its color from black to light grey after the heating process. The color change may happened due to the reduced carbon content during the burning process [19]. Next, POFA was sieved using a 0.30 mm (300μm) mesh according to [24] to remove the coarse particles that were incompletely combusted in the boiler.

Figure 2. The heated POFA had turned from (a) black to (b) light grey.

2.3. Fine and coarse aggregates

The sand was used as a fine aggregate with a particle size of less than 5 mm. It was dried in the oven at 110 °C ± 5 for 24 hr to remove moisture content. Meanwhile, the coarse aggregate used was gravel with a size range from 5 mm to a maximum size of 20 mm. All aggregates used have complied with [25]

3. Methodology

3.1. The preparation of concrete sleeper

The general requirement of concrete sleepers was conformed to [26], where the minimum compressive strength was 55MPa at 28 days. The proportion of the control sample was derived based on the design of the experiment (DOE) method with a water-cement ratio (w/c) was 0.35. Three design mixes with 0%, 20%, and 40% of POFA content were tested and assigned as OPC, POFA20, and POFA40, respectively. A mixture with 0% of POFA was assigned as a control sample. Table 1 shows the details of the concrete mix design.

Table 1. Mix design using DOE Method.

Materials	POFA	OPC	POFA	Aggregates		Water
	(%)	(kg)	(kg)	Fine (kg)	Coarse (kg)	(kg or liters)
CONTROL	0	577.14	0	389.77	1285.33	195
POFA20	20	461.71	115.43	389.77	1285.33	195
POFA40	40	346.28	230.86	389.77	1285.33	195

3.2. Water absorption test

The water absorption test indicates the porosity of the concrete. The test would be carried out following the [26] Part 1: Annex C: Method for measuring the water absorption of concrete sleepers at atmospheric pressure at the age of 7 and 28 days. The sequences of the test were as follows:

1) After removing the concrete from the water curing tank, it would be dried for 48 hours at 105 °C ± 2 °C in the oven and referred to as M_1.
2) After drying, the samples were set up in a container partially filled with potable water at a temperature between 15 °C and 20 °C for a period of 24 hr.
3) After 24 hr, the water level has been increased to 5 mm above the top of the samples over a minimum of 15 min.
4) After 48 h, the samples were weighted in the water (hydrostatic weight) to obtain the mass M_2. Figure 3 shows the hydrostatic weight test method
5) The samples were wiped to eliminate any surface water and weighed to get the mass M_3 (saturated weight).

The porosity of the sample has been calculated using equation (1). The requirement for porosity shall be less or equal to 12%.

$$V_p = \frac{M_3 - M_1}{M_3 - M_2} x100\% \quad (1)$$

Where V_p is the porosity, and M is mass in kg.

Figure 3. Hydrostatic weight method.

3.3. Compressive strength test

The compressive strength test has been conducted at 7 days and 28 days of curing age with reference to [27] Part 3: Method for Compressive Strength. Three 100 mm^3 cube samples per mixture would be used in the test and the strength value would be the average of the samples. The concrete sample was cleaned from surface water before being placed on the compression machine as shown in figure 4. The loading rate was set at 0.6 ± 0.2 MPa/s ($N/mm^2 \cdot s$) and continuously increase at the rate of ± 10% until the sample fails.

Figure 4. The compression machine.

4. Results and analysis

4.1. Water absorption test

The general requirement of the water absorption test was performed to determine the porosity of the concrete sleepers at a different percentage of POFA content. The test was conducted at 7 and 28 days of curing age. Figure 5 shows the result of the water absorption test at 7 and 28 days of curing age. From the result, we can observe that the water absorption was higher at an early age, and then reduced at 28 days of curing age. The water absorption was reduced by 14%, 22%, and 18% for OPC, POFA20, and POFA40, respectively. A similar result was obtained by [22] which recorded lower water absorption with longer curing age. A lower value of water absorption indicated that the concrete sleepers have a lower porosity. This may due to the pozzolanic reactivity between the POFA and OPC which improves the microstructure of the structure and produced denser concrete [19][28][29][30].

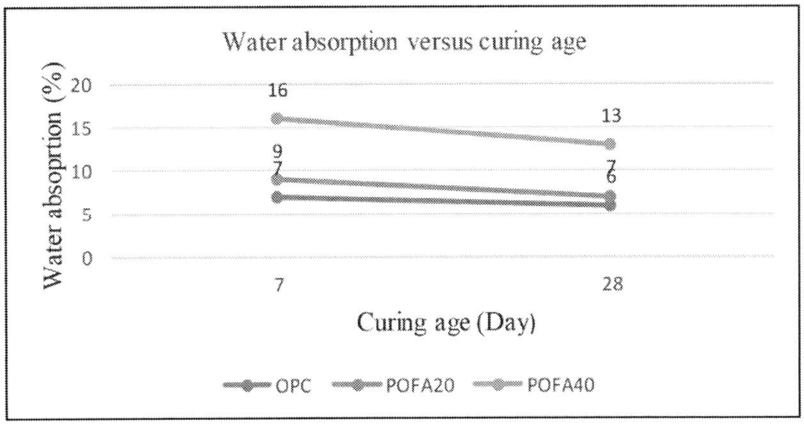

Figure 5. Water absorption (%) versus curing age (Day).

However, the results show that the water absorption of the concrete sleepers increases as the POFA replacement increased. As seen from figure 6, the water absorption of concrete sleepers increases as the POFA content increase at all curing age. On day 28 of curing age, the water absorption of the concrete sleepers was 6%, 7%, and 13% for OPC, POFA20, and POFA40, respectively. 20% of POFA increases water absorption by 14% compared to OPC. The highest water absorption recorded by the sample with 40% of POFA replacement, which increases the water absorption by 54% compared to the OPC sample. This shows that at a higher POFA replacement level, the permeability porosity of concrete sleepers was higher than the OPC. Similar results were observed from a study conducted by [20] where higher POFA content causes higher water absorption. This may due to ungrounded POFA which consists of larger and highly porous particles that tend to absorb more water [21][31]. Besides, larger particles of POFA produce a higher void in the concrete structure, thus resulted in higher permeable porosity.

According to [26], the porosity of the concrete sleepers should be lower than 12% in 28 days. However, POFA40 exceeds the maximum limit of 12%. Therefore it is inappropriate to be used in the railway operation. Hence, the optimum percentage of POFA replacement that fulfills the water absorption test was 20%.

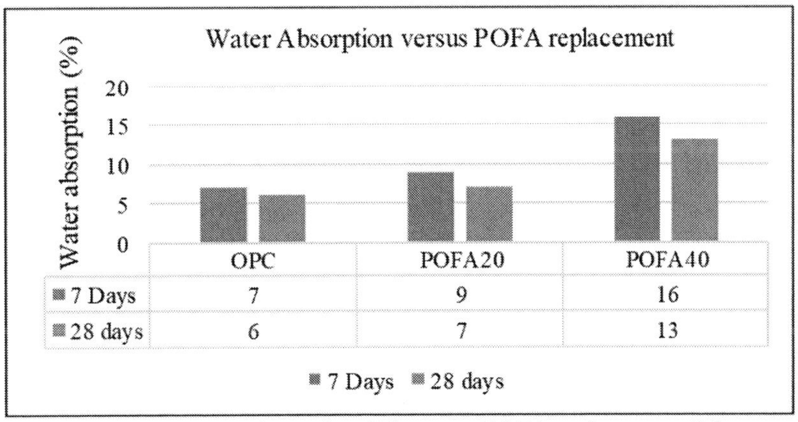

	OPC	POFA20	POFA40
7 Days	7	9	16
28 days	6	7	13

Figure 6. Water absorption (%) versus POFA replacement (%).

4.2. Compressive strength test

The compressive strength of concrete sleepers at 7 and 28 days of curing age is illustrated in figure 7. For the OPC sleepers, the compressive strength at day 7 and 28 of curing age was recorded as 44.04 MPa and 58.24 MPa, respectively. Since the compressive strength of OPC at 28 days was higher than 55 MPa, the design mix thus meets the [26] requirement and therefore, can be used as a reference sample of concrete sleepers. At an early age of 7 days, for concrete sleepers containing POFA replacement, all the compressive strength was lower than the OPC sample. This may due to the delayed formation of C-S-H gel due to high silica content over cement. A similar lower early strength trend recorded by [22,30–32]. At the early age of 7 days, the compressive strength was recorded at 31.11 MPa and 14.89 MPa for POFA20 and POFA40, respectively. However, at the later age of 28 days, the compressive strength of POFA20 and POFA40 increased to 35.52 MPa and 19.28 MPa, respectively. This shows that along with the curing age, a further pozzolanic reaction takes place between the silica in POFA and calcium hydroxide in cement to produce extra secondary calcium-silicate-hydrate (C-S-H), thus improved the microstructure of the concrete and therefore increase the compressive strength at a later age [19][21][33][34][35].

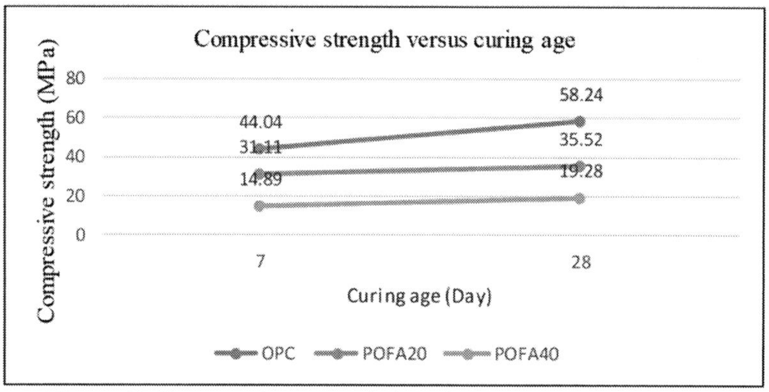

Figure 7. Compressive strength (MPa) versus curing age (Day).

Although the compressive strength of the concrete sleepers increases with the increased curing age, the compressive strength had reduced with a further increment of POFA replacement. Figure 8 shows the relationship between the compressive strength of the concrete sleepers and the POFA replacement content. From the results, the compressive strength was inversely proportional to POFA replacement content. As the percentage of POFA increase, the compressive strength reduced by 39% for POFA20 and 67% for POFA40. A similar trend was reported by [19] where the compressive strength of HSC decreases gradually as the POFA content increase. This may be due to the use of untreated POFA in this study that has low silica content due to its higher unburned carbon content and LOI value [20][29][36][37]. Thus, higher POFA content in POFA20 and POFA40 along with reduced cement content, resulting in the limitation of the reaction to cement and produce lower strength concrete than OPC [19][28]. Furthermore, a higher LOI value in POFA would cause excessive water absorption during the hydration process. As the result, the concrete sleepers with higher POFA content would losses their strength [19]. In terms of its physical properties, the POFA may have low pozzolanic properties due to its large particle size and porous structure [20][37][38][39]. However, although the compressive strength of POFA20 was 35.52 MPa and does not meet the minimum compressive strength of 55 MPa to be used as a concrete sleeper, it can be used in a slab track where the minimum compressive strength requirement was 27 MPa [4].

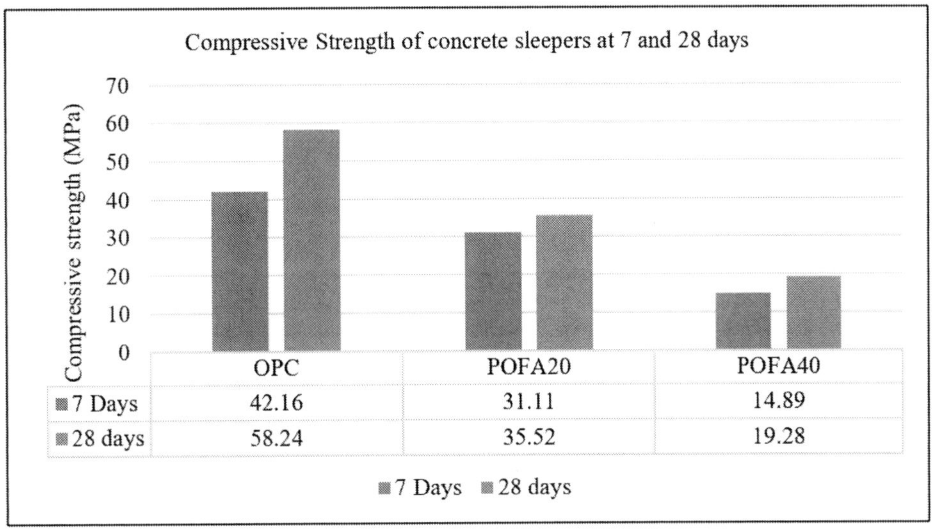

Figure 8. Compressive strength (MPa) versus the percentage of POFA replacement (%).

5. Conclusions

- The water absorption increases along with POFA content, indicating higher porosity. However, POFA40 recorded the highest water absorption with more than the allowable limit of 12%, therefore it is unsuitable to be used in the railway operation.
- The compressive strength of the concrete sleepers containing POFA increase with increasing curing age due to enhanced pozzolanic reaction which produced stronger concrete. However, POFA was not effective in improving the compressive strength of concrete sleepers. As 20% and 40% of POFA were utilized, the compressive strength of the concrete sleepers reduced below the minimum requirement of 55MPa, hence, they were not applicable as railway's concrete sleepers. However, 20% of POFA can be used in slab tracks since the compressive strength exceeds 27 MPa.

6. References

[1] Bezgin N O 2017 High performance concrete requirements for prefabricated high speed railway sleepers, *Constr. Build. Mater.* **138** 340–351

[2] Ferdous W, Manalo A, Van Erp G, Aravinthan T, Kaewunruen S and Remennikov A 2015 Composite railway sleepers - Recent developments, challenges and future prospects, *Compos. Struct.* **134** 158–168

[3] Anand S, Vrat P and Dahiya R P 2006 Application of a system dynamics approach for assessment and mitigation of CO 2 emissions from the cement industry, *Environmental Management.* **79** 383–398

[4] Li D and Kaewunruen S 2019 Mechanical Properties of Concrete with Recycled Composite and Plastic Aggregates, *Adv. Mater. Res.* **17** 231–238

[5] Kaewunruen S, Meesit R and Mondal P 2017 Early-age dynamic moduli of crumbed rubber concrete for compliant railway structures, *J. Sustain. Cem. Mater.* **6** 281–292

[6] Meesit R and Kaewunruen S 2017 Vibration characteristics of micro-engineered crumb rubber concrete for railway sleeper applications, *J. Adv. Concr. Technol.* **15** 55–66

[7] Katare V D and Madurwar M V 2018 Application of solid wastes for the production of sustainable concrete, *Lect. Notes Civ. Eng.* **7** 108–121

[8] Alsubari B, Shafigh P and Jumaat M Z 2015 Development of self-consolidating high strength concrete incorporating treated palm oil fuel ash, *Materials (Basel).* **8** 2154–2173

[9] Hoenig V, Hoppe H and Emberger B 2007 *Carbon Capture Technology - Options and Potentials for the Cement Industry* Technical Report No. TR 044 (Germany: European Cement Research Academy GmbH) pp 11-15
[10] Kaewunruen S, Li D, Chen Y and Xiang Z 2018 Enhancement of dynamic damping in eco-friendly railway concrete sleepers using waste-tyre crumb rubber, *Materials (Basel).* **11** 1169
[11] Huang X, Ge J, Kaewunruen S and Su Q 2020 The self-sealing capacity of environmentally friendly, highly damped, fibre-reinforced concrete, *Materials (Basel).* **13** 1–19
[12] Kim H, Hong J and Pyo S, Acoustic characteristics of sound absorbable high performance concrete, *Appl. Acoust.* **138** 171–178
[13] Agensi Inovasi Malaysia (AIM) 2012 National Biomass Strategy 2020, *Renew. Sustain. Energy Rev.* **68** 907–915
[14] Awal A S M A and Hussin M W 1997 The effectiveness of palm oil fuel ash in preventing expansion due to alkali-silica reaction, *Cem. Concr. Compos.* **19** 367–372
[15] Muthusamy K, Mirza J, Zamri N A, Hussin M W, Abdul Majeed A P P , Kusbiantoro A and Albshir Budiea A M 2019 Properties of high strength palm oil clinker lightweight concrete containing palm oil fuel ash in tropical climate, *Constr. Build. Mater.* **199** 163–177
[16] Tay J H and Show K Y 1995 Use of ash derived from oil-palm waste incineration as a cement replacement material, *Resour. Conserv. Recycl.* **13** 27–36
[17] Sata V, Jaturapitakkul C and Kiattikomol K 2007 Influence of pozzolan from various by-product materials on mechanical properties of high-strength concrete, *Constr. Build. Mater.* **21** 1589–1598
[18] Elbasir O M M, Megat Johari M A and Ahmad Z A 2019 Effect of fineness of palm oil fuel ash on compressive strength and microstructure of alkaline activated mortar, *Eur. J. Environ. Civ. Eng.* **23** 136–152
[19] Sumesh M, Alengaram U J, Jumaat M Z and Mo K H 2019 Microstructural and strength characteristics of high-strength mortar using nontraditional supplementary cementitious materials, *J. Mater. Civ. Eng.* **31** 04019017
[20] Le Tang W, Lee H S, Vimonsatit V, Htut T, Singh J K, Hassan W N F W, Ismail M A, Seikh A H and Alharthi N 2019 Optimization of micro and nano palm oil fuel ash to determine the carbonation resistance of the concrete in accelerated condition, *Materials (Basel).* **12** 130
[21] Zeyad A, Johari M A M, Tayeh B A and Yusuf M O 2017 Pozzolanic reactivity of ultrafine palm oil fuel ash waste on strength and durability performances of high strength concrete, *J. Clean. Prod.* **144** 511–522
[22] Zeyad A M, Tayeh B A, Saba A M and Johari M A M, Workability, Setting Time and Strength of High-Strength Concrete Containing High Volume of Palm Oil Fuel Ash, *Open Civ. Eng. J.* **12** 35–46
[23] BS EN 197-1 2019 *Composition, specifications and conformity criteria for common cements* (London: British Standard) pp 9-11
[24] C136-01 2005 *Standard Test Method for Sieve Analysis of Fine and Coarse Aggregates* (London: British Standard) pp 85–86
[25] BS EN 12620 2013 *Aggregates for concrete* (London: British Standard) pp 38
[26] BS EN 13230-1 2016 *Railway applications: Track, Concrete sleepers and bearers Part 1 : General requirements* (London: British Standard) pp 13-15
[27] BS EN 12390-3 2019 *Testing hardened concrete Part 3: Compressive strength of test specimens* (London: British Standard) pp 4-10
[28] Ismail M A and Feng C B 2019 The effect of acidic environment on the concrete utilizing palm oil fuel ash, *IOP Conf. Ser. Mater. Sci. Eng.* **495** 012104
[29] Alsubari B, Shafigh P, Ibrahim Z, Alnahhal M F and Jumaat M Z 2018 Properties of eco-friendly self-compacting concrete containing modified treated palm oil fuel ash, *Constr. Build. Mater.* **158** 742–754

[30] Philip J, Ismail M A K and Alsubari B 2019 Strength and sulphate resistance of high performance concrete containing different fineness of Palm oil fuel ash, *IOP Conf. Ser. Mater. Sci. Eng.* **495** 012100

[31] Zeyad A M, Megat Johari M A, Tayeh B A and Yusuf M O 2018 Efficiency of treated and untreated palm oil fuel ash as a supplementary binder on engineering and fluid transport properties of high-strength concrete, *Constr. Build. Mater.* **125** 1066–1079

[32] Megat Johari M A, Zeyad A M, Muhamad Bunnori N and Ariffin K S 2012 Engineering and transport properties of high-strength green concrete containing high volume of ultrafine palm oil fuel ash, *Constr. Build. Mater.* **30** 281–288

[33] Alsubari B, Shafigh P and Jumaat M Z 2016 Utilization of high-volume treated palm oil fuel ash to produce sustainable self-compacting concrete, *J. Clean. Prod.* **137** 982–996

[34] Sata V, Jaturapitakkul C and Kraiwood K 2004 Utilization of Palm Oil Fuel Ash in High-Strength Concrete, *J. Mater. Civ. Eng.* **16** 623–628

[35] Zeyad A, Johari M, Tayeh B A and Alshaikh I M 2019 Influence of Palm Oil Fuel Ash on Properties of High-Strength Green Concrete, *Sci. J. King Faisal Univ: Basic Appl. Sci.* **20**

[36] Karim M R, Hossain M M, Elahi M M A and Mohd Zain M F 2020 Effects of source materials, fineness and curing methods on the strength development of alkali-activated binder, *Building Engineering* **29** 101147

[37] Hassan W N F W, Ismail M A, Lee H S, Hussin M W and Ismail M 2019 Engineering Properties of High Strength Blended Concrete Enhanced with Nano POFA, *Mater. Sci. Eng* **495** 012105

[38] Hossain M M, Karim M R, Elahi M M A, Islam M N and Zain M F M 2020 Long-term durability properties of alkali-activated binders containing slag, fly ash, palm oil fuel ash and rice husk ash, *Constr. Build. Mater.* **251** 119094

[39] Alsubari B, Shafigh P, Ibrahim Z, Jumaat M Z 2018 Heat-treated palm oil fuel ash as an effective supplementary cementitious material originating from agriculture waste, *Constr. Build. Mater.* **167** 44–54

Effects of Different Expansive Agents on the Properties of Expansive Cementitious Materials

Abir Mahmood[1], A. B. M. Amrul Kaish[1], Sudharshan N. Raman[2], M. Jamil[3], Roszilah Hamid[1]

[1] Department of Civil Engineering, Faculty of Engineering and Built Environment, Universiti Kebangsaan Malaysia, 43600 UKM Bangi, Selangor, Malaysia
[2] Civil Engineering Discipline, School of Engineering Monash University Malaysia, Jalan Lagoon Selatan, 47500 Bandar Sunway, Selangor, Malaysia
[3] Department of Architecture and Built Environment, Universiti Kebangsaan Malaysia, 43600 UKM Bangi, Selangor, Malaysia
Corresponding author: P108539@siswa.ukm.edu.my

Abstract. Cracking happens in the case of restricted volumetric shrinking in cementitious materials which has a negative influence on the mechanical characteristics and longevity of concrete materials and hence, reduces the life of concrete structures. Many techniques have been developed to reduce the shrinkage cracking of concrete among which, usage of expansive agents (EA) has been utilised for decades. Different types of EA creates divergence due to its chemical characteristics. In this paper, three main categories (CaO, MgO & Sulphoaluminate based) EA have been reviewed based on four criteria of concrete structures, such as strength, expansibility, durability and flowability. The review clearly indicates that CaO-based EA boosts the strength but unable to control the temperature rise, which results in thermal cracking in the long run. While MgO-based EA is vastly used in China and Sulphoaluminate based EA have been industrially used worldwide for decades, both of these agents can successfully compensate thermal shrinkage while maintaining adequate mechanical strength and durability. Beside all this differences, all types of EA have been reported to decelerate the flowability of the concrete.

1. Introduction

Exposure to unsaturated environment by a cement-based system leads to a drying shrinkage due to water evaporation [1]. One major reason for mortar and concrete defects such as curling, cracking and de-bonding is shrinkage. These cracks are mentioned as the largest drawback of concrete structures, as they lower the durability due to the water entering to trigger the corrosion of steel bars, allow the leakage of rain, and spoil the aesthetic appearance [2]. It is thus critical for the concrete industry to manage shrinkage deformation as a cause of cracks in concrete in order to extend the structural applications of concrete [3]. To overcome this weakness, different techniques have been adopted such as surface treatments, e.g. prestressed technology, admixture surface treatment [4], shrinkage reducing admixtures (SRA) [5], temperature and humidity in curing condition [6], hybrid fibers [7] and expansive agents. Among the various solutions to control shrinkage, expansive agents can increase the volume of concrete due to specific chemical reactions and can be partially replaced with Ordinary Portland Cement (OPC) [8]. Expansive cement is an additive, industrially sold as powder, which is generally used to reduce concrete drying shrinkage and avoid the formation of cracks while enhancing high early strength and concrete hardening process. A concrete, consisting of expansive agent (EA) is also known as expansive concrete or shrinkage compensated concrete. Different EA that causes expanding hydrates after reaction occurs with water can be used to introduce internal expansion. It is important to note that the majority of these reactions lead to less voluminous products than reactants. An EA-containing concrete has an initial bulging bigger than the identical concrete that does not contain any EA. During the water curing of concrete this expansion happens. The maximum expansion is attained between the 2nd and 7th day of water curing, which varies on the type of expansion agent. Then, as soon as the water curing is done, drying shrinkage occurs.

Although EA are massively applied for compensating the drying shrinkage of concrete, it triggers thermal shrinkage and cracking. Because the heat of cement hydration is poorly conducted, its temperature may greatly rise and hence induce substantial thermal shrinking during cooling, which is critical for the shrinkage crack of subsequent ages. The temperature of the concrete may grow dramatically. Preventing thermal shrinkage cracking in mass cement through traditional methods such as the use of cement with low hydration heat, the use of mixtures with high volumes of additional cement materials as cement replacements, precooling raw materials and/or aftercooling concrete by incorporating cooling pipes are all widely used. However, these methods are both expensive and time consuming. [9]. Moreover, despite of the increment of mechanical strength and concrete volume, the harmful side-effects are not considered in the past or present researches. This paper will accumulate the primary categories of EA and find out the important factors that has not been addressed for future investigations.

2. Mechanism and usage of EA

During early concrete hardening the expansion agent is employed to provide a volumetric expansion that entirely or partially mitigates the volumetric contraction generated by various kinds of shrinkage according to distinct chemical processes [10]. Indifference to the type of EA, the expanding mechanism is the same.

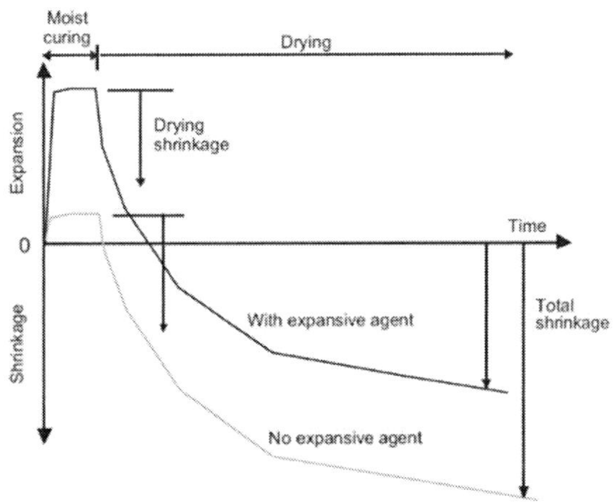

Figure 1. Illustration of the expansion and shrinkage in respect with moist and dry curing period of two concretes, one with an expanding agent and another without [10].

Gagné [10] has illustrated the difference in principle of concrete containing EA and vice versa. In figure 1, it can be analyzed that although the cement without EA has little expansion during the moist curing period, there is a drastic increase in shrinkage that stabilizes after a long period of time. Whereas, the cement with EA dramatically expands during wet curing condition and results in less shrinkage in dry curing condition.

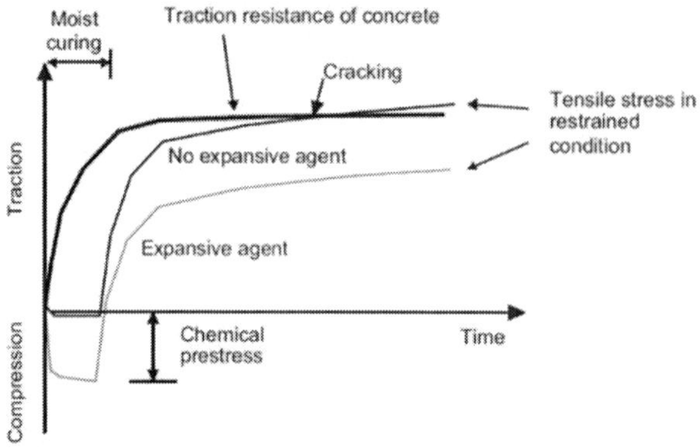

Figure 2. Illustrated the development of concrete strains with or without an expanding agent when volumetric fluctuations are limited [10]

Figure 2 portrays the stress variations in a concrete that is restrained to volumetric variations. For concrete with no EA, a restricted shrinking drying effect leads to tensile tension while shrinkage is developing. Crack occurs if the restrained shrinkage tensile stress is larger than the tensile strength of concrete [11]. Concrete expansion, which contains an EA, is also restricted, so that during moist curing, it produces internal tension of compression (chemical prestressing). The internal compression is reduced when the drying shrinkage begins to form after the curing phase. Lastly, the final tensile stress is less than the former scenario and can greatly minimise or reduce the likelihood of cracking (figure 2).

EA-containing concrete is mostly utilised for building bridge decks and ground decks [12][13]. Repairs and thin bonded concrete overlays are potential sectors where EA's can also be employed [10]. The decrease of the shrinkage in bridge decks is especially significant since the deck is attached to the main or secondary beams. Friction on the ground is the main cause of shrinkage for the plates. The lack of shrinking cracks is particularly crucial for long-term endurance for all these constructions.

3. Different types of EA

Though various EA have been produced over time, the major components of expansive cement include: CaO-based, MgO-based and Sulphoaluminate-type.

The expansive CaO-based agent contains dead-burner calcium granules (CaO) which, on entering into contact with water, create portlandite crystals ($Ca(OH)_2$). Commercial products are poured straight into the mixer; Portland cement, where they are frequently dissolved. According to the ACI 223 Committee report, these EA are Type G.

MgO-based expansive grout contains MgO from the clinker (calcined at 1450 °C), creating chemically stable hydration product, named $Mg(OH)_2$, comparatively less water required for MgO hydration, and specifically its designable expansion properties. Experiments from both experimental and industrial applications showed MgO's benefits in neutralizing concrete shrinkage and improving concrete durability.

Sulphoaluminate-type EA consists largely of a calcium sulphoaluminate (C_4A_3S) and dead-burn lime (CaO) combination which results in the quick development of ettringitis during water curing. This type of expansion is also applied in United States cements of Type K. A type K component is immediately added to concrete during batching and marketed in the form of a powder. The EA is mixed in some Portland Cement, which is to be removed from the cement dose [10]. Nevertheless, the neglible amount effects greatly in terms of expansibility and mechanical strength.

4. Properties of expansive grouts

Expansive cement can cause volume expansion and does not reduce but expand to some amount in the process of hydration. The usage of expansive cement can help to overcome and strengthen some of the flaws in regular cement concrete (typically used cement shrinks during the hardening process), allowing the structure to endanger permeability and create cracks), can enhance flexural, compressive and tensile strength and also the concentration of cement concrete structures and the integrity of concrete. The authors can be classified as follows.

Table 1. Main categories of Expansive Agent.

Different types of EA	Authors
CaO-based	[7][15][17]
MgO-based	[9][19][20][21][24][26][27][28]
Sulphoaluminate-based	[22][23][29][30][31]

4.1. Effects on Flowability

An EA can lead to a quick reduction in concrete workability due to the substantial water consumption initially involved [12]. To cater the decreased workability, additional superplasticizer is utilized. Anshuang et al. [14] conducted a research on Ultra High Performance Concrete using CaO, MgO and Sulphoaluminate based Chinese Expansive Grout. The results indicated that the fluidity decreased with the addition of EA. But very few research have been conducted examining the flowability of fresh grout thus, proper examination and methodology should be followed in order to examine the flowability of expansive mortar.

4.2. Effects on Strength

CaO-based expansive grouts are industrially produced in powder form so that it can be added in a small amount (3-10% of the cement mass) to the total mixture. Corinaldesi & Nardinocchi [15] investigated the effects of CaO-based EA in compliance with hybrid fibre reinforcement. He found that although different amounts of EA did not influenced compressive strength values, increased dosages of CaO-based EA were able to significantly increase flexural strength of cement-based composites reinforced by metallic fibre. Usage of Shrinkage Reducing Agent (SRA) alone decreases the compressive strength of the mixture [16]. On the other hand, Nardinnochi and Corinaldesi [17] findings show that addition of polyethylene glycol as Shrinkage Reducing Agent (SRA) to the mixture of CaO-based EA and fibre reinforced concrete penalizes the flexural strength of the concrete. The mechanism may be explained as follows: the cement matrix tends to be expanded because of the use of higher dosages of CaO-based EA and because of the adherence at the interface and the high elastic module, the metallic fibre refinement [18] may be used to prevent this expansion.

The EA from MgO causes slower expansion than the EA from CaO due to the comparatively slower MgO hydration. However, Mo et al. [9] have demonstrated, by modifying calcination and residence duration, the expansion characteristics of an EA based on MgO can be engineered. Additionally, Shen et al. [19] showed that the expansion of MgO caused densification of the concrete under restricted circumstances and, in turn, contributed to the improvement in mechanical strength.In even restricted conditions, nevertheless, the microstructure of the concrete can always be destructed when excessive expansion is achieved by an excessive addition of MgO and then mechanical strength and strength drop [20]. The use of a relevant kind and the right additional dosage of MgO-based EA to achieve the necessary expansion is therefore important to avoid degradation of the microstructure or loss of mechanical strength in concrete. MgO-based expansive grout is known to increase the tensile strength. Qian et al. [21] found out that the ultimate tensile strength increases to a greater extent compared with the compressive strength. This effect is considered as a "scattered prestress effect".

This EA is comprised primarily of a blend of calcium sulphoaluminate (C_4A_3S) and dead-burn lime (CaO) that quickly generates ettringite, also known as Type K cement. The amount of aluminate creates different types of Sulphoaluminate based expansive cement. Concrete's compressive strength is unaffected by Type K cement. but it significantly increases the tensile strength of the mixture when fully replaced for Ordinary Portland Cement (OPC) [22]. The presence of expansive cement in Fibre Reinforced Concrete (FRC) resulted in a better bond between fibers and cement matrix over time due to self-stressing. He also found that, although the flexural strength is same with the OPC cement, addition of steel fibre creates a radical increase to the flexural strength. A high mortar ratio of sulphoaluminate cement, aluminate cement, and gypsum results in strong early compressive strength while the late strength of the grouts grows slowly, according to Li & Hao [23].But it lacks the practicality of usage on site.

4.3. Effects on expansibility

CaO-based expansive grouts have early shrinking resisting effect due to hydration with water. Increment of CaO dosage causes decrement in total porosity. Sun et al. [7] found that the fibres prevented fracture initiation and propagation, decreased crack source size and quantity, and compensated with each other at different scales, improving the pore structure of concrete matrix. A self-prestress effect occurs due to the stress exchange between expanding cementitious paste and steel fibre, allowing a 30% decrease in 56-day drying shrinkage strains [17].

Moreover, Zhang et al. [24] conducted his research on industrially produced MgO-based expansive agent (MEA) and found out that the curing temperature of the expansive grout effects the expansion rate of the mixture. The combination of SRA and MEA cured at 850°C to 950°C could successfully counteract the cement paste shrinkage [24]. The presence of SRA lowered MEA hydration, and as cure age progressed, MEA hydration with or without SRA tended to be the same. It was because SRA breakdown decreases the polarity of water and SRA molecules absorb on MEA particles surface. Mo [25] said that MEA with increased responsiveness results in faster and bigger concrete expansion at an early age, but slower and less late expansion.Mo et al. [9] long-term research on MgO concrete samples or in-build surveillance on dam concrete shows that, once MgO hydration has been finished, the extension curve of MgO concrete is being stabilised. This is due to $Mg(OH)_2$, which have extremely poor solubility and which is 200 times less than $Ca(OH)_2$'s stable chemical characteristics. According to Chen and Ma [26], Gao et al. [27] and Ye et al. [28], as fly ash has moderating effects on the expansion of MgO, increasing the amount of fly ash added to cement pastes will result in a reduction in the expansion of MgO in the paste.

The expansion caused by the ettringite agent is signicantly more than the regular grout [29]. Wu et al. [30] found a clear early expansion effect during standard curing, due to calcium sulphoaluminate hydrate-calcium hydroxide EA, compared to the basic cement mortar. But, CaO & MgO based EA have more expansive characteristics in comparsion with Sulphoaluminate Expansive Cement. The results found in Afroughsabet et al. [22], also demonstrate that the use of Type K cement in place of OPC may compensate for shrinkage in concrete, and the length change remain relatively constant even after 56 days of exposure.

China has developed sulphoaluminate expansive cement of their own, known as United Expansive Agent (UEA). According to Sun [7], integration of EA with fibre on high strength grout improved the interface between fiber-matrix and aggregate-matrix, allowing fibres of varying sizes and types to resist shrinkage and cracking on a range of different scales. As a result, the shrinkage resistance of cement and fibers functioned earlier. Dong et al. [31] experimented using a Chinese Expansive agent from Nanjing Tejian Co. which is Low-alkali UEA. He found out that mineral admixtures remarkably influence the expansion characteristics of EA in grout. When the dosage of mineral admixtures is increased appropriately, the application impact of EA in grout is increased and the shrinkage effect of the grout is compensated for it. However, compared with curing conditions, mineral admixture has a minor influence on ethringite production.

4.4. Effects on durability

Durability of a compound is measured by examining its water permeability using different EA. Sun et al.[7] found out concrete reinforced with steel fiber has provided better improvement for the impermeability and durability. In fact, different sizes of steel fibers are providing different outcomes. Moreover, it had a considerably greater impact than that created by adding UEA or steel fibre alone when united expansive Agent (UEA) and steel fibre had. When three or two types of fibres are appropriately combined and integrated into the concrete, the relative coefficient of water permeability of the concrete might fall by 56% - 70% [7]. Usage of EA in hydraulic structures are less likely used in practical life due to lack of enough research, including chlorine attack. Unfortunately, many research have not been focused on the durability of structures and future probes are to be conducted to diversify the usage of EA.

5. Conclusions and recommendations

This study summarized all different types of EA that has been used around the world and correlated each other based on the intended outcomes. Although sulphoaluminate-based expansive cement have been widely used for the past six decades, China has over 40 years of experience in using MgO-based expansive cement in their structures. Furthermore, usage of EA alone is not sufficient due to its own inexpediency. Thus addition of external additives such as shrinkage reducing agent (SRA), steel fibre, hybrid fibre, polyvinyl alcohol (PVA) fibers, polyethylene terephthalate (PET) fibres, pozzolans etc. have shown better results in terms of material strength, expansibility, durability and flowability. The factors that effects the expansibility of different EA are dosage, curing condition and temperature.

However, very few research has been conducted on the vast usage of expansive cementitious composite in modern days. Despite of simple and easy usage of EA due to its powder form, the traditional expansive additives have many drawbacks. For example, their hydration and expansion rates are too fast to control. The expansive hydration components do not work well in water (a low thermal stability). Although atmospheric conditions can be controlled during the experiment, usage of this EA on site. Table 2 also shows that very few research focused on the flowability of the expansive cement which tends to be the matter of utmost importance during construction. In addition, the water content necessary for hydration processes is excessively high. But until now only MgO-based EA have partly succeeded to compensate this limitations. It is recommended that further research using new cementitious components are to be conducted to bring more versatility into the process, including research on fresh cement flowability and early high-strength.

Table 2. Research summary.

Parameters	Sun et al. [7]	Corinaldesi & Nardinocchi [15]	Mo et al. [9]	Li [20]	Qian et al. [21]	Afroughsabet et al. [22]	Wu et al. [30]	Anshuang et al. [14]
Strength	-	✓	-	✓	✓	✓	-	X
Expansibility	✓	✓	✓	-	-	✓	✓	✓
Durability	✓	-	-	-	-	-	-	-
Flowability	-	-	-	-	-	-	-	X

6. References

[1] Neubauer C M, Jennings H M and Garboczi E J 1997 Mapping drying shrinkage deformations in cement-based materials, *Cement and Concrete Research* **27** 1603–1612
[2] Nagataki S and Gomi H 1998 Expansive admixtures (mainly ettringite), *Cement and Concrete Composites* **20** 163–170
[3] Shah S P, Weiss W J and Yang W 1998 Shrinkage Cracking--Can It Be Prevented?, *Concrete International* **20** 51-55
[4] Zhao J L, Shen S L, Wang L B and Chen J 2011 A design approach for the interior anchorage zone of post-tensioned concrete structure, *KSCE Journal of Civil Engineering* **15** 487–495 3
[5] Nmai C K, Romita R, Hondo F and Buffenbarger J 1998 Shrinkage reducing admixtures, *Concrete International* **20** 31–37
[6] Jiang Z, Sun Z and Wang P, 2005 Autogenous relative humidity change and autogenous shrinkage of high-performance cement pastes, *Cement & Concrete Research* **35** 1539–1545

[7] Sun W, Chen H, Luo X and Qian H 2001 The effect of hybrid fibers and expansive agent on the shrinkage and permeability of high-performance concrete, *Cement and Concrete Research*, **31** 595–601
[8] Liu F, Shen S L, Hou D W, Arulrajah A and Horpibulsuk S 2016 Enhancing behavior of large volume underground concrete structure using expansive agents, *Construction and Building Materials* **114** 49–55
[9] Mo L, Deng M, Tang M and Al-Tabbaa A 2014 MgO expansive cement and concrete in China: Past, present and future, *Cement and Concrete Research* **57** 1–12
[10] Gagné R 2016 Expansive agents. *Science and Technology of Concrete Admixtures* 441–456
[11] Bissonnette B, Jolin M, Gagné R, Certain P V and Perez F, 2014. Evaluation of the Robustness of Shrinkage-Compensating Concrete Repair Concretes Prepared with Expansive Components, *ACI C223 Special Publication* p 17
[12] Phillips M V, Ramey, G E and Pittman D W 1997 Bridge deck construction using type K cement, *Journal of Bridge Engineering* **2** 176-182
[13] Richardson D, Tung Y, Tobias D and Hindi R 2014 An experimental study of bridge deck cracking using type K-cement, *Construction and Building Materials* **52** 366-374
[14] Anshuang S, Ling Q, Shoujie Z, Jiayang Z and Zhaoyu L 2017 Effects of shrinkage reducing agent and expansive admixture on the volume deformation of ultrahigh performance concrete, *Advances in Materials Science and Engineering* **2017** 6384859
[15] Corinaldesi, V and Nardinocchi, A 2016 Mechanical characterization of Engineered Cement-based Composites prepared with hybrid fibres and expansive agent, *Composites Part B: Engineering* **98** 389-396
[16] Tittarelli F, Giosuè C and Monosi S 2017 Combined use of shrinkage reducing admixture and CaO in cement based materials, *IOP Conference Series: Materials Science and Engineering* **245** 022093
[17] Nardinocchi A and Corinaldesi V 2014 Study of Ultra High Performance Fiber Reinforced Concrete with Expansive and Shrinkage Reducing Agents. In *Advanced Materials Research* **980** 137-141
[18] Li C 2000 Mechanical Performance and Long Term Durability of MgO Concrete, *Shui Dian Ke Ji Jin Zhan* **20** 30–35
[19] Shen D, Deng M, Mo L 2012, Effect of restraint degree on the expansion and tensile splitting strength of concrete containing MgO-based expansive agent, *Concrete* **10** 14–17
[20] Li X 2012 *Mechanical Properties and Durability Performance of Reactive Magnesia Cement Concrete* PhD Thesis (Cambridge: University of Cambridge)
[21] Qian X, Jiang L, Song Z, Chu H, Yan X and Zhang Y 2020 Impact of elevated curing temperature on mechanical properties and microstructure of MgO-based expansive additive cement mortars, *Structural Concrete* **21** 1082-1092
[22] Afroughsabet V, Geng G, Lin A, Biolzi L, Ostertag C P and Monteiro P J 2019 The influence of expansive cement on the mechanical, physical, and microstructural properties of hybrid-fiber-reinforced concrete, *Cement and Concrete Composites* **96** 21-32
[23] Li X and Hao J 2017 Effects of mortar ratio on the properties of super early strength grouting materials by ternary complex system, *Construction and Building Materials* **152** 769-776
[24] Zhang Z B, Xu L L, Liu F and Tang M S 2012 Effect of Amount on Expansion Property of MgO-type Expansive Agent Used in Cement-based Materials, *Advanced Materials Research* **391** 803-806
[25] Mo L 2008 *Microstructure and Property of MgO-type Expansive Agent* PhD Thesis (Nanjing: Nanjing University of Technology)
[26] Chen H and Ma X 2010 Effects of fly ash on expansion of MgO expansive cement and its mechanism, *Journal of Material Science & Engineering* **28** 181–185
[27] Gao P, Xu S, Cheng X, Li J and Lu X 2013 Research on autogenous volume deformation of concrete with MgO, *Construction Building Materials* **40** 998–1001

[28] Ye H, Fu C and Yang G 2019 Influence of dolomite on the properties and microstructure of alkali-activated slag with and without pulverized fly ash, *Cement and Concrete Composites* **103** 224-232
[29] Maruyama T and Karasawa H 2017 Effect of Type of Cement and Expansive Agent on the Hardening of Steam-Cured Concrete, *Key Engineering Materials* **744** 105-113
[30] Wu Y Y, Peng Y, Liu Y D and Jia L L 2013 Expansion of cement mortar mixing expansive agent under different curing conditions and compressive strength in constrained conditions, *Applied Mechanics and Materials* **405** 2534-2537
[31] Dong R Z, Wei J, Qi Y X, Guo L T, Chen B, Liu Y and Wei R 2014 Compensation effect of expansive agent on shrinkage of self-compacting concrete, *Applied Mechanics and Materials* **507** 401-405

Acknowledgments

The authors acknowledge Universiti Kebangsaan Malaysia for providing the necessary opportunities & funding for this research through "Research Graduate Assistance" scheme (FRGS/1/2019/TK01/UKM/02/2) and all my family members for encouraging me throughout the entire journey.

An Overview of Eco-Friendly Alternatives as the Replacement of Cement in Concrete

Habiba Afrin[1], Nazmul Huda[1] and Rouzbeh Abbasi[1]

[1] School of Engineering, Macquarie University, Sydney, Australia
Corresponding author: habiba.afrin@students.mq.edu.au

Abstract. Due to the global urbanization, economic development, and increasing rate of the world's population, the construction of new buildings and infrastructure is increasing. The manufacture of concrete has become an essential part of our life all over the world. Emitting 5-8% of carbon dioxide (CO_2), the production of concrete becomes the main obstacle to reach global climate action under the Paris 2050 Agreement. Reuse of waste or recycled waste materials in concrete as an environmentally friendly construction material has become highlighted as a feature of achieving sustainability, because of its potential environmental and economic benefits. The focus of this paper is to assess the application of alternative eco- friendly substitutes of cement for an innovative, economically attractive, and environmentally friendly alternative, and the transition towards of circular economy by reducing the number of natural resources consumed. Based on the existing studies, waste material (fly ash, bottom ash, coal ash, tire, steel slag, construction and demolition waste, glass, ceramic) incorporated with concrete accepted performance in the environment and economic perspective by reducing energy consumption, greenhouse gas emissions, costs, and other indicators.

1. Introduction

Increasing awareness of the sustainable construction sector has given rise to using environmentally and economically friendly materials. The construction sector is important all over the world. Being the most frequently used man-made material on earth, it has widespread use in buildings, pavement, pipe, dam, and drain. The major reason behind the expanding demand for concrete are the increased construction of high-rise buildings and its superior high early strength advantage [1,2,3]. Each year more than 10 billion tons of concrete are used worldwide which requires a large amount of raw material [2,3]. Besides, energy consumption during cement production, water consumption, generation of waste during manufacture, demolition, and construction waste recycling difficulty in general concrete is not a sustainable material for future [3,4,5,6]. The main material limestone at the production of cement is the main key factor by releasing 55% of cement-related emissions [1]. Figure 1 shows the increasing rate of cement production globally. Because of the current climate condition and global warming phenomena, it has become indubitable for the construction and other industries from adoption and introduces eco-friendly materials. So, replacing fully or partially cement with recycled material or waste products can reduce the use of raw material also reducing environmental impact. This research aims to find the feasibility of using waste or by-products that can substitute for cement in concrete reducing pressure on raw material use, energy consumption, water requirement,and environmental impact by maximizing resource implementation in concrete through waste recycling system and the transition towards of circular economy by reducing the number of natural resources consumed. Figure 2 shows the waste generation rate in Australia.

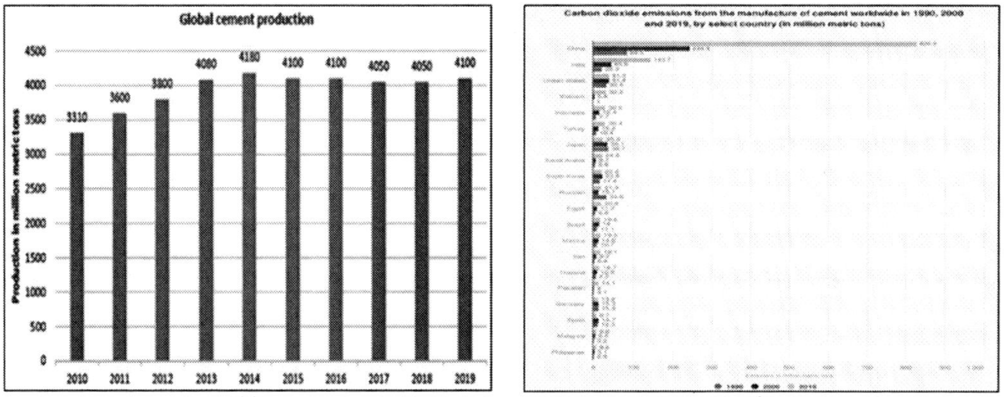

(a) (b)

Figure 1. (a) Cement production globally from 2010 to 2019 worldwide [2], (b) Globally carbondioxide emissions worldwide [3,4].

To find out which environmental impact, resource, and energy use could be reduced in the construction sector the qualitative techniques used from available official reports, legal documents, and scientific literature and the quantitative data gathered from literature and investigation of local manufacture of cement. This paper summarises the scope of using different eco-friendly materials as a replacement for cement to achieve sustainability.

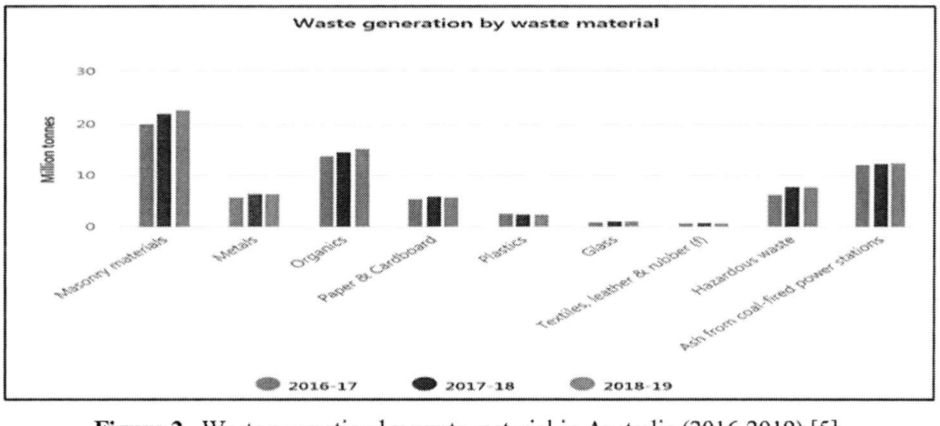

Figure 2. Waste generation by waste material in Australia (2016-2019) [5].

2. Cement, concrete, and environment

Nowadays, the construction sector is challenged for using many virgin materials and energy as well as environmental effects. Growing concern about the sustainable construction sector had risen awareness of the trend of utilization of waste materials. Green concrete is produced by utilizing waste or residual materials. So, it reduces the burden of landfilling and requires energy for processing [6,7]. Figure 2 shows waste generation by different waste materials in Australia [5]. Cement production contributes to greenhouse gases directly by emitting carbon dioxide when calcium carbonate is heated at 1500°C to 2000°C and indirectly through the use of energy, particularly if the energy is sourced from fossil fuels. As cement production requires a large amount of energy consumption, water requirement, and environmental impact, so finding a product that can substitute for cement makes good environmental sense. Partial replacement of energy-consuming cement with reusable materials is among the best way to gain sustainable construction material and finding an alternative way to reuse waste materials.

3. Feasible materials as a replacement of cement

At present, one of the main challenges in the concrete industry is to meet the demand for physical and economic awareness. Therefore, a growing awareness of sustainable construction has given the rise to recycling or reusing waste materials in concrete. Mostly from three sources i.e. industry, household/domestic activities, and construction and demolition site waste are generated. The industrialby-product is not a primary product nor separately produced in the industry. It is the remaining material i.e. fly ash, bottom ash, steel slag, foundry sand are common examples of industrial by- products used in concrete [8]. Plastic bags, glass containers, containers are some common examples of domestic waste [9]. Steel, wood products, and plaster, brick and clay tile, asphalt shingles, concreteare some common examples of construction and demolition waste.

3.1. Industrial by-product as a cement r,eplacement

3.1.1. Fly ash. A large volume of fly ash is produced and widely available in coal-burning power plants as a residual material and often ends up in landfills. But when used to make concrete which is a substitute for traditional concrete by utilizing recycled fly ash. Fly ash requires has less water requirement, improved paste properties behavior, and workability. It can reduce 20% of cement demand and 10% of water demand [2, 10]. To reduce the heat of hydration of concrete by replacing the same amount of cement with fly ash is one of the common uses. The main benefits of using fly ash, it can replace a large percentage of cement making it less costly than Portland cement [10], reduced bleeding, increased the strength of concrete, and reduced shrinkage when compared to traditional concrete [11]. Depend on various types of fly ash from different sources performance also get vary.

3.1.2. Silica fume. Silica fume is known as micro silica which is an ultrafine powder. From the condensation of Silicon dioxide in ferrosilicon alloy and silicon manufacture, this is produced as a residual material without any processing. Due to its excellent reactive pozzolan properties, it can replace around 7% - 12% cement in concrete [7]. This is the third most used cementitious substitute material in the concrete industry. The use of micro silica or silica fume has proven durability, less permeability, and increased compressive strength than traditional concrete [12]. Although its microparticle size makes it hard to operate, its cost-effectiveness puts it in high demand in the construction industry [13].

3.1.3. Granular-blast furnace slag. Granular-Blast Furnace Slag is one of the commonly used industrial by-products that can be recycled to make an environmentally friendly alternative to concrete [14]. It is a residual product of the steel industry and it is a glassy granular material. Theoretically, granular blast furnace slag can replace 100% of Portland cement but usually, 70% replacements are in common use. It can be used in concrete either as a binder or as an aggregate. As a binder, it can replace approximately 70–80 wt.% of cement [15]. The benefits of using blast furnace slag are to improve workability, gain higher long-term compressive and flexural strengths, reduced permeability, and improve durability and resilience like fly ash. The cost of using granular blast furnace slag is quite the same as Portland cement but this slag often poses toxic chemicals if it dumps in the landfill. But it can be utilized safely and beneficially in concrete. Here, figure 3 shows some of the commonly used cementitious materials [16].

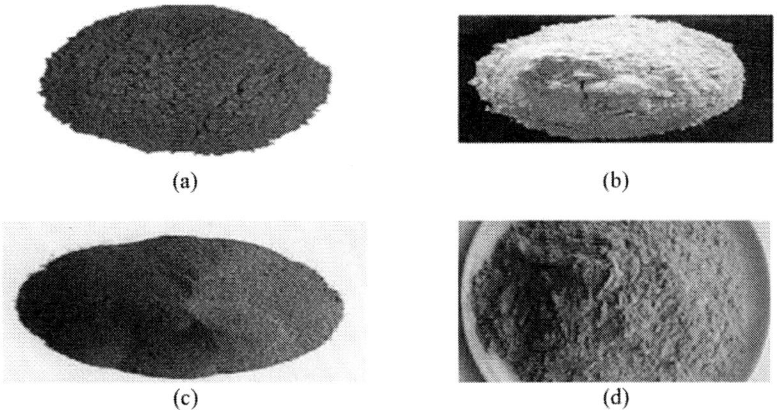

Figure 3. (a) Fly Ash, (b) GGBS, (c) Silica Fume, (d) Cement [17].

3.1.4. Bottom ash. Bottom Ash is a kind of non-combustible residue of burning in a power plant, boiler, furnace, or municipal solid waste incinerator. The electricity industry is being charged by producing a large amount of bottom ash and it has become an important economic and environmental objective [3]. Its main composition is silica, alumina, and iron with small amounts of calcium, magnesium, sulphate, etc [15]. Its chemical composition is varied by the source. It influences the workability, setting times, strength, porosity, the durability of hardened mass [18]. Here, strength reductions of concrete have mainly happened for higher porosity and higher water demand on the use of bottom ash in concrete. Bottom ash is heavier than fly ash and sometimes it remains toxic after recycling [18]. It has been effectively utilized in construction and railroad fill material, abrasive blasting grit, granules on asphalt roofing shingles, aggregate for concrete and masonry blocks [15].

3.1.5. Rice husk ash. Rice husk is the hard-protective coverings of rice grains that are separated from the grains during the rice production process [19]. Rice husk is a common waste material in all rice-producing countries. It is a pozzolanic reactive material that contains about 30–50% of organic carbon [20]. Traditionally, husks are removed from the raw grain to reveal whole brown rice which upon further milling to remove the bran layer will yield white rice. Current rice production in the world is estimated to be 700 million tons [21]. Rice husk obtains around 20% of the weight of rice. By 15% replacement of rice husk ash can increase compressive strength by about 20% in concrete. Strength and durability properties obtain by 20% replacement, beyond that is associated with a slight decrease in strength parameters by about 4.5% [21, 22]. Because of its high permeability, it is often required to maintainthe water-cement ratio perfectly [20].

3.1.6. Recycled waste tire. The tire is a non-biodegradable material and it has a very short life span of only 4 years. Waste tiresare one of the most challenging tricky and problematic waste streams, posing health and environmental concerns [23]. It creates a breeding ground for pests and presenting a significant fire hazard [24]. A two-stage process is used to make the particle into 0.075-0.475mm ground rubber i.e. magnetic separation and screening [25]. But, 20% of replacement may affect the compressive strength of tire rubber in concrete, so it is mostly used in low-strength require places like low rise buildings, parks, slabs, and farms to resist slip [26]. As a sound insulator, it has significant use in the construction sector [27]. Figures 4 and 5 show the different sizes of tire cut and processing.

IOP Conf. Series: Materials Science and Engineering 1200 (2021) 012003 doi:10.1088/1757-899X/1200/1/012003

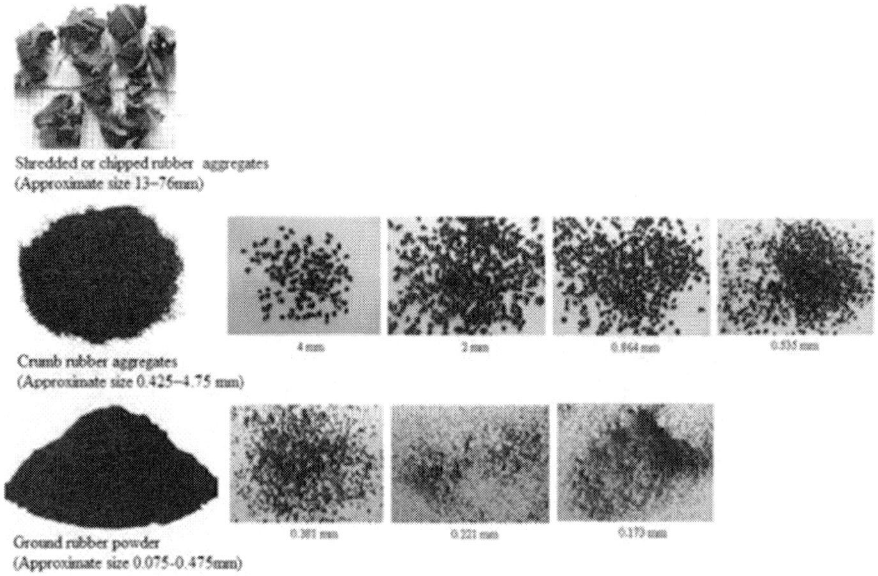

Figure 4. Different form of rubber aggregate [28,29].

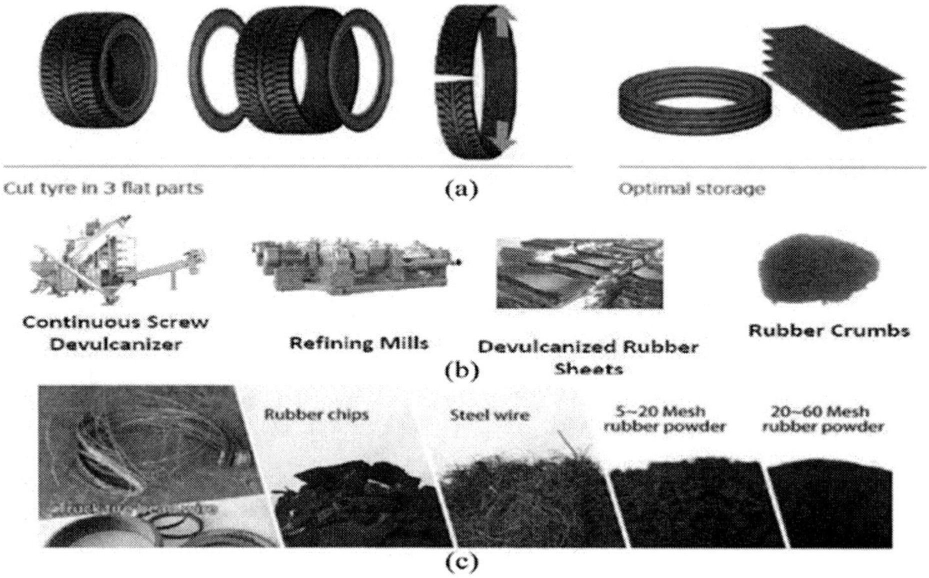

Figure 5. (a) Different size of tire cut [30], (b) powdering process, (c) forms of tire [31].

3.1.7. Waste ceramic powder. During the final polishing process, a large amount of ceramic waste is gathered [32]. Ceramic waste is a toxic powder that is harmful to soil, water, and air. But it has proven useful in concrete as a replacement for cement [9]. It also has a positive environmental impact by reducing the demand for raw materials. The ceramic waste powder has improved the workability and durability characteristics of concrete [33]. The ceramic waste powder can reduce compressive strength development and requiretime. Figure 6 shows the processing process of waste ceramic powder.

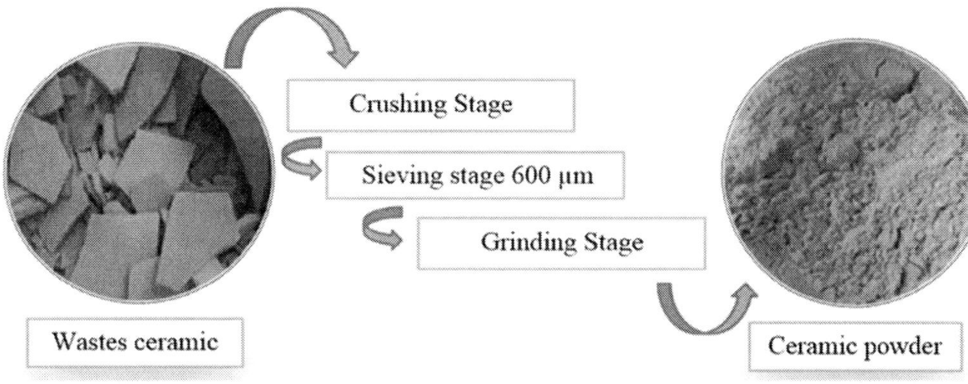

Figure 6. Waste ceramic powder processing [34].

3.2. *Domestic waste as a replacement of cement*

3.2.1. Recycled glass powder. Traditionally, a great amount of waste glass is produced every year in our domestic activities and frequently found untreated in landfills [6]. It can be used as a binder or aggregate in concrete as it has quite the same properties as sand. It becomes granulated by sieving employing sieves after they are crushed in the breaker machine and milled. Recycled glass powder commonly uses for surface treatment by blasting, reinforcement of synthetic resins, and footpath [6,35,36]. Figure 7 shows the processing process of recycled glass powder.

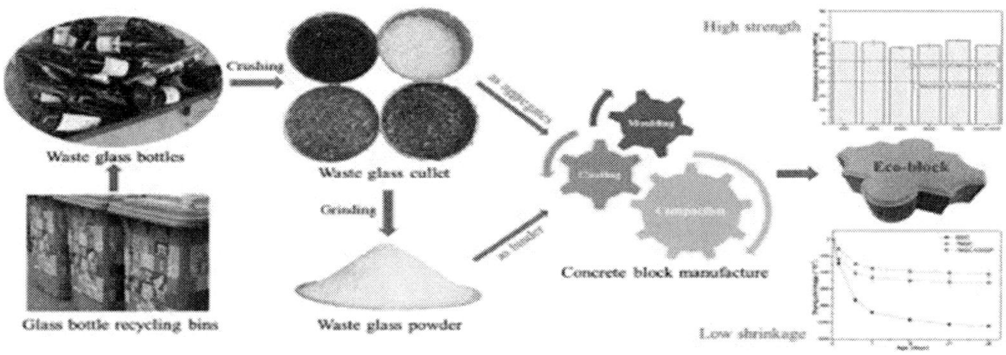

Figure 7. Recycling process of waste glass powder [36, 37].

3.2.2. Municipal Solid Waste (MSW) incineration ash. Because of the increasing population, a large volume of municipal waste is often gathered quickly and the lack of landfills also creates problems to dispose of waste. So, nowadays incineration is commonly used to reduce the volume of waste by 85%. But this abundant ash has great performance to increase the strength of concrete. Up to 5% replacement of cement was observed to give the same result as traditional concrete. Ash replacement more than 5%, there is a dramatic loss in strength regardless of curing age [38]. But with the mixture of calcium oxide to the raw mix with 15%, ash replacement can overcome the difficulties. Municipal solid waste incineration ash can replace cement raw mix under a well-conditioned situation [39]. Figure 8 shows the diagram of the MSW incineration process.

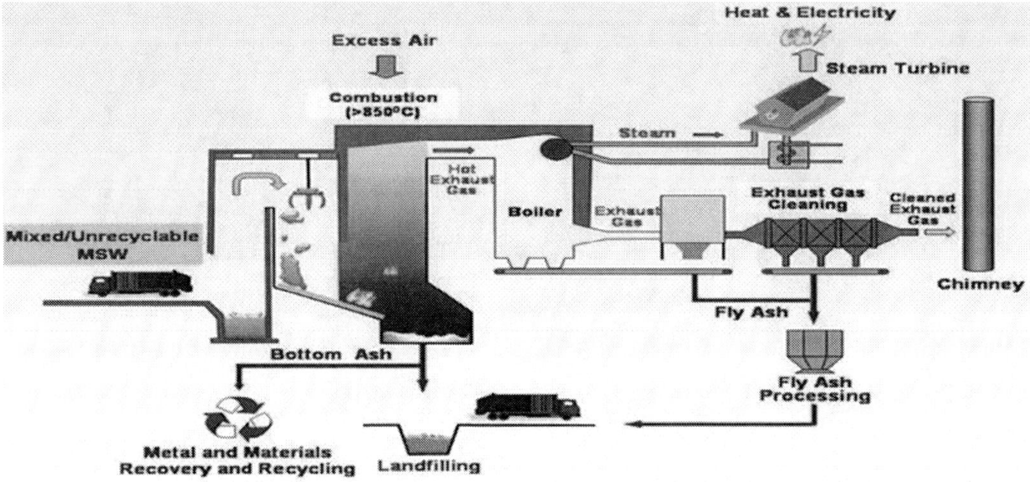

Figure 8. Schematic diagram of the MSW incineration process [39].

3.3. Construction and Demolition by-products as a replacement of cement

3.3.1. Recycled construction and demolition waste powder. A huge percentage of C&D waste is brick and concrete. It can bring a huge potential by using waste brick and concrete [13]. Recycled concrete and brick powders can be successfully used as cement replacements or binders in cement-based products. Replacing 5-7% of brick powder gives the same performance as conventional concrete [40]. The negative effect of concrete powder is mainly influenced by compression strength. Only a 10% cement replacement reduced compression strength by about 25% and a further decrease was almost the same [41]. Figure 9 shows the processing process of recycled construction and demolition in powder form.

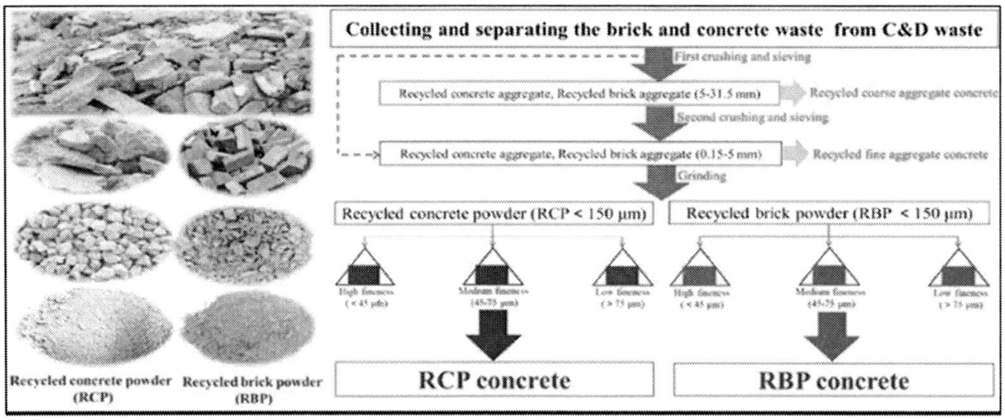

Figure 9. Concrete and demolition waste recycling process [41].

3.3.2. Recycled gypsum powder. Construction and demolition waste are the main source of waste gypsum and disposed of unsustainably untreated in landfills. But it has a good role to control the hydration in concrete and controlling the setting time [42]. Approximately 90% of the weight of a piece of drywall is gypsum, if the gypsum can be recovered from the drywall, the majority of the material can be recycled [7,43].

Improved workability, no shrinkage cracks, quick setting time, and high strength is the main added benefit of using gypsum [44]. Although it can replace 20% of cement and provide the same properties as control concrete it is not avoidable for its exceptional hydration properties [42].

3.3.3. Wood ash. Construction waste is the main source of wood waste. It is commonly utilized as burning fuel in various industries [45]. So, wood ash is often found as a residue in wood-burning mills or plants. Due to its excellent properties, wood ash can be used to replace cement partially in concrete production by 15-20% without hampering its strength than traditional concrete [2,46]. Hence, incorporating the usage of wood ash as a replacement for cement in cement is beneficial from the environmental pointof view as well as producing low-cost construction materials [2,45]. Thermal insulation properties also get improved in concrete by wood ash. Utilizing of wood ash partial replacement of cement adversely affects the slump of the concrete mix [46]. It is observed an increased rate of water abortion in wood ash content mix [2].

4. Conclusion

In recent years sustainable use of waste or by-products has been accelerated in the construction industry. Whereas the eco-friendly use of alternative waste recycled materials has shown a new path inthe concrete industry. The potential way of achieving sustainability can be gained by replacing asmuch Portland cement as possible with supplementary cementitious materials, especially those that areby-products of industrial processes. Foremost it is the increasing use of cementitious materials that canserve as partial substitutes for Portland cement, those materials that are by-products of industrial processes, such as fly ash and ground granulated blast furnace slag, wood ash, waste glass, waste tire rubber, and construction and demolition waste. But also, the substitution of various recycled materials for aggregate has made significant progress worldwide, thereby reducing the need to quarry virgin materials. However, transport distance of recycling waste, leaching with heavy metals, or recycling process may hinder the potential benefit.

5. References

[1] Siddika A, Mamun M A A, Alyousef R, Amran Y H M, Aslani F, and Alabduljabbar H 2019 Properties and utilizations of waste tire rubber in concrete: A review, *Construction and Building Materials* **224** 711-731

[2] Lazik P R and Garrecht H 2021 Wood ashes from electrostatic filter as a replacement for the fly ashes in concrete, *Journal of Construction Materials* **2** 2-6

[3] Andrew R M 2020 A comparison of estimates of global carbon dioxide emissions from fossil carbon sources, *Earth System Science Data* **12** 1437-1465

[4] Pierre F, Michael O, Matthew W J, Robbie M A, Judith H, Are O, Glen P P, Wouter P, Julia P, Stephen S, Corinne L Q, Josep G C, Philippe C, Robert B J, Simone A, Luiz E O C A, Almut A, Vivek A, Nicholas R B, Meike B, Alice B, Henry C B, Laurent B, Selma B, Naveen C, Frédéric C, Louise P C, Wiley E, Liesbeth F, Piers M F, Thomas G, Marion G, Dennis G, Thanos G, Luke G, Nicolas G, Ian H, Kerstin H, Vanessa H, Richard A H, Tatiana I, Atul K J, Emilie J, Koji K, Etsushi K, Vassilis K, Jan I K, Peter L, Nathalie L, Andrew L, Sebastian L, Zhu L, Danica L, Gregg M, Nicolas M, David R M, Julia E M S N, Shin-Ichiro N, Yosuke N, Kevin O, Tsuneo O, Paul I P, Denis P, Benjamin P, Laure R, Eddy R, Christian R, Jörg S, Roland S, Ingunn S, Adam J P S, Adrienne J S, Toste T, Pieter P T, Hanqin T, Bronte T, Guido V W, Nicolas V, Anthony P W, Rik W, Andrew J W, David W, Andrew J W, Wenping Y, Xu Y and Sönke Z 2020 Global carbon budget 2020, *Earth System Science Data* 2020 **12** 3269-3340

[5] Statistics A B S *Australia Bureau of Statistics* 2019-2020

[6] Tamanna N 2020 *Use of waste glass as aggregate and cement replacement in concrete* PhD Thesis (Queensland: James Thesis Cook University)

[7] Meyer C 2009 The greening of the concrete industry, *Cement and Concrete Composites* **31** 601-605

[8] Francesco C, Antonella P and Claudia B 2018 Life cycle assessment of recycled concretes: A case study in southern Italy. *Sci Total Environ* **615** 1506-1517

[9] El-Dieb A S and Kanaan D M 2018 Ceramic waste powder an alternative cement replacement – Characterization and evaluation. *Sustainable Materials and Technologies* **17** 00063

[10] Yifei S, Yue L, Yuzhou T, Xueliang Y, Qingsong W, Jinglan H and Jian Z 2019 Life cycle assessment of autoclaved aerated fly ash and concrete block production: a case study in China. *Environ Sci Pollut Res Int.* **26** 25432-25444

[11] Malhotra V 2000 Role of supplementary cementing materials in reducing greenhouse gas emissions, *Concrete technology for a sustainable development in the 21st century*, **5** 6

[12] Mindess S 2019 Developments in the Formulation and Reinforcement of Concrete, *Woodhead Publishing*

[13] Rocha S and Sousa-Coutinho J 2019 Construction and demolition waste as partial cement replacement. *Advances in Cement Research* **31** 411-422

[14] Kandiri A and Fotouhi F 2021 Prediction of the module of elasticity of green concretes containing ground granulated blast furnace slag using hybridized multi-objective ANN and Salp swarm algorithm, *Journal of Construction Materials* **2** 2-2

[15] Grubeša I N, Barišić I and Samitinjay S B 2016 *Characteristics and uses of steel slag in building construction* (Sawston: Woodhead Publishing)

[16] Akram M M, Ghasan F H, Ahmad R Z A and Mohammad I 2020 Long-term mechanical and durable properties of waste tires rubber crumbs replaced GBFS modified concretes. *Construction and Building Materials* **256** 119505

[17] Jonalagadda K B, Jagarapu D C K and Eluru A J M T P 2020 Experimental study on mechanical properties of supplementary cementitious materials, *Materials Today:*Proceeding **27** 1099-1103

[18] Maneerung T, Kawi S and Wang C H 2015 Biomass gasification bottom ash as a source of CaO catalyst for biodiesel production via transesterification of palm oil, *Energy Conversion and Management* **92** 234-243

[19[Della V P, Kühn I and Hotza D J M l 2002 Rice husk ash as an alternate source for active silica production, *Materials Letter* **57** 818-821

[20] Hwang C L and Chandra S 1996 The use of rice husk ash in concrete *Waste materials used in concrete manufacturing* ed S Chandra Elsevier (New York: William Andrew Publishing) pp 184-234

[21] Siddique R 2007 *Waste materials and by-products in concrete* (Berlin: Springer Science & Business Media)

[22] Yuan S W 2013 *Energy, Resources and Environment: Papers Presented at the First US-China Conference on Energy* (Elsevier)

[23] Feraldi R, Cashman S, Huff M and Raahauge L 2013 Comparative LCA of treatment options for US scrap tires: material recycling and tire-derived fuel combustion, *The International Journal of Life Cycle Assessment* **18** 613-625

[24] Mucsi G, Szenczi Á and Nagy S 2018 Fiber reinforced geopolymer from synergetic utilizationof fly ash and waste tire, *Journal of Cleaner Production* **178** 429-440

[25] Thomas B S and Gupta R C J R 2016 A comprehensive review on the applicationsof waste tire rubber in cement concrete, *Renewable & Sustainable Energy Revire* **54** 1323-1333

[26] Siddika A, Mamun A A A, Alyousef R, Mugahed A Y H, Aslani F, Alabduljabar H 2019 Properties and utilizations of waste tire rubber in concrete: A review, *Construction and Building Materials* **224** 711-731

[27] Zhang G, Chen F, Zhang Y, Zhao L, Chen J, Cao L and Xu C 2021 Properties and utilization of waste tire pyrolysis oil: A mini review, *Fuel Processing Technology* **211** 106582

[29] Yang L 2010 *Exploration on Microstructure and Structural Theory of Crumb Rubber Concrete Tianjin* PhD Thesis (Tianjin: Tianjin) p 64

[30] Li L, Ruan S and Zeng L 2014 Mechanical properties and constitutive equations of concrete containing a low volume of tire rubber particles, *Construction and Building Materials* **70** 291-308

[31] Valente M and Sibai A 2019 Rubber/crete: Mechanical properties of scrap to reuse tire-derived rubber in concrete; A review, *J Appl Biomater Funct Mater.* **17** 2280800019835486

[32] Siringi G, Abolmaali A, and Aswath P B J T S W J 2015 Properties of concrete with tire derived aggregate partially replacing coarse aggregates, *The Scientific World Journal* **2015** 863706

[33] Aswin S and Mohanalakshmi V 2018 Effects of Ceramic Tile Powder on Properties of Concrete and Paver Block, *GRD Journals- Global Research and Development Journal for Engineering* **3** 84-87

[34] Mohammadhosseini H, NorHasanah A S L, Tahir M M, Alyousef R, Samadi M, Alabduljabbar H, Mohamed A M 2020 Effects of Waste Ceramic as Cement and Fine Aggregate on Durability Performance of Sustainable Mortar, *Arabian Journal for Science and Engineering* **45** 3623-3634
[35] Faridmehr I, Fahim Huseien G and Hajmohammadian Baghban M 2020 Evaluation of Mechanical and Environmental Properties of Engineered Alkali-Activated Green Mortar,*Materials* **13** 4098
[36] Saribiyik M, Piskin A and Saribiyik A 2013 The effects of waste glass powder usage on polymer concrete properties, *Construction and Building Materials* **47** 840-844
[37] Islam G M S, Rahman M H., and Kazi N 2017 Waste glass powder as partial replacement of cement for sustainable concrete practice, *International Journal of Sustainable Built Environment* **6** 37-44
[38] Harbec D, Zidol A, Tagnit- Hamou A and Gitzhofer F 2017 Mechanical and durability properties of high performance glass fume concrete and mortars, *Construction and Building Materials* **134** 142-156
[39] Shih P H, Chang J E and Hiang L C C 2003 Replacement of raw mix in cement production by municipal solid waste incineration ash, *Cement and Concrete Research* **33** 1831-1836
[40] Charles H K L, Alvin W M I, John P B and Gordon M 2010 Use of Incineration MSW Ash: A Review, *Sustainability,* **2** 1943-1968
[41] Shahidan S, Mohamad A M A, Kupusamy K, Salwa S M Z andNoorwirdawati S 2017 Utilizing Construction and Demolition (C&D) Waste as Recycled Aggregates (RA) in Concrete, *Procedia Engineering* **174** 1028-1035
[42] Tang Q, Zhiming M, Huixia W, Wang W 2020 The utilization of eco-friendly recycled powder from concrete and brick waste in new concrete: A critical review, *Cement and Concrete Composites* **114** 103807
[43] Hansen S and Sadeghian P 2020 Recycled gypsum powder from waste drywalls combined with fly ash for partial cement replacement in concrete, *Journal of Cleaner Production* **274** 122785
[44] Toghroli A, Shariati M, Sajedi F, Ibrahim Z, Koting S, Mohamad E T and Khorami M 2018 A review on pavement porous concrete using recycled waste materials, *Smart Struct. Syst*, **22** 433-440
[45] Pavlů T 2018 The Utilization of Recycled Materials for Concrete and Cement Production- A Review, *IOP Conference Series: Materials Science and Engineering* **442** 012014
[46] Chowdhury S, Mishra M and Suganya O 2015 The incorporation of wood waste ash as a partial cement replacement material for making structural grade concrete: An overview, *Ain Shams Engineering Journal* **6** 429-437
[47] Arunkumar K, Muthukannan M, Suresh K A, Chithambar G A 2021 Mitigation of waste rubber tire and waste wood ash by the productionof rubberized low calcium waste wood ash based geopolymer concrete and influence of waste rubber fibre in setting properties and mechanical behavior, *Environ Res.* **194** 110661

Performance of Eco Engineered Cementitious Composites Containing Supplementary Cementitious Materials as a Binder and Recycled Concrete Fines as Fine Aggregate

M R Md Zain[1], C L Oh[1] and L S Wee[2]

[1]School of Civil Engineering, College of Engineering, Universiti Teknologi MARA, 40450 Shah Alam, Selangor, Malaysia
[2]School of Civil Engineering, College of Engineering, Universiti Teknologi MARA, Johor Pasir Gudang Campus, 81750 Masai, Johor, Malaysia
Corresponding author: raizam@uitm.edu.my

Abstract. Engineered cementitious composites (ECC) mixtures demand a large cement content, which is detrimental to their sustainable development because mass cement production is hazardous to the environment and human health. Thus, this paper investigates the mechanical performance of eco engineered cementitious composites (ECC) under axial compressive loading and direct tensile strength tests. The eco ECC used in this investigation was comprised of cement, superplasticizer, fly ash (FA) or ground granulated blast furnace slag (GGBS), polypropylene (PP) fibre, water and recycled concrete fines (RCF). Two (2) eco ECC mixture series were designed and prepared. GGBS70 (70 percent GGBS + 30 percent cement), FA70 (70 percent Fly Ash + 30 percent cement), GGBS80 (80 percent GGBS + 20 percent cement), and FA80 (80 percent Fly Ash + 20 percent cement) are the four Cement-GGBS and Cement-Fly Ash combinations examined in this study. Also every combination had two different RCF percentages, R0.2 (0.2 percent RCF) and R0.4 (0.4 percent RCF). The main objective of this research is to determine the optimum mix design for eco ECC that contains supplementary Cementitious Materials (SCMs) such as GGBS or FA. Additionally, recycled concrete fines (RCF) were used as a substitute for sand. The influence of different cement replacement materials and RCF content on compressive and tensile strength was experimentally investigated. The inclusion of GGBS as a partial replacement of cement in the eco concrete mixture results in greater compressive strength than Fly Ash (FA). The test results revealed that increasing the RCF content in the ECC mixture resulted in higher compressive and tensile strength. When the sand to binder ratio was adjusted between 0.2 and 0.4, the compressive and tensile strength of the ECC mixture increased.

1. Introduction

Construction began with the intention of reaching human needs for shelter, as well as protecting themselves from inclement weather and wildlife. Humans begin to refine their construction practises over time, shifting from natural sources to synthetic materials, which boosts the durability of building structures. As a result of the widespread use of synthetic materials, particularly concrete, as the primary element, the demand for cement industries has grown. Ordinary Portland cement (OPC) has been used in engineering for hundreds of years as an essential component of concrete. Cement manufacturing necessitates extensive use of fossil and biomass fuels [1] as well as a vast amount of raw materials and energy [2]. It is reasonable to say that raw material depletion has worsened in tandem with rising cement demand. Besides that, the industry is reported to contribute approximately 5% to 8% of global CO2 and greenhouse gas emissions due to the cement manufacturing process [3]. It was revealed to have resulted in poor air quality and noise pollution [4], including such particulate matter releases from raw mills, kiln systems, clinker coolers, and cement mills (air pollution) and noise pollution during the raw material preparation process, storage of materials, and the heavy machinery used [1,5,6]. According to research findings, cement manufacturing emits approximately 0.8 tonnes of carbon dioxide into the environment for every tonne of cement produced [4]. If particulate matter is suspended in the atmosphere for an extended period of time, it can form haze, which can endanger human health, such as severe respiratory and cardiovascular problems, which can lead to death.

Concrete is a composite material with multiple components (cement, water coarse aggregates and fine aggregates). Because it has only 0.1 percent strain capacity, conventional concrete is reported to be particularly brittle and rigid [7]. Concrete fractures are generally triggered by the material's inability to withstand tensile force [8], which reduces the toughness and durability of concrete. Steel reinforcement is thus required to withstand high tensile forces. Reinforced concrete structures, on the other hand, are facing serious problems as a result of extreme environmental and mechanical conditions. As a result, concrete must be modified to ensure long-life structures while keeping future generations in mind.

Throughout the modern era, the concrete production process has been improvised to meet the evolving industrial and construction needs. Multiple studies were conducted in order to develop a solution for creating a more sustainable and quality way of building, so that the performance and durability of the buildings would improve. The incorporation of engineered cementitious composite (ECC) into structures as a substitute for conventional concrete has become one of the studies that were conducted to resolve this concern. The characteristics of ECC are very different from those of conventional concrete. ECC is very ductile, as opposed to conventional concrete, which is brittle. ECC ingredients are specifically developed to withstand various cracking and regulate crack width [9]. ECC has been shown to be more flexible than conventional concrete because it has more than a 3 percent of strain capacity [7]. This is due to the ECC's fibres, which aid in increasing the impact resistance of concrete through bridging action [8]. Nevertheless, ECC necessitates a high cement usage to compensate for the lack of strength caused by the absence of coarse aggregates. As a result, there is an urgent need to consider using green or eco-friendly construction materials.

Growing issues of sustainability between many of the construction material resources, as well as public awareness of the problem, pave the way for green or eco construction as a new, innovative method to replace the conventional method. Among the alternative solutions are the use of Supplementary Cementitious Materials (SCMs), recycled concrete aggregates, and many other industrial wastes. SCMs are frequently used as a supplementary cementitious material in attempt to lessen the cement content of concrete. SCMs also improves the workability of hardened concrete, as well as its strength and durability. Green components such as Ground Granulated Blast-furnace Slag (GGBS), fly ash (FA), silica fume, eggshell and recycled concrete fines (RCF) have been widely used in ECC studies to find a way towards sustainable development [10][11][12].

The blend of SCM and RCF in a mixture is regarded as eco Engineered Cementitious Composites (ECC). Due to the fineness of particles, ground granulated blast-furnace slag (GGBS) mix has greater mobility characteristics. Earlier studies have shown that increasing the GGBS content in the concrete mix improves workability and reduces water absorption [13], thereby increasing the durability of the concrete. The performance of concrete in terms of strength by varying the percentage of fine aggregates substitution and cement replacement in their mixture have been performed by numerous numbers of researchers. Fly ash is often used as a supplementary cementitious material, and mechanical properties has been examined utilising diverse proportions of fly ash varying from 0% to 60% [14]. Dhanya et al. [15] investigated the strength characteristics of concrete employing steel fibre and GGBS as partial substitutes for cement. The findings of the study demonstrate that partial substitution by both GGBS and steel fibre enhances the performance of the concrete. Besides that, researchers [16,17] signify that incorporating an appropriate amount of fly ash, can strengthen the performance of concrete comprising recycled concrete aggregate (RCA).

ECC is made up of cement, water, fly ash, sand, fibre, and water reducer. Whereas, conventional concrete is only composed of water, cement, coarse and fine aggregates, and it cannot withstand various cracking owing to the lack of fiber-matrix interaction. Polypropylene (PP) fibres and polyvinyl alcohol (PVA) fibres are two examples of fibres which could be used in ECC. Because PP fibre is widely available in Malaysia, it was used in this study's ECC mix for comparisons in terms of compressive strength, tensile strength, and failure mechanisms, alongside GGBS or fly ash as partial cement replacement and RCF as sand replacement in ECC mix.

2. Methods

The aim of this study is to investigate the mechanical behaviour of ECC in terms of compressive and tensile strengths using GGBS and Fly Ash, respectively. The Universal Testing Machine (UTM) was used to conduct laboratory testing on ECC specimens containing SCM as a binder and RCF as fine aggregates. Laboratory assessment was performed to evaluate the ultimate tensile strength and compressive strength of eco ECC subjected to direct tensile and compression tests, as well as to observe the failure generation mechanism of the eco ECC. The following sub-topics provide detailed explanations of the materials, mix design proportions, specimen preparation, and test methods.

2.1. Materials

The raw materials used in the concrete were Ordinary Portland Cement (OPC), recycled concrete fines (RCF), water, FA or GGBS, PP fibres, and SP as shown in figure 1. The RCF employed in ECC was taken from the demolished building which act as fine aggregates replacement and it was crushed and sieved to a size below 600 μm. The PP fibre used had a diameter of 19.5 m and a length of 6 mm. In all ECC mixtures where the binder is a cement-GGBS mixture, the water binder ratio was set at 0.27. Moreover, the amount of SP was kept under control at around 10 kg/m^3 to ensure the workability of the ECC mortar. Table 1 show the elemental compositions of OPC, GGBS and FA.

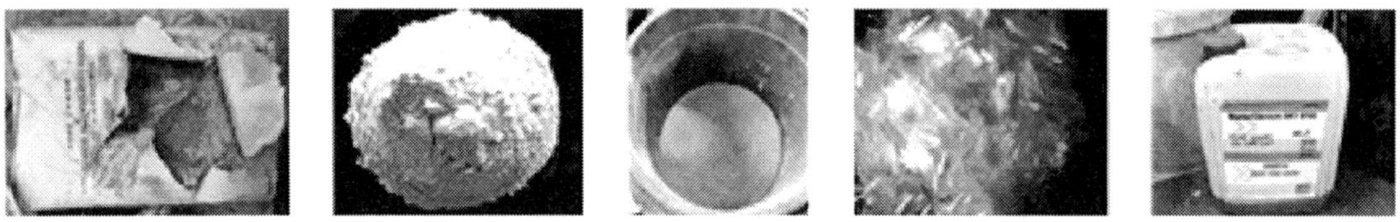

(a) (b) (c) (d) (e)

Figure 1. Materials used in the production of ECC (a) OPC, (b) GGBS, (c) RCF, (d) PP fibre, (e) SP

Table 1. Chemical compositions of cement, GGBS and FA in units of weight percent.

Chemical compositions	MgO	Al_2O_3	SiO_2	P_2O_5	SO_3	K_2O	CaO	TiO_2	MnO	Fe_2O_3
OPC	0.96	2.13	12.80	<LOD	2.69	0.53	77.53	0.18	0.06	3.12
GGBS	4.03	10.15	35.67	<LOD	3.80	0.29	43.68	0.58	0.19	1.61
FA	1.29	16.11	72.76	0.23	0.37	1.06	1.41	0.99	0.09	5.69

* LOD embodies Low on density

Figure 2 depicts the particle size distribution or grading curves of sand and RCF. As observed in figure 2, sand has larger particle sizes than RCF. Based on these results, it can be indicated that RCF, rather than sand, should be used in ECC matrix since RCF has a smaller particle size distribution and aids in the development of ductility in ECC samples.

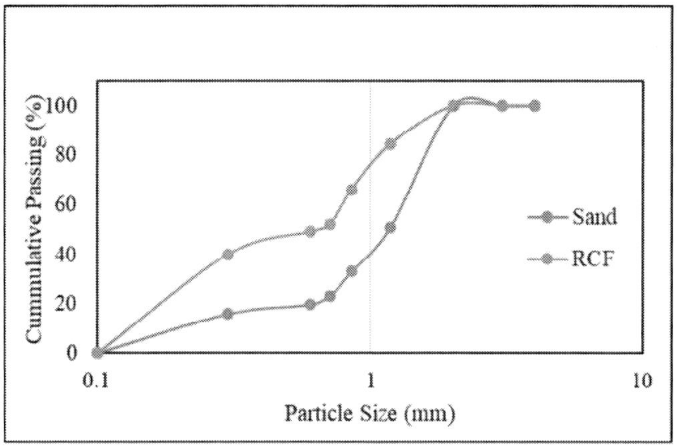

Figure 2. Particle size distribution of sand and RCF.

2.2. Mix design proportions and specimen preparation

For the experimental investigations, three (3) series of concrete mixtures were designed and prepared. Among the three series of concrete mixtures, series 1 and 2 were created to investigate the effects of different percentage variations of GGBS and RCF. In this study, four Cement-GGBS and Cement-Fly Ash combinations are investigated. Every combination had two different RCF percentages, R0.2 (0.2 percent RCF) and R0.4 (0.4 percent RCF) as depicted in table 3. For instance, mixture ID GGBS70R0.2 indicates that the mixture contains 30 percent cement, 70 percent GGBS and 0.2 percent RCF. The other remaining mixture (series 3) acts as control mixture, in which no SCMs or RCF substitution (water/cement ratio is equal to 0.46). The mix proportion of normal concrete and ECC mixtures are presented in table 2 and table 3.

Table 2. ECC mix proportion.

Specimen designation	Unit weight (kg/m^3)						
	OPC	Water	FA	GGBS	RCF	SP	PP Fiber
GGBS80R0.2	286	358	-	1143	286	10	23
FA80R0.2	286	358	1143	-	286		
GGBS80R0.4	257	350	-	1028	518		
FA80R0.4	257	350	1028	-	518		
GGBS70R0.2	429	386	-	1001	286		
FA70R0.2	429	386	1001	-	286		
GGBS70R0.4	385	350	-	900	518		
FA70R0.4	385	350	900	-	518		

Table 3. Normal concrete mixture proportion (kg/m^3)

OPC	Water	Coarse Agg.	Sand
415	195	1135	605

Normal concrete mixtures were made by mixing cement, water, fine aggregates, and coarse aggregates in a concrete mixer and left them to mix uniformly. The mixture was then poured with water and mixed for 10 minutes. When preparing ECC specimens, similar procedures were followed. Cement, GGBS or FA, and RCF were mixed uniformly in a concrete mixer for 3 to 5 minutes. In order to prepare an ECC mix containing PP fibre, combine the superplasticizer, water, and PP fibre in a clean bucket. After thoroughly mixing the PP fibre and SP in water, they were slowly and evenly poured into the concrete mixer, where all binders were thoroughly mixed. Following the completion of the mixing process, fresh concrete was moved and carefully placed at cylindrical (for compression) and dog bone (for tensile) formworks, as can be seen in figures 3(a) and 3(b). All of the fresh concrete in the formworks was left for 24 hours to harden before going through the curing process for 7 days and 28 days.

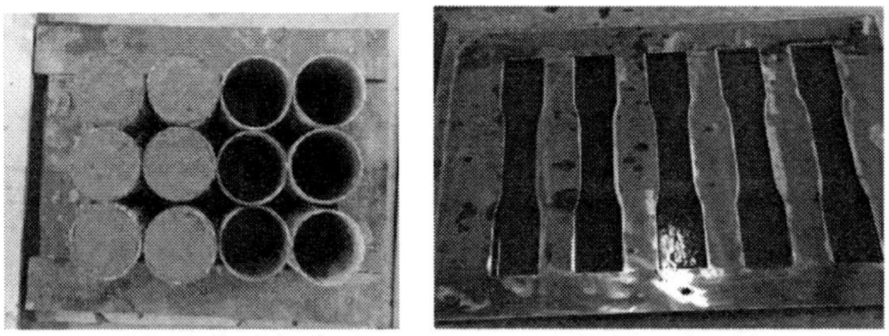

(a) Cylindrical (b) Dog bone

Figure 3. Cylindrical and dog bone formworks.

2.3. Test methods

Using a 3000 kN compression machine, five (5) cylindrical ECC samples of 50 mm in diameter and 100 mm in height from each mix, along with five (5) cylindrical normal concretes, were tested up to failure at its 7 days and 28 days compressive strength. During the test, ECC specimen was loaded at 0.5 mm/minute (constant rate). Then, eco ECC plate samples in the shape of a dogbone with a cross sectional dimension of 330 x 40 x 13 mm were prepared, characterized, and tested subjected to direct tensile test. Figure 4 depicts the tensile test performed on a Shimadzu Machine.

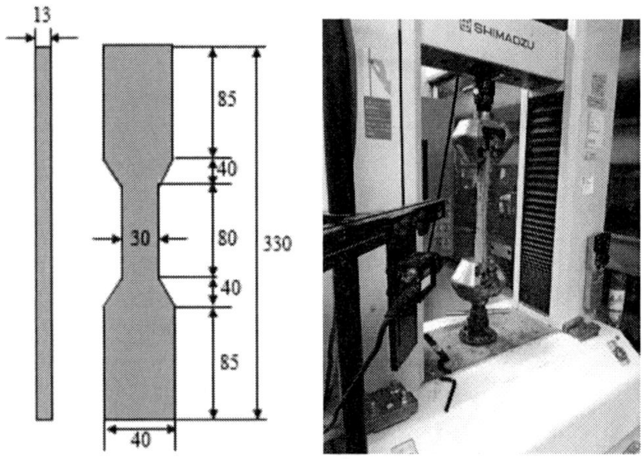

Figure 4. Dog bone specimen and experimental setup.

3. Results and discussion

The experimental results, such as compressive strength, tensile strength, and failure mechanism, were elaborated and discussed in detail in the subsequent sections.

3.1. Axially compressive strength of concrete specimens

The average results of compressive strengths for all specimens subjected to compression test at 28 days were expressed as demonstrated in table 4.

Table 4. Compressive strength of concrete specimens under axial compressive loading.

Series	Specimen designation	Ave. Compressive Strength 28 days (MPa)	Enhancement Factor (%)
1	GGBS80R0.4	30.70	0.88
	GGBS80R0.2	28.25	0.81
	FA80R0.4	12.59	0.36
	FA80R0.2	6.72	0.19
2	GGBS70R0.4	40.84	1.16
	GGBS70R0.2	35.94	1.03
	FA70R0.4	22.45	0.64
	FA70R0.2	12.86	0.37
3	Normal concrete	35.00	-

As can be seen in table 4, the mixture ID delineates GGBS70R0.0.4 had the highest compressive strength at 28 days of curing age with 40.84 MPa, followed by GGBS70R0.2 with 35.94 MPa. The increment ratio of the ECC specimen from the normal concrete specimen is implied by the enhancement factor, as shown in table 4. Use of 70% GGBS replacement of cement with PP fibre in ECC specimens enhances the compressive strength of ECC specimens by 1.03 and 1.16, respectively, when compared to conventional concrete. The increment could be attributed to the use of eco materials, GGBS, and the addition of fibres such as PP fibre, which have been shown to improve the microstructure of the concrete. The addition of GGBS to ECC concrete improves its mechanical properties. Because of the higher workability of GGBS, the ECC mixture bonds well, improving compressive strength. The action of the fibres contained also made a significant contribution to the ECC mixtures' higher compressive strength. But since normal concrete cannot withstand tensile force, fibres are added primarily to assist in carrying tension within the concrete matrix. Table 4 also demonstrates that certain ECC samples containing 70% or 80% fly ash had a significant reduction in compressive strength when compared to normal concrete specimens. These outcomes may be possible as a result of the addition of RCF content in ECC samples, which may contribute in concrete strength degradation attributed to the prevalence of impurities in RCF [17].

3.2. Tensile strength of ECC concrete specimens

The average crack width, ultimate tensile strength and ultimate tensile strain of ECC tested based on the direct tensile test are shown in table 5.

Table 5. Tensile Performance of ECC specimens.

Specimen designation	Ave. Crack Width (mm)			Ultimate Tensile Strength (MPa)	Ultimate Tensile Strain (%)
	Min.	Max.	Ave.		
Normal Concrete	-	-	-	3.33	0.88
GGBS80R0.4	0.10	0.44	0.27	6.07	1.27
GGBS80R0.2	0.16	0.40	0.28	3.13	0.49
FA80R0.4	0.20	0.48	0.34	2.19	0.30
FA80R0.2	0.12	0.40	0.26	2.79	0.35
GGBS70R0.4	0.14	0.50	0.32	7.99	0.10
GGBS70R0.2	0.16	0.40	0.28	5.26	0.45
FA70R0.4	0.16	0.36	0.26	3.06	0.52
FA70R0.2	0.14	0.60	0.37	2.47	0.48

The first cracking strength is typically defined as its first response to the decline at which the response of tensile stress deviates from linearity. In general, all ECC specimens have higher first cracking strength than normal concrete. With a tensile strength of 7.99 MPa, GGBS70R0.4 has the highest tensile strength. It demonstrates that using GGBS as a cement replacement improved the tensile behaviour. Meanwhile, ECC specimens denotes as FA80R0.4 has the lowest tensile strength of 2.19 MPa. The mixture ID GGBS80R0.4 has the highest tensile strain value of 1.27 percent, and the mixture ID denotes as GGBS70R0.4 has the lowest tensile strain value of 0.10 percent.

Figure 5 depicts a graph of tensile strength versus displacement for a mixture containing 70% cement replacement and 0.2 percent RCF when compared to normal concrete. It demonstrates that the GGBS-cement combination (GGBS70R0.2) has a higher tensile strength value of 5.26 MPa when compared to the Fly Ash-cement combination (FA70R0.2) and normal concrete, which have lower values of 2.47 MPa and 3.33 MPa, respectively. This demonstrates that the tensile strength of the mixture containing 0.2 percent RCF with 70 percent GGBS-cement combination (GGBS70R0.2) is greater than the tensile strength of the mixture containing 70 percent Fly Ash-cement combination (FA70R0.2) and normal concrete. Furthermore, when the mixture contains 80 percent cement replacement with 0.2 percent RCF is compared to normal concrete, the normal concrete has a higher value of tensile strength with 3.33 MPa compared to the GGBS-cement combination (GGBS80R0.2) and Fly Ash-cement combination (FA80R0.2), which have lower values of 3.13 MPa and 2.79 MPa, respectively. In terms of tensile strength, this demonstrates that the normal concrete mixture outperforms the mixture containing 0.2 percent RCF and 80 percent cement replacement.

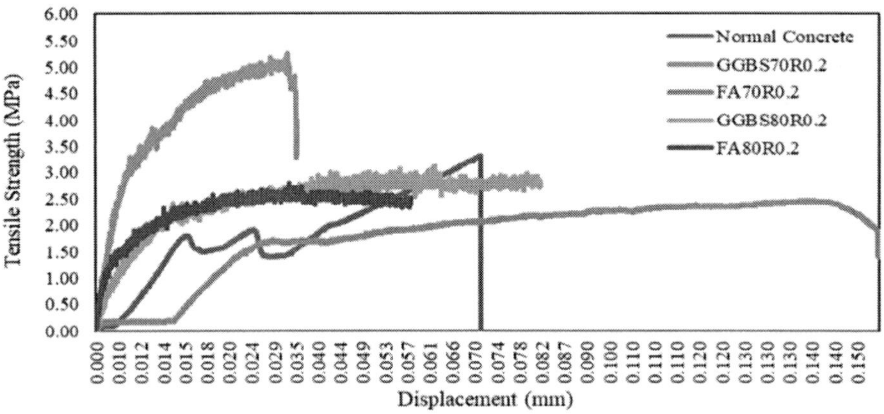

Figure 5. Tensile strength-displacement curve between 70 % and 80% of GGBS and Fly Ash replacement for 0.2 % RCF.

Figure 6 represents the tensile strength-displacement curve for a mixture containing 70% cement replacement and 0.4 percent RCF when compared to normal concrete. It indicates that the GGBS-cement combination (GGBS70R0.4) seems to have a higher tensile strength value of 7.99 MPa when compared to the Fly Ash-cement combination (FA70R0.4) and normal concrete, which have lower values of 3.06 MPa and 3.33 MPa, respectively. This shows that the tensile strength of the mixture containing 0.4 percent RCF with 70 percent GGBS-cement combination (GGBS70R0.4) is significantly greater than the tensile strength of the mixture containing 70 percent Fly Ash-cement combination (FA70R0.4) and normal concrete. In addition, as can be seen in figure 6, the tensile strength of the mixture containing 0.4 percent RCF with 80 percent GGBS-cement combination (GGBS80R0.4) is greater than the tensile strength of the mixture containing 0.4 percent RCF with 80 percent Fly Ash-cement combination (FA80R0.4) and normal concrete. In this study, the tensile performance of eco ECC (GGBS and Fly Ash as partial cement replacement) were measured through laboratory testing to assess its strength and failure behavior. Tensile strength test yields consistent results in the eco ECC containing Fly Ash, with the value of its strength remaining in the small range of 2.19 MPa to 3.09 MPa. While the green ECC containing GGBS exhibits inconsistency in the results, ranging from 3.13 MPa to 7.99 MPa. This can be explained by the physical properties of

GGBS and Fly Ash. The GGBS is irregular in shape, whereas the Fly Ash is regular. The irregular shape has less workability but a higher bonding strength, whereas the regular shape has better workability but not strength due to poor interlocking behavior and a weak bond strength.

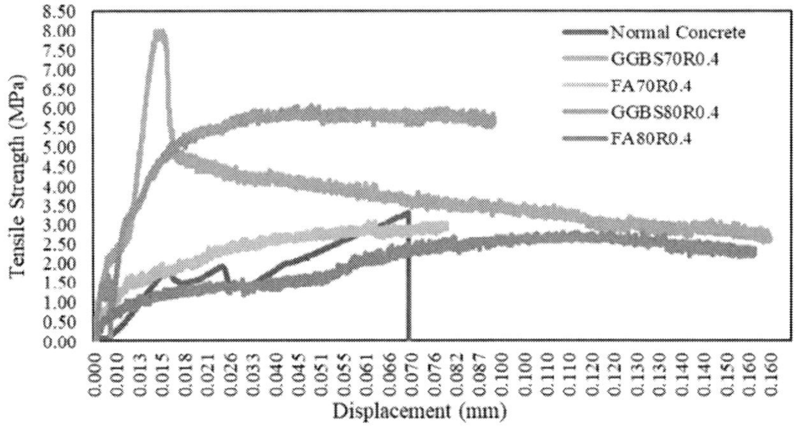

Figure 6. Tensile strength-displacement curve between 70 % and 80% of GGBS and Fly Ash replacement for 0.4 % RCF.

3.3. Failure behavior of ECC concrete specimens

The dog bone shape specimen is used to regulate the location of the cracking in order to ensure that the cracking occurs in the middle section of the specimen. The shape is in the form of a dog bone because of the stress concentration at the narrow neck, which yields a more accurate result. Pure tension occurs at the narrow neck area [18]. When the sample is loaded with a tensile force, the shoulders are wider than the gauge section, causing a stress concentration in the middle. This stress concentration increases the likelihood of the sample rupturing away from the ends. This indicates that cracking during the tensile strength test was mostly observed in the middle of the dog bone specimen, as shown in figure 7. Nonetheless, cracks in the specimen can also be seen about one-third from the middle section of the specimen.

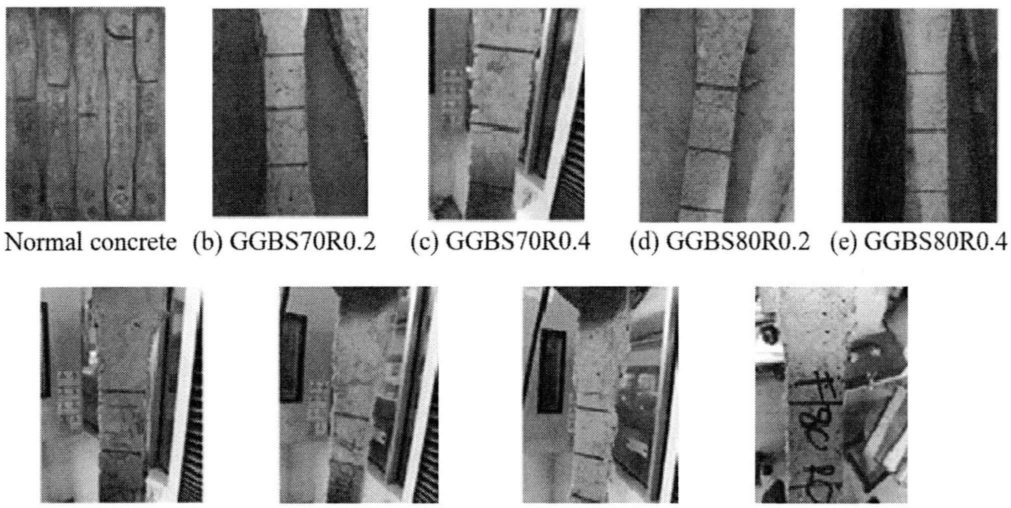

(a) Normal concrete (b) GGBS70R0.2 (c) GGBS70R0.4 (d) GGBS80R0.2 (e) GGBS80R0.4

(f) FA70R0.2 (g) FA70R0.4 (h) FA80R0.2 (i) FA80R0.4

Figure 7. Mode of failure of eco ECC specimens under direct tensile strength test.

4. Conclusions

The following conclusions can be drawn based on the findings of experimental tests on ECC specimens incorporating SCMs (GGBS or FA) as a binder and RCF as fine aggregates under compressive and direct tensile strength tests.

i) The inclusion of 70% GGBS to ECC concrete specimens improves its compressive strength performance compared to normal concrete. The greater workability of GGBS allows ECC mixture to bond well thus boosting compressive strength. The action of the fibres contained, which support in carrying the tension within the concrete matrix, also contributed to the higher compressive strength attained by the ECC specimens.

ii) Meanwhile, compressive tests on eco ECC specimens containing 70% and 80% fly ash reveal a significant reduction in strength compared to the control concrete specimens. This could be due to the inclusion of RCF content in ECC samples, which might also contribute to concrete strength degradation caused by the presence of impurities in RCF.

iii) Tensile strength tests signify that the eco ECC containing Fly Ash produces consistent results, with the value of its strength remaining in the low range, as opposed to the green ECC containing GGBS, which yields inconsistent results. The GGBS has an irregular shape (less workability but a higher bonding strength), whereas the Fly Ash has a regular shape (has more workability but less strength due to poor interlocking behaviour and a weak bond strength).

iv) The cracking behavioural pattern has mostly been detected in the middle of the dog bone specimen during the tensile strength test, as well as some crack can also be seen about one-third from the middle section of the specimen.

5. References

[1] Stajanca M and Estokova A 2012 *Environmental Impacts of Cement Production* Lviv Polytechnic National University of Košice Institutional Repository **138** 296-302

[2] Hayati U, Huseyin Y A and Osman Gencel 2016 Investigation on characteristics of blended cements containing pumice, *Construction and Building Materials* **118** 11-19

[3] Claisse P A 2015 *Civil Engineering Materials* (United Kingdom: Butterworth-Heinemann) pp 163-184

[4] Ziming H, Aiqin S, Zhenghua L, Yue L, Hansong W and Wenzhen W 2020 Effect of wollastonite microfibers as cement replacement on the properties of cementitious composites: A review, *Construction and Building Materials* **261** 119920

[5] Nabilla M, Khairunisa M, Rahimah E, Andri K and Mohd H H 2021 Environmental impact of cement production and Solutions: A review, *Materials Today: Proceedings* 1-6

[6] Gao W 2003 *Noise control engineering* (Jiangsu Shen: Wu Hang University of Science and Technology Press)

[7] Uttamraj S, Ashwanth K and Rafeeq M 2016 A Comparative Study on Conventional Concrete and Engineered Cementitious Composites (ECC-PVA) Review, *IOSR Journal of Mechanical and Civil Engineering* 19-25

[8] Sukumar A and John E 2014 Fiber Addition and Its Effect on Concrete Strength, *International Journal of Innovative Research on Advanced Engineering* **1** 144-149

[9] Li V C 2008 Engineered cementitious composite (ECC): Material, structural, and durability performance, *Concrete Construction Engineering Handbook.* Second Edition, 1001–1048

[10] Li V C, Michael L, Shuxin W, Martin W and Gregory K 2004 Development of Green ECC for Sustainable Infrastructure Systems *Proc. Int'l Workshop on Sustainable Development and Concrete Technology* (Beijing, China) 181-192

[11] Md Zain M R, Oh C L, and Lee S W 2021 Investigations on rheological and mechanical properties of self-compacting concrete (SCC) containing 0.6 μm eggshell as partial replacement of cement, *Construction and Building Materials* **303** 124539

[12] Lee S W, Oh C L and Md Zain M R 2019 Mechanical properties of engineered cementitious composites using local ingredients, *Journal of Mechanical Engineering* **16** 145-157
[13] Khan A N, Memon, D A, Rizvi S H, Bhanbhro Q and Bheel N 2018 Fresh and Hardened Properties of Ground Granulated Blast Furnace Slag Made Concrete, *International Journal of Modern Research in Engineering and Management* **1** 01-07
[14] Alvin H, Vikas S and Arpan H 2014 Effect of fly ash on compressive strength of Portland Pozzolona cement concrete, *Journal of Academia and Industrial Research* **2** 479-479
[15] Dhanya R, Arasan G V and Ganapathy R 2017 Study on strength properties of concrete using GGBS and steel fiber as partial replacement of cement, *Journal of Industrial Pollution Control* 1255-1259
[16] Kurda R, de Brito J and Silvestre J D 2017 Combined influence of recycled concrete aggregates and high contents of fly ash on concrete properties, *Construction and Building Materials* **157** 554–572
[17] Verian K P, Ashraf W and Cao Y 2018 Properties of recycled concrete aggregate and their influence in new concrete production, *Resources, Conservation and Recycling* **133** 30–49
[18] Ke-Quan Y, Zhou-Dao L, Jian-Gua D and Surendra P S 2020 Direct tensile properties and stress-strain model of UHP-ECC, *J. Mater. Civ. Eng.* **32** 04019334

Acknowledgments
The authors are thankful the Ministry of Higher Education for funding this research via the Fundamental Research Scheme (FRGS) (600-IRMI/FRGS 5/3 (378/2019). The authors would like to express their gratitude to Muhammad Syazwan Obaidah Md Obaidah, who was directly involved in the experimental work. Also, thanks to all technical staff at the Concrete Laboratory, School of Civil Engineering, College of Engineering, Universiti Teknologi MARA (UiTM), Shah Alam for their unwavering support during these studies.

IOP Conf. Series: Materials Science and Engineering 1200 (2021) 012005 doi:10.1088/1757-899X/1200/1/012005

Study on Mechanical Properties of Foamed Concrete Incorporating Palm Oil Fuel Ash and Mussel Shell as Partial Cement Replacement

A F Rahman[1], W I Goh[1], N H Othman[1] and M S Kamaruddin[1]

[1] Department of Structural and Materials Engineering, Faculty of Civil Engineering and Built Environment
Universiti Tun Hussein Onn Malaysia, 86400 Batu Pahat, Johor, Malaysia
Corresponding author: wigoh@uthm.edu.my

Abstract. Concept of sustainable construction has gradually become one of the concern issues in our construction industry in recent years. Concrete which acts as an important construction material has contributed to excessive consumption of natural resources. Simultaneously, tonnes of waste materials were produced from agricultural activity in form of palm oil fuel ash (POFA) while mussel shell from marine hatchery. Utilization of agricultural waste as cement replacement is an option to reduce the environmental impact from the construction industry. In this study, these waste materials were used as partial cement replacement to produce foamed concrete in wet density of 1800 kg/m^3. The main purpose of this research is to study the workability and mechanical properties of foamed concrete which contain uniform 20% of POFA combined with 5% to 10% of mussel shell powder (MSP) and mussel shell ash (MSA) respectively. The cube specimens were cast in dimension 100 mm x 100 mm x100 mm to test the compressive strength at 7th and 28th. The cylinder specimens were cast in 100 mm diameter x 200 mm diameter for split tensile test to determine the tensile strength and compression test to determine modulus of elasticity at 28th day. The result showed workability of foamed concrete decreased as more cement was replaced by POFA combined with MSP and MSA. Foamed concrete mixture with 20% POFA and 5% MSP was selected as optimum percentage of cement replacement due to reduction less than 5% compromised performance in compressive strength at 16.52MPa while tensile strength at 1.83MPa.

1. Introduction

Foamed concrete is a type of lightweight concrete with different physical and mechanical properties compares to normal concrete. Foamed concrete can be designed ranging from 300 – 1850 kg/m^3 as compared to normal concrete at 2400 kg/m^3 [1]. The foamed concrete which also known as cellular concrete without coarse aggregates was created by void in form of small bubble that utilize air molecules [2]. Advantages of foamed concrete in construction works due to its lightweight density, fire resistance, thermal conductivity and acoustical properties [3]. Due to the presence of air bubbles and micro-structural cells in the concrete which make it porous in nature, the foamed concrete had shown unexpectedly good thermal insulation and sound absorbance [4]. The amount of waste produced by humans are increased significantly and many researchers integrated the waste materials into construction materials technology. Improvement in term of raw materials, manufacturing process, packaging and transportation are necessary to minimize environmental hazards of cement products that may help optimize reserved natural resources.

Meanwhile, palm oil fuel ash (POFA) and seashell are among those waste materials that can be used as partial cement replacement in the concrete production. These waste materials have caused land and air pollution and will become a threat for human life. In this project, combination of POFA and mussel shell were used as partial cement replacement in foamed concrete to determine the workability and mechanical properties of foamed concrete. POFA contains silica oxide content while mussel shell has high calcium oxide (CaO) content to be utilized as cement replacement materials. The silica oxide content in POFA will react with calcium hydroxide ($Ca(OH)_2$) from the hydration process which was deteriorated to concrete and the pozzolanic reactions produce more calcium silicate hydrate (C-S-H) which was a gel compound which reduce the amount of calcium hydroxide. This property helped to produce stronger and denser concrete as well as enhanced the durability of concrete [5].

2. Materials

2.1. Fine Aggregate

Natural river sand accordance to the standards BS EN 12620 was used as fine aggregate. Fine aggregate was sieved passing 5mm mesh and clear from contaminant. The collected sand was stored damp area without free standing water to ensure materials kept in saturated surface dry condition.

2.2. Palm oil fuel ash (POFA)

Palm oil fuel ash was produced by burning of palm oil shell and husk in palm oil mill at 800 to 1000°C in order to use the steam produced to generate electricity in biomass thermal power plants [6]. There were two main physical properties of POFA which are specific gravity and grain size distribution. These two parameters were related with the fineness of POFA which affected the performance of concrete. Specific gravity of POFA was varying which depends on few factors such as treatment process and burning temperature, but according to many researchers, the specific gravity of POFA never exceeded 3.0.

Study conducted by Sata et al. [7] highlight that when the POFA particles size was finer, the specific gravity of POFA increased up to 2.78 due the reduction of porosity. In general, POFA had the median particle size range from 30.80 to 62.5μm with 34.80 to 70.00% by weight retained on a 45μm sieve [8]. The fineness of POFA can affect the pozzolanic nature of material and hydration rate. The pozzolanic reactivity increased due to the increase in fineness of POFA after the grinding process [9]. table 1 shown the physical properties of POFA in Malaysia.

Table 1. Physical properties of POFA in Malaysia [10,11].

Physical Properties	Central	North	South
Specific gravity	2.48	2.42-2.56	2.22-2.42
% Passing through 45 μm sieve (%)	95.0	N.A	90.00-95.02

POFA has been used as supplementary cementitious materials (SCM) by previous researchers [12-16] to determine POFA can be used to replace cement in concrete without affecting the strength of concrete produced. From the study of Awal et al [17], the researchers had observed that the compressive strength of concrete that contains 20% POFA was almost same with OPC concrete. According to study of Alhadooh et. al [18] which had used POFA as pozzolan in concrete with 10% to 50% of POFA, the results shown that the concrete mixture with 20% of POFA replacement had achieve the highest compressive strength of 35 MPa at the 28-days of curing process. In terms of workability, the study of Foti & Cavallo [19] stated that increase in percentage of POFA in concrete mixture had lowered the workability and decreased the slump value. On the other hand, the research of Mo et. al [20] had concluded that the using of ultra-fine POFA particles will increased the workability of concrete. Ultra-fine POFA had low carbon content and concrete which containing unburned carbon had caused decrease in workability.

2.3. *Mussel shell*

Mussel shell (bivalve mollusk) was one type of seashell which have triangular shape and elongates with round edges. Commercial fisheries industry in Malaysia had recorded more than 50 thousand tonnes metric of seashell class of seafood so does the waste produced. Shellfish farming had becoming one of the world's fastest food sector and it had increased sharply going from 1 million tonnes in 1950 to about 14.6 million tonnes in 2010 [21]. The chemical composition of seashell contained higher than 90% of calcium carbonate ($CaCO_3$) by weight and this composition was similar to limestone powder which use to produce Portland-limestone cement (PLC).

Two physical properties, specific gravity and grain size distribution of any SCM are important as these have significant influence on the properties of concrete. According to the study of Skariah et. Al [22], the specific gravity of mussel shell powder was 2.86 with generic value does not exceed 3.0. Mussel shell primarily comprises $CaCO_3$ which means its specific gravity was closer with calcium carbonate. The mussel shell powder was obtained after crushing and grinding and the size of powder was varied as this depends on the grinding process. The average particle size of mussel shell powder was 29.87 µm and its Blaine surface area was determine as being 6186 cm^2/g. According to Momeen et al. [23], the split tensile strength and modulus of elasticity (MOE) reduce in the range of 10 to 30% when seashell powder was added into the mixture at high portion. Previous study of Olivia et al. [24] shown by replacing 2, 4, 6 and 8% of seashell powder, tensile strength of concrete composed of OPC and seashell powder was higher than the normal concrete.

In this study, mussel shell was cleaned from residual organic then pulverised into small particles. Processed mussel shell was diversified into two form which are mussel shell powder (MSP) that not subjected to heat treatment while mussel shell ash (MSA) was burned at 1000 °C for 1 hour in furnace.

2.4. *Foaming agent*

Foaming agent was prepared with water in the foam generator. Foaming agent Sika AER-50/50 used in the study with dilution ratio 1:20. The foam created microbubbles thus allowing air entrained subsequently reducing the density of concrete. Pre-foaming method as shown in figure 1 was produced stable foam before mixed with mortar to produce foamed concrete.

Figure 1. Pre- foaming method using foaming generator.

3. Methodology

3.1. *Mix proportions*

Design mixture is the calculation process to determine the right quantities of each raw materials to achieve desired properties. In this study, design for the foamed concrete materials calculation was referring to previous equation proposed by Kearsley et al. [25] as in equation 1.

$$p_m = x + x\left(\frac{w}{c}\right) + x\left(\frac{s}{c}\right) + RD_f \times V_f \quad (1)$$

Where;

p_m = target casting density, kg/m^3

x = cement content, kg

w/c = water/ cement ratio

s/c = sand / cement ratio

RD_f = Relative density of foaming agent

V_f = volume of foam, m^3

Meanwhile, the foamed concrete wet density was selected at 1800 kg/m^3 with controlled water / cement ratio at 0.50 respectively and fine aggregate content used twice quantity of cement. Table 2 summarized type of foamed concrete mix with respect to percentage of partial cement replacement.

Table 2. Types of concrete mixture.

Types of concrete mixes (%)	Specific dimension		
	Cube (Compressive) 100mm x 100mm x 100mm	Cylinder (Split tensile) 100mm dia. x 200mm height	Cylinder (Modulus of elasticity) 100mm dia. x 200mm height
Control	6 nos.	3 nos.	3 nos.
20 POFA	6 nos.	3 nos.	3 nos.
20 POFA + 5 MSP	6 nos.	3 nos.	3 nos.
20 POFA + 5 MSA	6 nos.	3 nos.	3 nos.
20 POFA + 10 MSP	6 nos.	3 nos.	3 nos.
20 POFA + 10 MSA	6 nos.	3 nos.	3 nos.

Foamed concrete mixture were fixed with 20% POFA and varies percentages of mussel shell type at 5% and 10% respectively. Specimens consists of cube with dimension 100mm x 100mm x 100mm for compressive strength test while cylinder with dimension 100 mm diameter x 200 mm height for split tensile strength test and modulus of elasticity.

3.2. Fabrication

Raw materials were dry mixed in a rotary mixer for 5 minutes to ensure consistency between each matrix and binder. Next, potable water was added into the dry materials and mixed for 1 minute. Specific volume of foam was measured and added into the wet mixture until uniform consistency with density of 1800 kg/m^3 achieved. Finally, the fresh foamed concrete was ready to be poured into the mould. J-ring test with aided of a slump cone was carried out to determine workability of foamed concrete in accordance to ASTM C1621. Foamed concrete specimens were left for air curing with curing periods of 7 days and 28 days.

4. Result

4.1. Specific gravity

The physical properties of POFA and MSA were examined in the specific gravity test using Le Chatelier flask method. The density of POFA, MSP and MSA compared with ordinary Portland cement. The testing conducted accordance to the standard of specification BS 4550-3.4:1978 Standard Test Method for Density of Cement. Specific gravity of material is tabulated in table 3.

Table 3. Specific gravity of OPC, POFA, MSP and MSA.

Material	Specific gravity
OPC	3.16
POFA	1.96
MSP	2.59
MSA	2.54

Based on the result, OPC had the highest specific gravity value at 3.16 which in normal range between 3.0 and 3.20. POFA had the lowest specific gravity value at 1.96. This was associated to the fine size of POFA particles. Previous studies by [26] highlighted the value of specific gravity directly influenced by grinding process to increase contact surface area. Next, specific gravity value of MSP which is 2.59 and MSA 2.54. Heat treatment process on mussel shell particle slightly lower the specific gravity values by eliminating organic materials subjected to high temperature.

4.2. Grain size distribution

Figure 2 shows distribution of particles analysis OPC, POFA, MSP and MSA. Based on the result, POFA was observed has highest fineness percentage with D60 value 7.89μm. MSP shows better size distribution with recorded lowest 10.55μm and largest D60 value at 36.19μm. Value of D10, D30 and D60 of MSA were largest which were 10.55μm, 25.35μm and 36.19μm respectively.

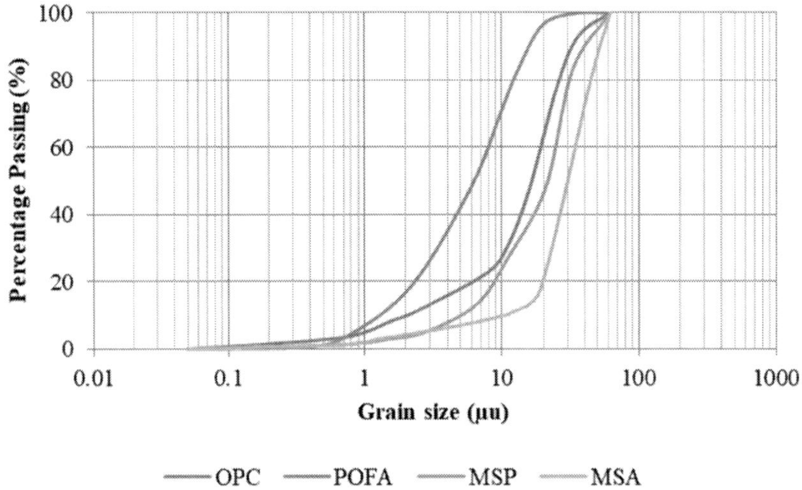

Figure 2. Grain size distribution of materials.

4.3. Workability

Testing on workability of foamed concrete was conducted using J-ring flow test assisted by inverted slump cone. Based on the result in table 4, control sample had the highest workability which flow and spread fluently without any visible blocking. The J-ring flow value decreased as more cement was replaced by POFA and MSP and the variation of flow with respected on control sample was ranged between 14 to 16%. Addition of cement replacement materials mainly from POFA absorbs the water surrounding grain particles. The porous nature of POFA particles had led to high water demand and thus decreasing the concrete workability [27]. Combination of POFA and MSP reduced the flow to 372.5mm and 367.5mm as compared to control specimen of flow 435mm. On the other hand, mixture of POFA and MSA have resulted 375mm and 370mm which are slightly better than unburned mussel shell. Workability decreased as flow decreased due to the lower specific gravity and higher porosity of POFA and MSA as compared to OPC. The increased in porosity absorbed more water in the production of foamed concrete mix.

Table 4. Workability result of fresh foamed concrete.

Mix Proportion	J-ring Flow (mm)	Variation of flow with respected on control sample (%)
Control	435	-
20% POFA + 0% MSP	377.5	-14.1
20% POFA + 5% MSP	372.5	-15.4
20% POFA + 5% MSA	375	-14.8
20% POFA + 10% MSP	367.5	-16.8
20% POFA + 10% MSA	370	-16.1

4.4. Compressive strength

General trend of decline can be observed in figure 3 when percentage of cement replacement materials increases. Foamed concrete specimens with combination of POFA+MSP and POFA+MSA show strength reduction to 10% than control mix. At 7th day, mixture with 20POFA+5MSA has the highest compressive strength compared to other two mix proportion at 15.57 MPa which 3%% less than the control. Foamed concrete with 20POFA+10MSP had the lowest compressive strength at 14.74 MPa. However, the presence of MSP and MSA combined with POFA in foamed concrete shows potential in early development in compressive strength. due to high amount of silica dioxide (SiO_2) in POFA and MSA which enhanced the pozzolanic reaction in concrete mix. Besides, MSA contains high amount of calcium oxide (CaO) which reduced the time of hydration process leading to reduction of strength at early age. Presence of mussel shell as calcium oxide contributed in tricalcium silicate (C_3S) which is responsible for early strength development in cement matrix as a result of the increased lime content [28]

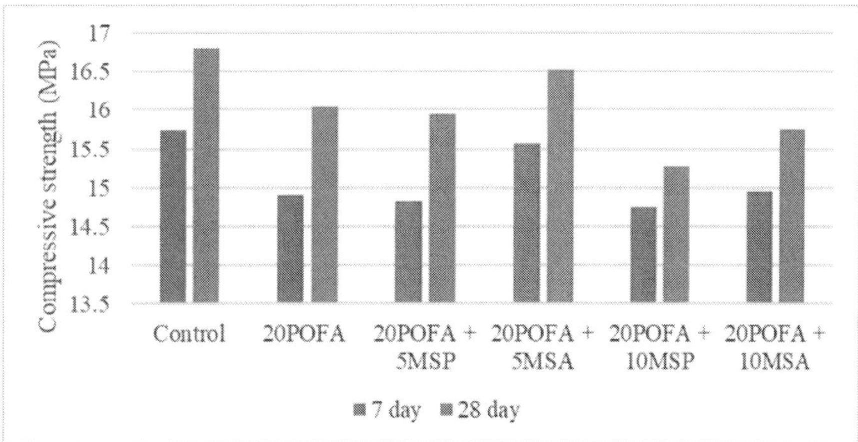

Figure 3. Compressive strength with various mix.

In term of 28 days compressive strength, mixture 20POFA+5MSA has the highest value at 16.52 MPa which 2% reduction than control. Heat treatment MSA provided better pozzolanic reactivity when combined with POFA as partial cement replacement. POFA, higher compressive strength was obtained in the long term, due to their pozzolanic reaction. Compressive strength reduction was observed in foamed concrete mixture 20POFA+10MSP and 20POFA+10MSA respectively. This result agrees with previous experimental studies conducted by other researchers [29,30] that partial cement materials up to 30% can reduce up to 20% the compressive strength of the concrete. This is mostly due to the high water absorption from cement replacement materials which distorted hydration process during C-S-H formation.

4.5. Split tensile strength

Split tensile test of foamed concrete specimen was conducted to determine the effect on tensile strength of POFA combined with MSP and MSA. Based on figure 4, similar trend of strength reduction was observed when the percentage of cement replacement increases. Optimum percentage was shown in mixture of 20POFA+5MSA at 1.83MPa with negligible difference as compared to control specimen. Similar tensile strength was also recorded with mixture 20POFA which suggest presence of 5% MSA will produce similar performance without reducing tensile capability of foamed concrete. However, both foamed concrete mixture at 20POFA+10MSP and 20POFA+10MSA show strength reduction to 11%. Foamed concrete mixture at 20POFA+10MSP has the lowest tensile strength at 1.65MPa. Variance of strength result was observed between two different types of mussel shell preparation. Unburned mussel shell contained residual organic materials originally attached to the prismatic layer that influence hydration process in foamed concrete.

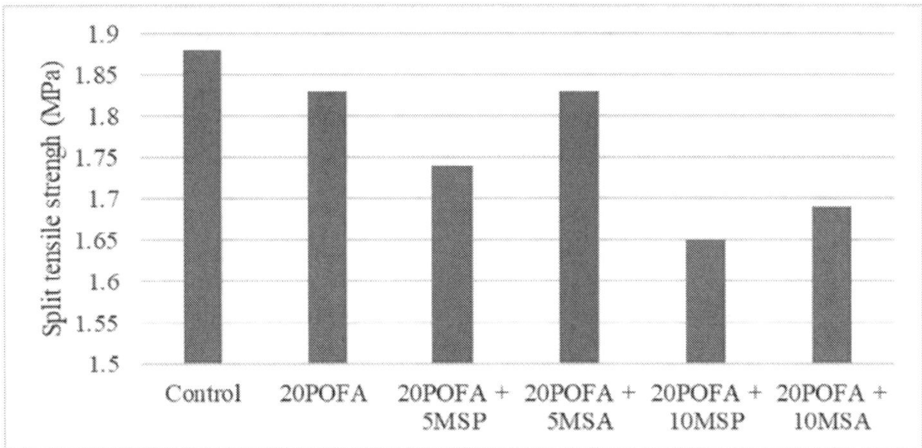

Figure 4. Split tensile strength with various mix.

Additionally, high percentage of cement replacement materials distorted bonding at the interface between binder paste and fine aggregates which produce lower tensile strength concrete. Incorporation of POFA was not effective in improving the interfacial zone between aggregate and cement particle [31]. POFA combined with MSP and MSA mixes were observed not provide additional tensile strength but optimum proportion between both cement replacement materials can be considered with acceptable split tensile strength reduction.

4.6. Modulus of elasticity

Modulus of elasticity are the main mechanical properties will contribute to the flexural and shear stiffness of masonry structures. Data were obtained from the compression test of cylinder specimen with attached strain gauges were analyzed using stress-strain curve. Effect on modulus of elasticity due to combination of POFA with MSP and MSA as partial cement replacement in foamed concrete mixtures are shown figure 5.

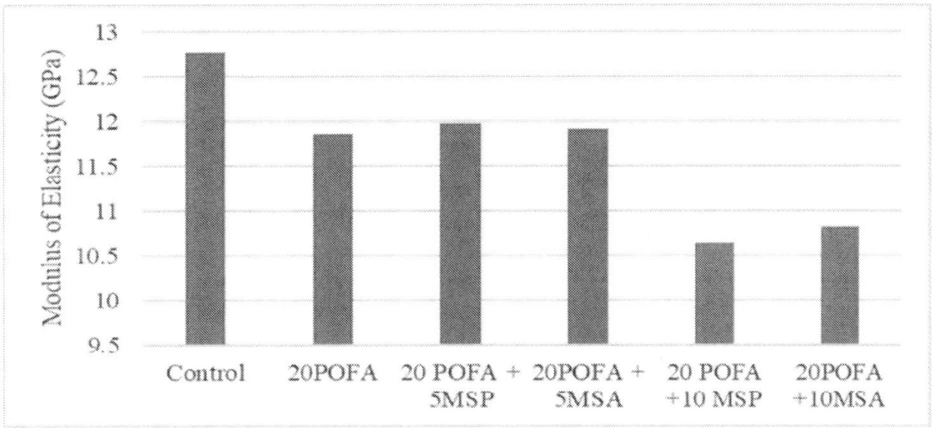

Figure 5. Modulus of elasticity with various mix.

Based on the result, it was observed that foamed concrete with 20%POFA show less variance in term of modulus of elasticity when combined with 5% of MSA and MSP respectively. The trend of reduction in was similar within 8% which indicates addition of mussel shell materials as partial cement replacement will not enhance stiffness of foamed concrete specimen. Further reduction in modulus of elasticity was shown when the total percentage of cement replacement at 30% with value recorded at 18% lower than control specimen.

Previous researcher Lim et al. [32] highlighted cement replacement materials in foamed concrete simultaneously reduced their stiffness due to higher ductility than OPC. As the percentages of POFA and MSA increases, high water absorption from the fine particles disrupts reaction of C-S-H gel resulting lesser cylinder compressive strength which subsequently further reducing modulus of elasticity than normal foamed concrete.

5. Conclusion

This paper utilize combined matter on workability and mechanical properties of foamed concrete incorporating POFA combine with MSP and MSA. The combination of 20% POFA with 5% MSA shows potential as sustainable materials to be implemented as partial cement replacement in construction sector. Based on the data acquired, the findings of this study are summarized as follows:

i. Specific gravity of POFA has the lowest values which correspond to grain size distribution highest fineness percentage with D60 value 7.89μm.
ii. Workability of foamed concrete significantly reduced to 370mm for 20POFA+ 10MSA while 367.5mm for 20POFA+10MSP compared to control at 435mm. High water absorption from cement replacement materials directly lower the flow ability in foamed concrete.
iii. The compressive strength of foamed concrete mixture was reduced when the percentage of cement replacement increases. Presence of MSA combine with POFA has significance in term of early strength development. Mixture 20POFA+5MSA has potential due to slight compressive strength reduction as compared to control.
iv. Variance of strength result was observed between two different types of mussel shell preparation. Optimum percentage was shown in mixture of 20POFA+5MSA at 1.83MPa.
v. Reduction in modulus of elasticity was shown 18% lower than control specimen when the total percentage of cement replacement at 30%. Higher ductility from of foamed concrete resulting less stiffness produced by the binder materials.

6. References

[1] Jhatial A A, Inn G W, Mohamad N, Johnson Alengaram, Hung Mo, K, and Abdullah, R. 2017. Influence of polypropylene fibres on the tensile strength and thermal properties of various densities of foamed concrete *IOP Conference Series: Materials Science and Engineering* **271** 1–7

[2] Richard A O and Ramli M 2013 Experimental Production of Sustainable Lightweight Foamed Concrete, *Br. J. Appl. Sci. Technol.* **3** 994–1005

[3] Munir A, Abdullah Huzaim, Sofyan Irfandi and Safwan 2015 Utilization of palm oil fuel ash (POFA) in producing lightweight foamed concrete for non-structural building material, *Procedia Eng.* **125** 739–746

[4] Mugahed Y H, Farzadnia N and Abang Ali A A 2015 Properties and application of foamed concrete: A review, *Constr. Build. Mater.* **101** 990-1005

[5] Jhatial A A, Mohamad N, Mo K H and Sohu S 2020 Incorporation of palm oil fuel ash and egg shell powder as supplementary cementitious materials in sustainable foamed concrete. *Tehnički vjesnik* **27** 1394-1402

[6] Ahmad M H, Omar R C, Malek M A, Noor M 2008 Compressive Strength of Palm Oil Fuel Ash Concrete, *J. Prev. Interv. Community* **(A) 27** 297–306

[7] Sata V, Jaturapitakkul C and Kiattikomol K 2007 Influence of pozzolan from various by-product materials on mechanical properties of high-strength concrete, *Constr. Build. Mater.* **21** 1589–1598

[8] Khankhaje E, Hussin M W and Mirza J 2016 On blended cement and geopolymer concretes containing palm oil fuel ash, *Mater. Design* **89** 385–398

[9] Tangchirapat W, Jaturapitakkul C and Chindaprasirt P 2010 Use of palm oil fuel ash as a supplementary cementitious material for producing high-strength concrete, *Constr. Build. Mater.* **23** 2641–2646

[10] Chindaprasirt P, Homwuttiwong S and Jaturapitakkul C 2007 Strength and water permeability of concrete containing palm oil fuel ash and rice husk-bark ash, *Constr. Build. Mater.* **21** 1492–1499

[11] Muthusamy K, Zamri N and Amirulkhairi M 2015 Effect of mixing ingredient on compressive strength of oil palm shell lightweight aggregate concrete containing palm oil fuel ash, *Procedia Eng.* **125** 804–810

[12] Jhatial A A, Goh W I, Mastoi. K, Rahman A F and Kamaruddin S 2021 Thermo-mechanical properties and sustainability analysis of eco-friendly structural foamed concrete by reusing palm oil fuel ash and eggshell powder as supplementary cementitious materials, *Environ. Sci. Pollut. Res.* **9** 1-22

[13] Kamaruddin S, Goh W I, Mutalib N A, Jhatial A A, Mohamad N and Rahman A F 2021 Effect of Combined Supplementary Cementitious Materials on the Fresh and Mechanical Properties of Eco-Efficient Self-Compacting Concrete, *Arab. J. Sci. Eng.* **57** 1-21

[14] Lee W K, Lim H S, Song S, Hussin M W and Ismail M A 2018 Use of an agricultural by-product, nano sized Palm Oil Fuel Ash as a supplementary cementitious material, *Constr. Build. Mater.* **183** 139-149

[15] Kroehong W, Damrongwiriyanupap N, Sinsiri T and Jaturapitakkul C 2016 The effect of palm oil fuel ash as a supplementary cementitious material on chloride penetration and microstructure of blended cement paste, *Arab. J. Sci. Eng.* **41** 4799-4808

[16] Bamaga O, Hussin M W and Ismail M A 2013 Palm oil fuel ash: promising supplementary cementing materials, *KSCE J. Civ. Eng.* **17** 1708-1713

[17] Awal A S & Shehu I A 2013 Evaluation of heat of hydration of concrete containing high volume palm oil fuel ash, *Fuel* **105** 728–731

[18] Aldahdooh M A, Bunnori N M and Johari A M 2013 Development of green ultra-high performance fiber reinforced concrete containing ultrafine palm oil fuel ash, *Constr. Build. Mater.* **48** 379–389

[19] Foti D and Cavallo 2018 Mechanical behavior of concretes made with non-conventional organic origin calcareous aggregates, *Constr. Build. Mater* **179** 100–106
[20] Mo K H, Alengaram U J, Jumaat M Z, Lee S C, Goh W I and Yuen C W 2018 Recycling of seashell waste in concrete: A review, *Constr. Build. Mater* **162** 751–764
[21] Lertwattanaruk P, Makul N and Siripattarapravat 2012 Utilization of ground waste seashells in cement mortars for masonry and plastering, *J. Environ. Managmnt* **111** 133–141
[22] Skariah B, S Kumar and H Sahan 2017 Sustainable concrete containing palm oil fuel ash as a supplementary cementitious material, *Renew. Sustain. Energy Rev.* **80** 550–561
[23] Momeen M, Islam U, Mo K H and Alengaram U J 2016 Durability properties of sustainable concrete containing high volume palm oil waste materials, *J. Clean. Prod.* **137** 167–177
[24] Olivia M, Mifshella A A and Darmayanti L 2015 Mechanical properties of seashell concrete, *Procedia Engineering* **125** 760–764
[25] Kearsley E P 2005 Designing mix composition of foamed concrete with high fly ash contents *Cemnt. Conc. Resrc.* **32** 233-239
[26] Soltanzadeh F, M Emam-Jomeh, Edalat-Behbahani A and Soltan-Zadeh Z 2018 Development and characterization of blended cements containing seashell powder, *Constr. Build. Mater.* **161** 292–304
[27] Muthusamy K, Mirza J and Zamri A 2019 Properties of high strength palm oil clinker lightweight concrete containing palm oil fuel ash in tropical climate, *Constr. Build. Mater.* **199** 163-177
[28] Adesina P A and Olutoge F A 2019 Structural properties of sustainable concrete developed using rice husk ash and hydrated lime, *J. of Build Eng.* **25** 100804
[29] Alsubari B, Shafigh P, Ibrahim Z, Alnahhal M F and Jumaat M Z 2018 Properties of eco-friendly self-compacting concrete containing modified treated palm oil fuel ash, *Constr and Build Mat.* **112** 723-730
[30] Zeyad A M, Megat Johari, Tayeh M A and Yusuf M O 2017 Pozzolanic reactivity of ultrafine palm oil fuel ash waste on strength and durability performances of high strength concrete, *J. of Clean Prod* . **144** 471-488
[31] Luhar S and Ta Wui Cheng 2019 Incorporation of natural waste from agricultural and aquacultural farming as supplementary materials with green concrete: A review, *Compst. Part B* **175** 107-176
[32] Lim S K, Tan C S, Lim O Y and Lee Y L 2013 Fresh and hardened properties of lightweight foamed concrete with palm oil fuel ash as filler, *Constr. Build. Mater.* **46** 39–47

Acknowledgments
This research was supported by Ministry of Higher Education Malaysia (MOHE) through Fundamental Research Grant Scheme for Research Acculturation of Early Career Researchers (RACER/1/2019/TK06/UTHM/1).

Study on Workability High Strength Concrete Containing Pineapple Leaf Fiber (PALF)

J. Hadipramana[1], F. V. Riza[1], T. Amirsyah[1], S.N. Mokhatar[2] and M.Ardiansyah[1]

[1] Civil Engineering Department, Faculty of Engineering, Universitas Muhammadiyah Sumatera Utara. Jl. Kapten Muchtar Basri #3, 20238 Medan, Indonesia.
[2] Structural Dynamics and Computational Engineering Focus Group, Universiti Tun Hussein Onn Malaysia, 86400Parit Raja, Johor, Malaysia.
Corresponding author: fetravenny@umsu.ac.id

Abstract. The addition of fibers in concrete increases bridging force in interfacial transition zones inside the concrete matrix. Pineapple Leaf Fiber (PALF) is a natural fiber that has the potential to replace artificial fibers as reinforcement on concrete. As reinforcement fiber in concrete, PALF will undergo fibrillation and water absorption in concrete mixture and will change the mechanical properties of fresh concrete. So, the purpose of this study is to study the workability characteristics of fresh state concrete given PALF. Some variations of PALF composition are 0.04, 0.09, and 0.15 % wt. of cement is mixed into fresh concrete mixture with water-cement (w/c) ratio variations of 0.35, 0.38, and 0.41. The planned concrete is high-strength concrete, then the provisions of determination of coarse and fine aggregates using sieve analysis according to high-strength concrete material standards and concrete mixtures are given superplasticizer 0.09% wt. of cement constantly for all mixtures. A slump test method is still an option in this study, where this method becomes the standard to determine workability effectively. The results showed that an increase in concrete strength was achieved in the addition of PALF composition in concrete compared to control (normal concrete), in addition to the tendency of the relationship between w/c ratio, slump value, and compressive strength in all PALF variations achieved in this study.

1. Introduction

Concrete has a higher compressive strength property than tensile strength. Factually, the concrete will withstand tensile when weighed down, especially on distressed rear surfaces. Concrete is brittle, when a crack occurs, then stress cannot be transferred anymore and concrete will failure instantly. The strength of concrete will depend on the bond between the aggregates in the concrete matrix [1], then thecrack that occurs arises from microcrack or shear stress in the interfacial transition zone (ITZ) betweenthe aggregates where this area is the weakest in the concrete matrix.

Many studies were conducted to improve the weaknesses in this area, one of the efforts is the addition of fibers into the concrete to increase the strength of concrete [2][3][4]. Fibers in concrete will increase theinterfacial between aggregates. However, in reality, the presence of fiber in concrete will affect the workability performance of concrete. Fiber will occur fibrillation [4][5] and affect the moisture content in the concrete mixture due to absorption by fiber thus affecting the slump value [6]. The presence of fiber in the concrete mixture creates friction between the fiber and the concrete particles and inhibits the flow of fresh concrete. Therefore. large quantities are not recommended and can result in shafts in concrete. But in certain cases in lightweight concrete, fiber can increase workability since it effectively restrains the segregation, although fiber lowers the slump value [7]. In the meantime, the effect of absorption can be reduced by adding extra water following the absorption of fiber.

In this study, the fiber used is Pineapple Leaf Fiber (PALF) which is abundant and wasted. PALF has good properties [8] and can be used as reinforced fiber in concrete. PALF is a natural fiber, which can absorb water and affect the water-cement (w/c) ratio of concrete and it is easy to have fibrillation. Therefore, this study investigated the influence of PALF on the workability of high-strength concrete, further exploring the slump character of high-strength concrete mixtures containing PALF with w/c cement variations.

2. Materials and methods

2.1. Materials

The materials used to develop high-strength concrete in this study, through several requirements of the Indonesian National Standard and ASTM. Coarse and fine aggregates through the test stages as shown in table 1. This is very important because aggregate is the basic material of concrete and must have good quality before it becomes high-quality concrete. Sieve analysis can determine Fine Modulus (FM)coarse and fine aggregate [9], and several other tests to determine aggregate and cement properties suchas density, bulk density, specific gravity, and absorption [10].

PALF used as concrete fiber has a composition of 0.04%, 0.09%, and 0.15% wt. of cement. Pineapple leaf fibers are produced by manual [11]. The fibers are obtained by separating the fibers in the leaves with the outermost layer of the leaves, then the fibers are dried and the result is as in Figure 1. The chemical properties of PALF had determined experimentally by [12] and [8]. The physical and mechanical properties of PALF based on its density were reviewed and reported by [11] and [8]. The properties of PALF as in Table 2, which in this study, the PALF used has a diameter of 12 micrometers.

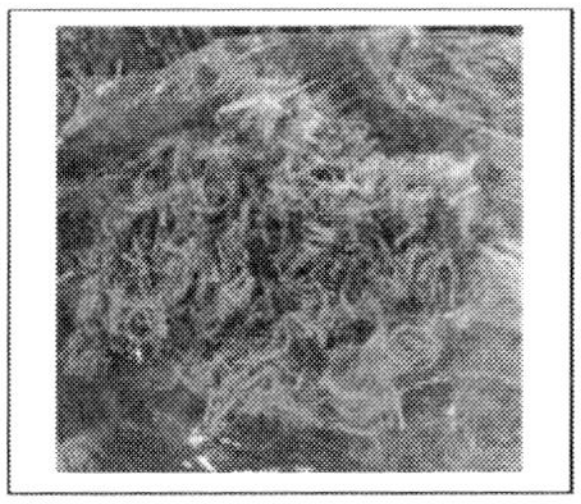

Figure 1. PALF used in the study taken from pineapple leaves.

Table 1. Properties of the constituent material of concrete.

	Coarse Aggregate	Fine Aggregate	Cement	PALF
Fine Modulus (FM)	7.068	2.775	-	-
Density (gr/cm^3)	2.716	2.571	1.439	-
Specific Gravity (gr/cm^3)	-	-	3.102	-
Bulk Density (gr/cm^3)	1.511	1.165	-	-
Water content (%)	0.604	2.145	-	-
Absorption (%)	0.752	1.730	-	5.27
Mud Content (%)	0.767	3.3	-	-
Maximum diameter (mm)	40	2.36	-	-
Abrasion (%)	16.460	-	-	-

Table 2. Physical properties of PALF [8].

Fiber diameter (Micrometer)	Tensile Strength (Mpa)	Young's Modulus (Gpa)	Elongation (%)
5.0–30.0	170	6.26	0.8 – 1.6
105–300	293.08	18.934	1.41
50–91	210–695	15–53	-
60-6	413-8	6.2	1.6
30–60	413	6.5	1.6
20–8	413–1627	34.5–82.5	1.6

2.2. Mix Design of Concrete and Experimental methods
Concrete mix planning based on Indonesian National Standard [13] as in table 3. The samples of concrete were produced in form of cubes with a size of 15cm x 15cm x 15cm, with a total of 24 sampleseach for three treatment ages: 7 days, 14 days, and 28 days.

Table 3. Mix proportions of concrete.

PALF (% wt. cement)	w/c	Cement (Kg/m^3)	Fine Aggregate (Kg/m^3)	Coarse Aggregate (Kg/m^3)	Superplasticizer (% wt. cement)
0	0.35	485.00	574.85	1213.65	0.09
0	0.38	447.37	605.04	1221.47	0.09
0	0.41	414.63	625.22	1234.11	0.09
0.04	0.35	485.00	574.85	1213.65	0.09
0.04	0.38	447.37	605.04	1221.47	0.09
0.04	0.41	414.63	625.22	1234.11	0.09
0.09	0.35	485.00	574.85	1213.65	0.09
0.09	0.38	447.37	605.04	1221.47	0.09
0.09	0.41	414.63	625.22	1234.11	0.09
0.15	0.35	485.00	574.85	1213.65	0.09
0.15	0.38	447.37	605.04	1221.47	0.09
0.15	0.41	414.63	625.22	1234.11	0.09

3. Results and discussion

3.1. Sieve analysis
Aggregate gradation plays an important role in forming high-quality concrete [14]. In figure 2 and Figure 3, it is shown that the aggregate is within the required range (sufficient) for coarse and fine aggregate respectively [15]. Usually, fine modulus (FM) is determined for fine aggregate analysis, but in this study, fine and coarse aggregate combined although in the analysis determine FM separated [16].The merger of overall gradations [17] resulting in very small stage zones and using only a nominal maximum aggregate of 20 mm. By using ASTM C33[16], the aggregate ratio will be wider to determinethe slump and varying strength.

Figures 2 and 3 use a logarithmic scale, showing that the percentage distribution passes the maximum sieve of coarse aggregate is the sieve pass number 38.1 mm [18], and the normal distribution of the percentage passes fine aggregate sieve is rather rough sand [18] with an FM of 2,775. In the combinationof these two aggregates, it has a moderate mixed gradation and has a mixed FM of 5.5. All aggregate sizes can certainly fill the space on the concrete matrix evenly. Furthermore, the sieve analysis can determine the quality of concrete to be produced.

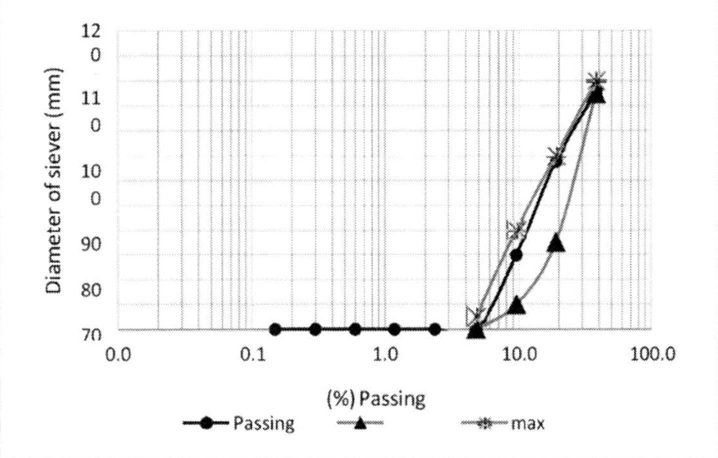

Figure 2. Coarse aggregate gradation.

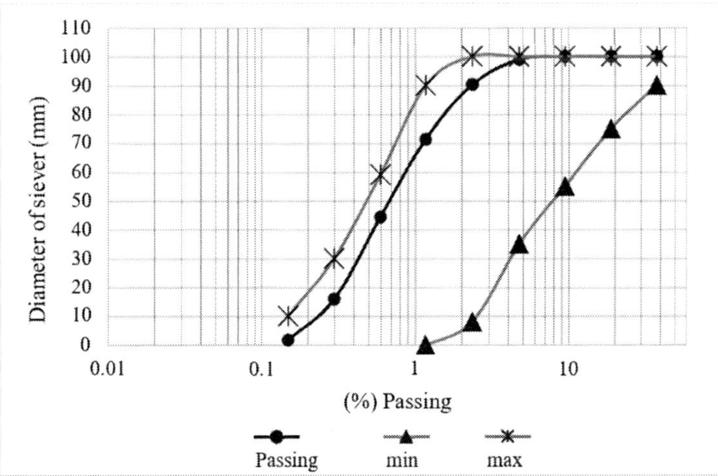

Figure 3. Fine aggregate gradation.

3.2. Workability performance and w/c Ratio

The workability mix with various w/c ratios of PALF concrete as shown in table 4 indicates the slump mix of the concrete mixes ranges from 30 to 45 mm. This slump is generally called a low slump, where these results are influenced by the w/c ratio, PALF presence, and superplasticizer in concrete. However, the composition of the superplasticizer on the overall concrete mix is constant.

The results showed the lowest w/c ratio of 0.35 had the lowest slump value and at a high w/c ratio of 0.41 on normal concrete had a high slump. Figure 4 shows the tendency of slump value influenced by w/c ratio in each variation of concrete PALF. At w/c ratio 0.35 for all concrete mixes have a low value,but it can be seen that in concrete PALF variation is lower than control. This is a similar finding by theprevious experiment by using fiber material as an addition to concrete [18,19,20], here it is clear that the use of fiber on concrete will reduce the slump value. The effect of fiber in a mixture is seen in figure 4,where the slump value in the mixture contains lower fiber than concrete without fiber. The presence offiber will affect the viscosity of fresh concrete as found by Aydin [21], that the addition of carbon fiber changes the mix of concrete, the more fiber the viscosity will decrease, even for artificial fiber such as steel fiber will affect the value of slump [21].

Table 4. Slump test results.

Composition of PALF (%) wt. of cement	0	0.04	0.09	0.15
Avg. Slump Height (mm) w/c Ratio 0.35	35	33	30	33
Avg. Slump Height (mm) w/c Ratio 0.38	40	43	38	35
Avg. Slump Height (mm) w/c Ratio 0.41	45	45	40	40

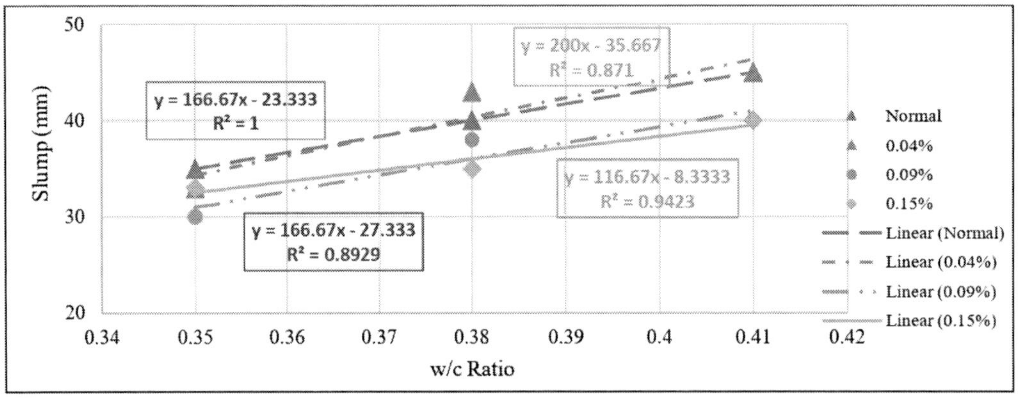

Figure 4. The tendency of workability mix with various w/c ratios of PALF Concrete.

3.3. Compressive strength vs concrete ages

Figures 5 to 7 show compressive strength of concrete with PALF variations and w/c ratios of 0.35, 0.38,and 0.41 respectively. In all w/c ratios, concrete with PALF variation of 0.09% has the highest strengththen decreases at PALF 0.15%. The presence of fiber helps in bridging microcracks on concrete [5][22],thus increasing the strength of concrete, however, when the fiber content is not appropriate or excessiveit will decrease the strength of the concrete [21][22][23].

At w/c the ratio of 0.35 has the highest concrete strength compared to other w/c ratios. It agrees withthe previous study [24], that the strength of concrete is expressed as a function of time and w/c ratio, decreasing the w/c ratio will be increasing of concrete strength, as well as increasing strength of concretealong with the length of time.

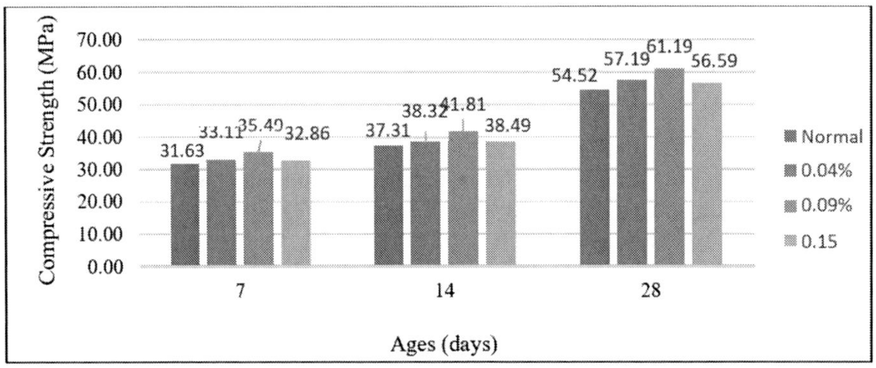

Figure 5. Compressive Strength of various PALF in concrete with w/c ratio 0.35 according to the ageof concrete.

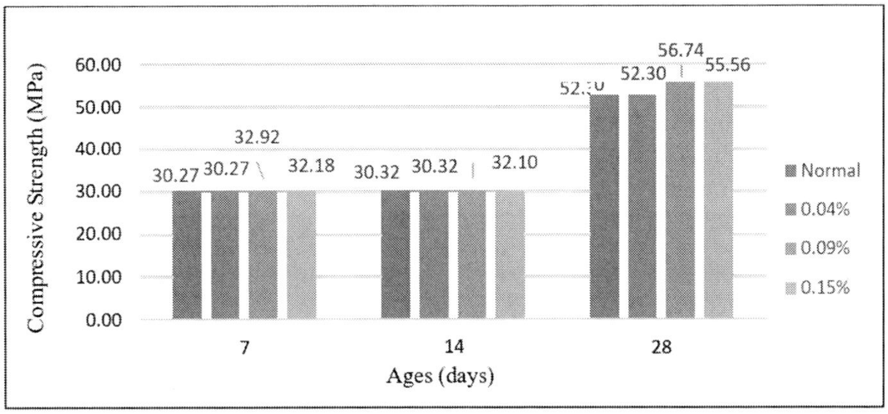

Figure 6. Compressive Strength of various PALF in concrete with w/c ratio 0.38 according to the ageof concrete.

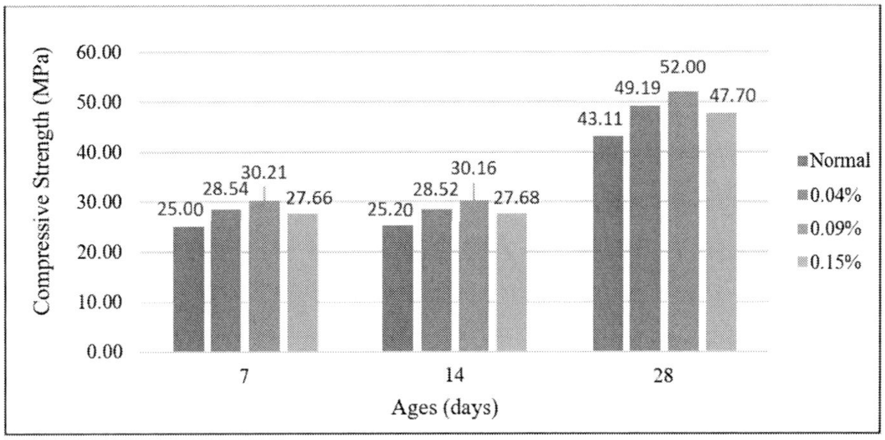

Figure 7. Compressive Strength of various PALF in concrete with w/c ratio 0.41 according to the age of concrete.

3.4. Relation between compressive strength, w/c ratio, and slump for variation of PALF

Figures 8 to 11 show PALF variations in concrete: control, 0.04%, 0.09%, and 0.15% respectively indicating a relation between a decrease in concrete strength to an increase in w/c ratio, and also a relation between the tendency of increased slump value to the increase in w/c ratio. In each variation of PALF tendency can be mapped, where the w /c ratio decreases then the strength of concrete will riseand the slump value is small. In the variation of concrete control, 0.04%, and 0.15% have the same curvetrend while 0.09% tend that at w/c ratio 0.38 concrete strength is high, so the slump value is quite small.This agrees with previous works, that the presence of fiber in the concrete mixture will affect the slump value, further affecting the strength of the concrete [25]. Along with that, the increasing w/c ratiowill increase the slump loss and resulting in a decrease in concrete strength [26]. Nonetheless, thoseresults cannot be used for a very low w /c ratio, where there are restrictions on use, because concrete will lack water and shafts occur in concrete thus lowering the strength [27].

Figure 8 to 11 shows that the strength of concrete and the slump value to the w/c ratio have the same tendency in all PALF variations, where the relationship of concrete strength and cement ratio is expressed with the values y = (-3872x2) + (2752.7x) – 434.59 and y = 166.67x -23,333 for the value ofconcrete strength to slump values.

IOP Conf. Series: Materials Science and Engineering 1200 (2021) 012006 doi:10.1088/1757-899X/1200/1/012006

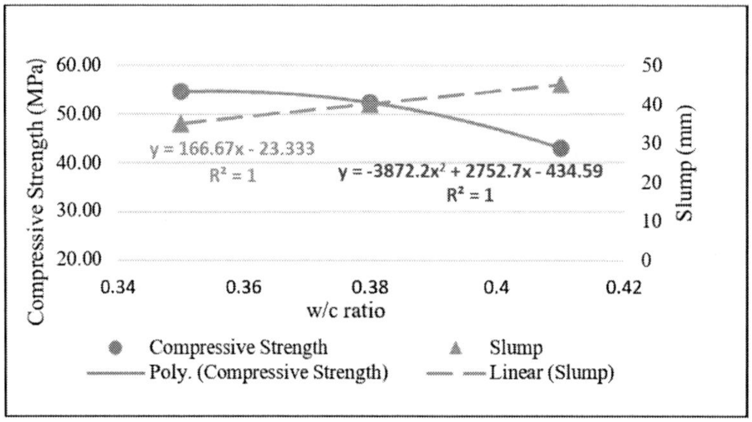

Figure 8. Relation between compressive strength, w/c ratio, and slump for control concrete.

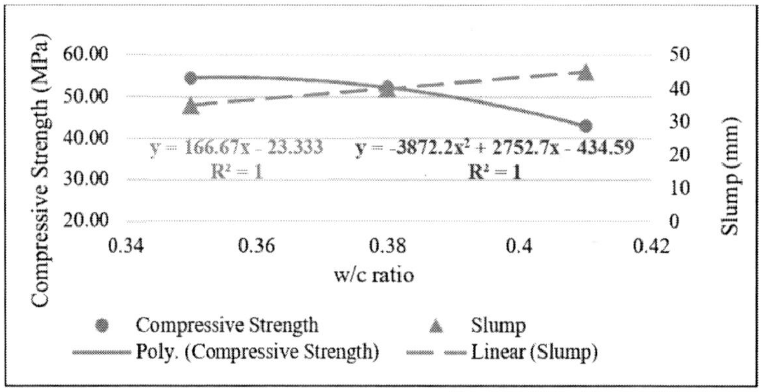

Figure 9. Relation between compressive strength, w/c ratio, and slump for 0.04% PALF Concrete.

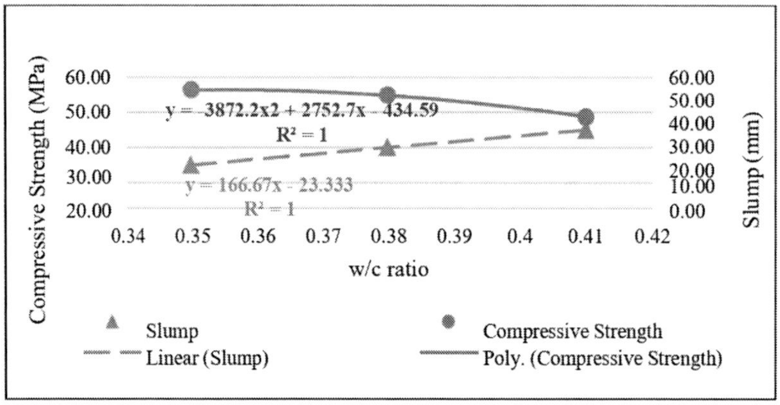

Figure 10. Relation between compressive strength, w/c ratio, and slump for 0.09% PALF Concrete.

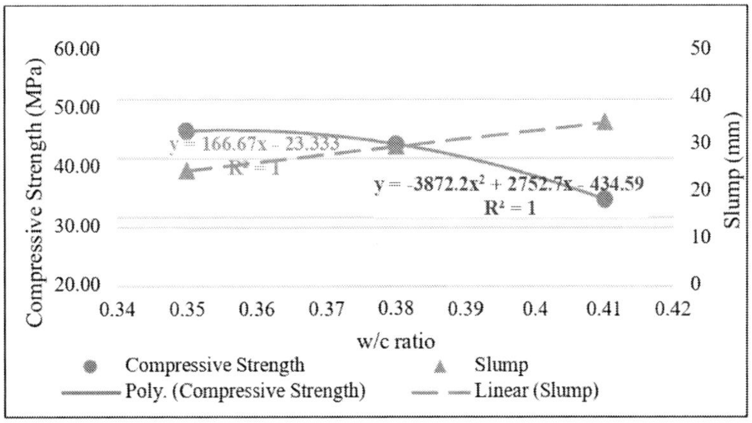

Figure 11. Relation between compressive strength, w/c ratio, and slump for 0.15% PALF Concrete.

4. Conclusion

The addition of fiber to high-strength of concrete affects the slump value, but with w/c ratio and PALF percentage amount reduces the effect of the reduction. Lack of water due to absorption by fiber can be overcome by adding water following the absorption capacity of water by fiber. The w/c low ratio increases concrete strength despite PALF presence in concrete. The presence of PALF in concrete helpsin bridging microcrack on concrete thus increasing the compressive strength on concrete. The relationship between concrete strength and water ratio has the same tendency for concrete with all PALFvariations.

5. References

[1] Xuan D X, Shui Z H, and Wu S P 2009 Influence of silica fume on the interfacial bond between aggregate and matrix in near-surface layer of concrete, *Constr. Build. Mater.* **23**2631–2635

[2] Flores Medina N, Barluenga G, and Hernández-Olivares F 2015 Combined effect of Polypropylene fibers and Silica Fume to improve the durability of concrete with naturalPozzolans blended cement, *Constr. Build. Mater.* **96** 556–566

[3] Afroughsabet V and Ozbakkaloglu T 2015 Mechanical and durability properties of high-strength concrete containing steel and polypropylene fibers, *Constr. Build. Mater.* **94** 73–82

[4] Hadipramana J, Samad A A A, Zaidi A M A, Mohammad N and Ali N 2013 Contribution of polypropylene fibre in improving strength of foamed concrete, *Adv. Mater. Res.* **626** 762–768

[5] Blazy J and Blazy R 2021 Polypropylene fiber reinforced concrete and its application in creating architectural forms of public spaces, *Case Stud. Constr. Mater.* **14** e00549

[6] Ghanem S Y, Bowling J, and Sun Z 2021 Mechanical Properties of Hybrid Synthetic Fiber Reinforced Self- Consolidating Concrete, *Compos. Part C Open Access.* **5** 100154

[7] Chen B and Liu J 2019 Contribution of hybrid fibers on the properties of the high-strength lightweight concrete having good workability, *Cem. Concr. Res.* **35** 913–917

[8] Todkar S S and Patil S A 2019 Review on mechanical properties evaluation of pineapple leaffibre (PALF) reinforced polymer composites, *Compos. Part B Eng.* **174** 106927

[9] ASTM C136-01 2001 *Concrete Aggregates* (Pennsylvania: ASTM International)

[10] ASTM C128-01 2003 *Standard Test Method for Density, Relative Density (Specific Gravity), and Absorption of Fine Aggregate* (Pennsylvania: ASTM International)

[11] Asim M, Abdan K, Jawaid M, Nasir M, Dashtizadeh Z, Ishak M R, and Enamul Haque M 2015 A review on pineapple leaves fibre and its composites, *Int. J. Polym. Sci.* **2015** 1-16
[12] Bhaduri S K, Sen S K, and Dasgupta P C 1983 Structural studies of an acidic polysaccharideisolated from the leaf fibre of pineapple, Ananas comosus MERR. **121** 211–220
[13] SNI 03-2834 2000 *Tata cara pembuatan rencana campuran betonnormal* (Indonesia: Standar Nasional Indonesia) pp 1–34
[14] Rashid M A and Mansur M A 2009 Considerations in producinghigh strength concrete, *J. Civ. Eng.* **37** 53–63
[15] ASTM C 136 – 01 1997 *Standard Test Method for Sieve Analysis of Fine and Coarse Aggregates* (Pennsylvania: ASTM International) pp 1–6
[16] BS 882 1992 *Specification for aggregates from natural sources for concrete* (United Kingdom: British Standards Institution)
[17] Ashraf W B and Noor M A 2011 Performance-evaluation of concrete properties for different combined aggregate gradation approaches *Procedia Eng.* **14** 2627–2634
[18] BS EN 12620 2008 *Aggregates for Concrete* (United Kingdom: British Standards Institution)
[19] Q. Cao Q, Cheng Y, Cao M, and Gao Q 2017 Workability, strength and shrinkage of fiber reinforced expansive self-consolidating concrete, *Constr. Build. Mater.* **131**178–185
[20] Matar P and Assaad J J 2019 Concurrent effects of recycled aggregates and polypropylene fiberson workability and key strength properties of self-consolidating concrete, *Constr. Build. Mater.* **199** 492–500
[21] Aydin A C 2007 Self compactability of high volume hybrid fiber reinforced concrete, *Constr.Build. Mater.* **21** 1149–1154
[22] Hadipramana J, Samad A A A, Zaidi A M A, Mohammad N, and Wirdawati 2012 Influenceof Polypropylene Fiber in strength of foamed concrete, *Adv. Mater. Res.* **488–489** 253–257
[23] Yap S P, Bu B U, Alengaram U J, Mo K H, and Jumaat M Z 2014 Flexural toughness characteristics of steel-polypropylene hybrid fibre-reinforced oil palm shell concrete, *Mater.Des.* **57** 652–659
[24] Gavela S, Nikoloutsopoulos N, Papadakos G, Passa D, and Sotiropoulou A 2018 Multifactorial experimental analysis of concrete compressive strength as a function of time and water-to-cement ratio, *Procedia Struct. Integr.* **10** 135–140
[25] De Figueiredo A.D and Ceccato M R 2015 Workability analysis of steel fiber reinforcedconcrete using slump and ve-be test, *Mater. Res.* **18** 1284–1290
[26] Singh S B, Munjal P, and Thammishetti N 2015 Role of water/cement ratio on strengthdevelopment of cement mortar, *J. Build. Eng.* **4** 94–100
[27] Živica V 2009 Effects of the very low water/cement ratio, *Constr. Build. Mater.* **23** 3579–3582

Acknowledgments
The author expressed his gratitude and acknowledgment given by the Ministry of Education and Culture - Research and Technology (Kemendikbud-Ristek) for the PDUPT source of grants2021 and Higher Education Service Institute – Region 1 (LLDIKTI-1) as well as the Institute of Research and Community Service - Universitas Muhammadiyah Sumatera Utara (LPPM- UMSU) and Department of Civil Engineering, Faculty of Engineering, UMSU, Medan Indonesia for research and staffing resources.

New technology in 3D Concrete Printing by Using Ground Granulated Blast-Furnace Slag: A Review

Norhafizah Salleh[1],Nur Syahera Jamalulail[1], Noor Azlina Abdul Hamid[1], Zalipah Jamellodin[1], Masni A Majid[1] and Nurul Huda Suliman[2]

[1]Faculty of Civil Engineering and Built Environment, Universiti Tun Hussein Onn Malaysia, Batu Pahat, Johor, 86400, MALAYSIA
[2] Faculty of Civil Engineering, Universiti Teknologi MARA, 40450 Shah Alam, Selangor, Malaysia
Corresponding author: nhafizah@uthm.edu.my

Abstract. 3D building printing is a technology for producing 3D models of an object to build any shape or size in layers by using computer software. The development of 3D printing was going to be more famous and commercial in the future to reduce the construction cost and labor demands, sustainability, and to the greenest way. Concrete is the mixture that consists of the ingredients of water, binder (cement) and aggregates (rock, sand, gravel). The productions of Portland cement in construction leads to the emissions of carbon dioxide (CO2) gas into the air. Waste material has been used as cement replacement in this research study to reduce carbon dioxide (CO2) gas emissions. This research study was going to evaluate the viability of concrete for 3D printing and printing emphasizing the impact on potential opportunities of this innovative industry. The behaviour of 3D concrete printing and potential of modified mortar in 3D concrete mix design by using Ground Granulated Blast-Furnace Slag (GGBS) is used to evaluate the potential uses of GGBS in concrete mixture for 3D building printing. This research study involved the review of concrete compressive strength and workability of 3D concrete printing with the control aspect during process manufacturing. The result shows that the mix design of 3D concrete printing with 30% and 40% produced concrete strength of 47.33MPa and 47.67MPa respectively. Furthermore, control aspect requirements of concrete for 3D printing were discussed in the field extrudability, flowability, buildability, strength between layers, aggregates, and water-cement ratio. Throughout this study, the manufactures of 3D building printing materials using environmentally friendly elements can contribute effectively create a sustainable environment automatically.

1. Introduction

Adaptive technologies have opened up new possibilities for many rapidly emerging industrial manufacturing sectors. Thanks to the flexibility factor, 3D printing has taken technologies to a new level they are also commonly used in building. Special fine-grained concrete (mortars) are necessary for the construction of real buildings and structures using 3D printing, which can ensure the safe use, robustness and resilience of buildings [1]. Concrete was the essential materials used in building construction works. Ground Granulated Blast-Furnace Slag (GGBS) have potential to replace the Portland cement as reported in many researchers. GGBS is an alternative to Portland cement for sustainable building [2][3][4][5][6]. The substitution of GGBS with Portland cement can lead to a reduction in the production of carbon dioxide gas which can be used to substitute up to 80% of ordinary Portland cement [7]. Portland cement has usually been used in a concrete mixture as cement. The use of GGBS as waste material in concrete mixture in this research study will be carried out either for 3D building printing processes or not. Also, this research paper overlooks the participation of the numerous research institutes and concrete experts in 3D building printing. In this research reviewing the possible uses for the future use of 3D printing will be briefly studied. It explains the possibilities of innovation and advantages of mortar 3D construction printing to save money and time for future construction project technologies.

In 2014, a new chapter in construction technology began a real revolution in the construction industry as the first house was printed [8]. Today, the construction industry faces a variety of significant challenges, such as low labour productivity, a high statistical rate of construction site emergencies, and the complications of building process control. Besides, one of the largest users of non-renewable natural resources is the building industry. The use of Portland cement in the construction systems of concrete technology results in the release of carbon dioxide (CO^2) gas into the air [1]. The increase in emissions of CO^2 gas causes warming, degrading the quality of human life. Overall, the world's production of Portland cement adds 1.6 million tonnes, or around 7%, of CO^2 to the atmosphere. For every 1000 kg of cement made, 900 kg of carbon dioxide is released due to the high energy output of Portland cement [9]. The invention of green concrete materials is important to minimise the use of cement. In addition to the design of the structure, in order to process the 3D building printing, the Ground Granulated Blast-Furnace Slag (GGBS) will be substituted as a Portland cement in the concrete blend (mortar).

The objectives of this study are to study the behaviour of concrete on 3D printing and potential of modified mortar in 3D mix design by using Ground Granulated Blast-Furnace Slag (GGBS). Besides, to evaluate the viability of concrete for 3D printing and printing emphasizing the impact on potential.

2. Concrete 3D printing Mix Design

The amount of GGBS is an estimate based on mix design ratio before preparing the concrete mixture. The amount of cement in concrete mixture is mix according to the different percentage to get a good quality of concrete mixture for 3D printing. The strength and workability of the mix design should be relevant and follow the criteria. Table 1 shows the mixture of Portland cement with varying percentage of GGBS according to the different researchers.

Table 1. Percentage of Portland cement and GGBS mix design.

Researcher	Portland Cement (%)	GGBS (%)	Results
Shreyask [2]	100	0	30-40 % GGBS give the best performance
	90	10	
	80	20	
	70	30	
	60	40	
Zhou et al. [4]	100	0	30 % of GGBS give the higher compressive strength
	70	30	
	50	50	
	30	70	
Barnett et al [5]	100	0	30-40 % GGBS give the best performance
	80	20	
	65	35	
	50	50	
	30	70	
Talib Khalid et al. [8]	100	0	40 % of GGBS give the higher compressive strength
	70	30	
	60	40	
	50	50	
	90	10	40 % of GGBS give the higher compressive strength
	80	20	

Chandra	70	30
Thakur and	60	40
Kumar [9]	50	50
	40	60
	30	70

To get a good quality of concrete mixture, the consistency of cement should be checked through compressive strength and workability of the mix design [10]. The concrete mix must be included the characteristic of 3D printing of extrudability, flowability and buildability.

In mix design of this research study, it is focusing on the percentage of GGBS uses in the concrete mixture to build the 3D building printing [11]. Based on the literature review that has been interpreted, the percentage of GGBS uses in the concrete mixture to replace the Portland cement according to a past case study can be 10% until 90%. But, the researchers of the past case study mentioned that the 30% until 40% replacement is highly recommended in concrete mixture to achieve a good strength and workability in 3D building printing.

According to Chandra Thakur and Kumar [12], the observation compressive strength of concrete of partially replaced Portland cement with GGBS decreases at 3 days and 7 days while at 28 days, it increases giving the optimum at 40% replacement. Shreyask [7], and Talib Khalid et al. [13], believed that at 30% - 40% replacement by GGBS, the level of compressive strength is higher than other concrete mixture. Besides, Zhou et al. [9], found that 30% replacement by GGBS gives the maximum compressive strength value. It can be proved in the table 2 that the compressive strength at 40% replacement takes the higher strength compared to the others. Table 6 show the comparison compressive strength according to a percentage of GGBS replacement [14]. Figure 1 shows the graph of compressive strength through 3 days, 7 days, and 28 days of the concrete mixture respectively. It can be seen that the compressive strength of concrete mixture at 40% replacement was higher compared to the other concrete mixture.

Table 2. Comparison of compressive strength according to percentage of GGBS replacement [14].

Portland Cement %	GGBS %	Compressive strength (N/mm^2)		
		3 days	7 days	28 days
90	10	21.00	34.00	45.67
80	20	20.33	33.33	46.00
70	30	19.00	31.67	47.33
60	40	17.33	30.67	47.67
50	50	16.33	29.67	47.00
40	60	14.33	27.00	45.67
30	70	12.67	24.00	45.00

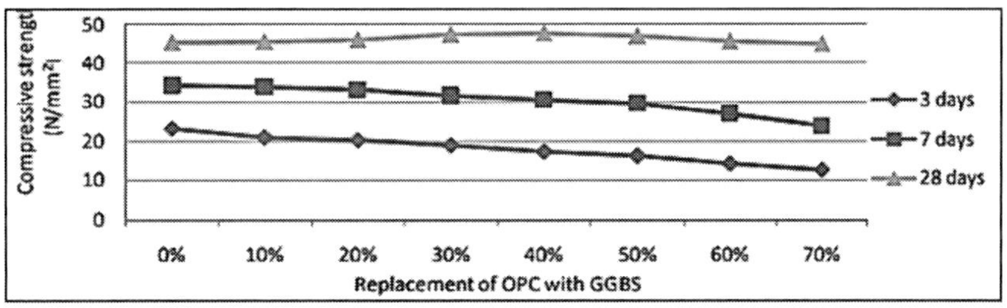

Figure 1. Compressive strength of Portland cement replacemnt with GGBS [14].

3. Concrete 3D Printing Laboratory Testing

Laboratory testing for 3D concrete printing experimented by previous researchers with various types of mix design suitable for concrete mixtures. Laboratory testing on concrete mixture includes compressive strength test and workability. In order to get good concrete mixture for 3D printing, the testing that needs to be highlight for concrete mixture needs to be analysed. The behaviour of concrete strength was usually is tested by using compressive strength test to find out the workability of concrete mixture in 3D printing. The compressive test is important to find out the characteristic of concrete mixture in accordance with the criteria. All the experiments need to figure out the behaviour of concrete mixture either it is suitable or not for 3D printing.

Workability of concrete is very important because the optimum strength cannot be attaining if the concrete is not placed properly. Moreover, improper compaction and due to improper placement also can be a reason for low workability of the concrete. The workability of concrete will increase with the increase of GGBS replacement level [6]. So that the GGBS may be used to replace Portland cement which will reduce the cost and also reduce the consumption of Portland cement. The workability of concrete mixture for 3D printing will determine by slump test, flow table, Vicat test, and extrusion test. Table 3 show compressive strength according to mix design regarding different researchers.

Table 3. Compressive strength according to mix design regarding to different researchers

Researcher	Mix design	Compressive strength (N/mm^2)		
		7 days	14 days	28 days
Shreyask [7]	40% GGBS + 60% Portland cement	15.70	18.25	26.45
Zhou et al. [9]	30% GGBS + 70% Portland cement	21.00	25.50	33.22
Barnett et al. [10]	35% GGBS + 65% Portland cement	16.72	23.44	48.00
Talib Khalid et al. [13]	30% GGBS + 70% Portland cement	14.23	30.06	45.10
Chandra Thakur and Kumar [14]	40% GGBS + 60% Portland cement	17.33	30.67	47.67

Based on previous studies on the use of GGBS in concrete mixture, Zhou et al. [19], have stated that GGBS possessed lower compressive strength than Portland cement at an early age but 30% of replacement GGBS in the mixture are slightly higher than Portland cement mixture. Another researcher, Talib Khalid et al. [13], stated that 30% until 40% replacement of GGBS got higher compressive strength in concrete mixture. However, the value of compressive strength getting lower when the replacement of GGBS in the mixture increase [13]. Figure 2 proved that the compressive strength of 30% replacement of GGBS is higher than the Portland cement mixture. Figure 2 show compressive strength GGBS mixture which (a) absolute strength; (b) relative strength [14].

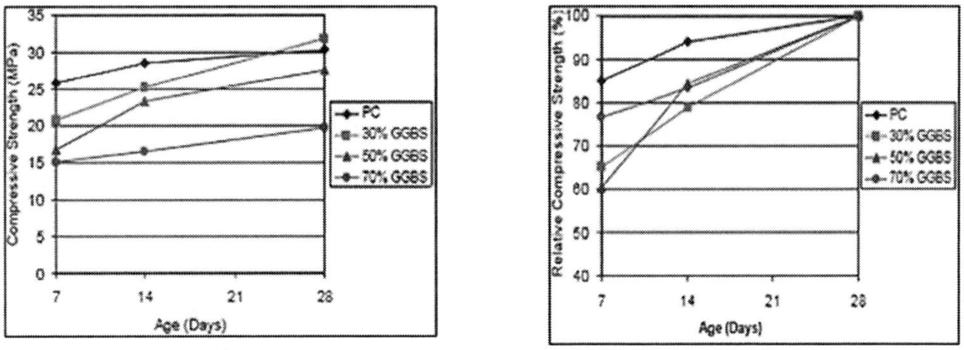

(a) (b)

Figure 2. Compressive strength GGBS mixture (a) absolute strength; (b) relative strength [4]

The extrusion test is usually conducted by using extrude machine which a prototype printing system with a chosen diameter nozzle, for linear extrusion as well as the base where the nozzle prints the layers of the mixture. By applying this method, the printability and buildability of the concrete mixture are going to determine as a characteristic of 3D printing process whether the mixture accepted or not.

4. Control aspect of 3D concrete printing

Control aspect is important to effectively print 3D printing structures, the main properties of concrete mixture needs to be considered are extrudability and flowability, buildability, strength between layers, aggregates, water-cement ratio. To achieve a stable foundation, the concrete needs to be bound within the layers when stacked on top of each other. The concrete should not be in a hardened condition, so as the concrete is placed on the surface of the previous layer, hydration of the concrete should be in progress. This factor is often linked to the workability and setting time of concrete for 3D building printing. Besides, the shape of layers of mortar should be considered to preserve the integrity of the structure and to ensure strong bonding between the layer [12]. Figure 3 shows the layer of 3D printing extrusion model [12].

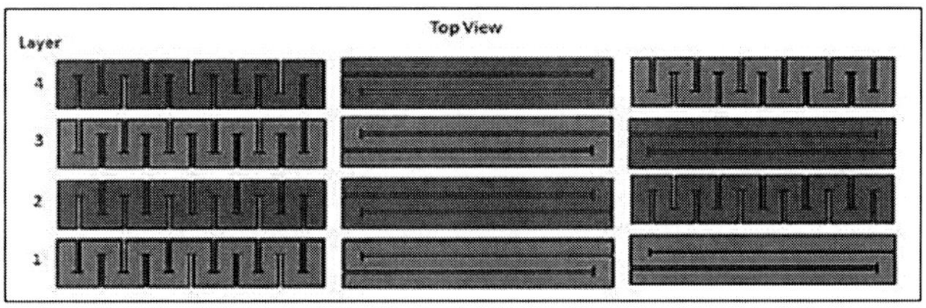

Figure 3. The layer of 3D printing extrusion model [12]

Talib Khalid et al. [13], mentioned that the mixture containing GGBS is required less water compared to mixture without GGBS. In a real application, the workability of 3D concrete printing is prone to even small variations of environmental conditions such as temperature, humidity, moisture of raw materials and others. Buswell et al. [15] mentioned that 3D construction printing a part that allows the extrusion to follow a finite path that places the material deposited, and is also repeated to create vertical height at each layer. Other than that, Garg and Kapoor [6] also mentioned that GGBS has good effects on the workability as the water binder ratio decrease. With the content of GGBS replacement in the mixture increase, the workability of concrete will be increasing too. Table 4 shows the criteria for accepting concrete mixture as printable or not [16].

Table 4. Criteria for accepting concrete mixture as printable or not [16].

Characteristic of 3D printing concrete	Accepted	Not accepted
Printability	1. The mixture is extruded through the nozzle. 2. Good printing quality meaning no voids, no dimensional variations of extruded material.	If 1 or 2 do not apply
Buildability	3. Five layers of printing material can be achieved without collapse. 4. Height of first layer versus height of fifth layer.	If 3 or 4 do not apply

The layered extrusion of concrete components allows the freshly printed material to have some unique rheological properties to successfully utilize the physical or mechanical properties of 3D printed concrete objects. The control aspect for extrudability, flowability and buildability are given in table 5 [16].

Table 5. Control aspect in concrete 3D printing [16].

Control aspect	Limitation
Extrudability	The capability to be extruded properly through the printing nozzle with a continuous material flow.
Flowability	The capacity to be worked and moved to the printing nozzle through a pumping system throughout a given time interval.
Buildability	The capacity to both remain stacked in layers after extrusion and sustain the weight of the subsequent layers that are deposited by the printing process.

5. Conclusion remarks

Through the case study, this research paper begins with an introduction to the assorted of 3D concrete printing technologies that are practice with waste materials which Ground Granulated Blast- Furnace Slag (GGBS). It can be inferred that the mix template for 3D printing uses 30% to 40% GGBS substitute for Portland cement. This is because many researchers typically use the compressive strength and workability of the concrete mixture with 30% up to 40% in 3D printing constructions. The technology has demonstrated its ability for 3D building printing and will become successful if the tools are applied further. Leading to the use of waste material as a cement substitute and the elimination of carbon dioxide gas emissions, the workability of the concrete mixture by using a partial percentage of GGBS is certainly recommended. In the future, the competition within contractors will increase due to demand in cheaper building and market will gradually grow the 3D building printing industry.

6. References

[1] Demyanenko O, Sorokina E, Kopanitsa N and Sarkisov Y 2018 Analysis of Strength Characteristics of Ggbs Concrete, *Int. J. Adv. Eng. Technol. E- Int J Adv Engg Tech V II* **12** 82–84

[2] Awasare V and Nagendra M V 2014 Analysis of Strength Characteristics of GGBS Concrete, *Int. J. Adv. Eng. Technol. E- Int J Adv Engg Tech V/Issue II* **12** 82–84

[3] Suhaini N A M 2020 Performance of modified epoxy mortar using composite cement containing Ground Granulated Blast Furnace Slag (GGBS) *IOP Conf. Ser. Mater. Sci. Eng.* **991** 012135

[4] Jamalulail N S and Salleh N 2021 Comprehensive Review of 3D Concrete Printing by Using Ground Granulated Blast-Furnace Slag, *Recent Trends in Civil Engineering and Built Environment* **2** 800–807

[5] Abdul Latip A, Ali N, Abdul Hamid N A, Salleh N, Abdullah S R and Shahidan S 2020 Performance of Modified Mortar Containing Epoxy *IOP Conference Series: Materials Science and Engineering*, **713(1)** 012004

[6] Garg E. K. and Kapoor E. K 2016 A Review on Ground Granulated Blast-Furnace Slag as a Cement replacing material, *Int. J. Eng. Res. Manag,* **3** 214–217

[7] Shreyask 2017 Characteristics of GGBS as an Alternate Material in Conventional Concrete **5** 3174-3184

[8] Hager I, Golonka A and Putanowicz R 2016 3D printing of buildings and building components as the future of sustainable construction, *Procedia Engineering* **151** 292–299

[9] Zhou X M, Slater J R, Wavell S E and Oladiran O 2012 Effects of PFA and GGBS on early-ages engineering properties of Portland cement systems, *Journal of Advanced Concrete Technology* **10**74–85

[10] Barnett S J, Soutsos M N, Millard S G and Bungey J H 2006 Strength development of mortars containing ground granulated blast-furnace slag: Effect of curing temperature and determination of apparent activation energies, *Cement and Concrete Research* **36** 434–440

[11] Yamgar S B and Takkalaki S R 2018 Study and Analysis of Strength of GGBS Concrete, *International Journal of Engineering and Management Research* **8** 28–4

[12] Nadarajah N 2018 *Development of concrete 3D printing* Master Dissertation (Espoo: Aalto University) 67

[13] Talib Khalid W T, Qazi O, Abdirizak A M and Rashid R S M 2019 Comparison of different waste materials used as cement replacement in concrete *IOP Conference Series: Earth and Environmental Science* **357** 012010

[14] Chandra Thakur and Kumar S 2007 Assessment of the Properties of Cement and amp; Mortar using GGBS, *International Journal of Innovative Research in Science, Engineering and Technology* **3297** 15224–15231

[15] Buswell R A, De Silva W R L, Jones S Z and Dirrenberger J 2018 Cement and Concrete Research 3D printing using concrete extrusion : A roadmap for research, *Cement and Concrete Research* **112** 37–49

[16] Garg E K and Kapoor E K 2016 A Review on Ground Granulated Blast-Furnace Slag as a Cement replacing material, *International Journal of Engineering Research And Management (IJERM)* **03** 214–217

Acknowledgement

Special acknowledgement to the Faculty of Civil Engineering and Built Environment, Universiti Tun Hussein Onn Malaysia (UTHM) for providing the technical support and facilities, which without the technical support and facilities, the study would not have been possibly completed.

Partial Replacement of Fine Aggregate Using Waste Materials in Concrete as Roof Tile: A Review

K Supar[1], F A A Rani[2], N L Mazlan[2] and M K Musa[2]

[1] Centre of Diploma Studies, Universiti Tun Hussein Onn Malaysia, 84600 Pagoh, Muar, Johor, Malaysia
[2] Faculty of Engineering Technology, Universiti Tun Hussein Onn Malaysia, 84600 Pagoh, Muar, Johor, Malaysia
Corresponding author: khairis@uthm.edu.my

Abstract. The use of waste material as a partial replacement has become popular in concrete mixture studies. Many research has utilized waste materials like cement, fine aggregate, coarse aggregate, and reinforcing materials substitute. The current paper focuses on some of the waste elements that are utilized in a concrete mortar (use in roof tile) as a partial replacement for fine aggregates such as rubber ash, sawdust, seashells, crumb rubber, pistachio shells, cinder sand, stone dust, and copper slag. There are many variations of mix proportion and water-cement ratio for every waste material. Compressive strength was compared and found that stone dust and the combination of seashell and coconut fiber shows an incensement when used to replacing fine aggregate. The suitable replacement level for stone dust is 25% and 50%. While the suitable replacement levels for the combination of sea shell and coconut fiber are 20% and 30%. Material from the rubber families such as rubber crumb and rubber ash is only suitable for replacement levels. Rubber families especially rubber crumbs have shown low water absorption value which is good in the production of roofing products. As we know, the roof should have waterproof properties to prevent any leaks from happening when it rains. Most of the waste materials added as fine aggregates in concrete have increased the amount of water absorption and found that sawdust is the most abundant material with a high percentage of water absorption compared to the others. Research on the partial replacement of fine aggregates replaced with waste materials is needed more extensively to provide more confidence about their use in concrete mortars, especially on roof tiles.

1. Introduction

Construction in the country is developing in this day and age of modernism and sophistication. As a result, the market for building materials increased. Concrete is a mixture of fine aggregate, coarse aggregate, and even cement. A great number of houses have been constructed as a result of the growing population. Shop construction, public building construction, and so into use have advanced significantly. This has resulted in the insufficiency of an article source of construction materials. Due to the lack of construction materials and also dumping of waste materials, researchers are considering methods to create useful waste materials which can reduce the number of resources and costs. Roof tile is a man-made component of durable material, such as ceramic, stone, metal, or glass that is commonly used to cover roofs, floors, and walls. The name comes from the French word "tuile," which comes from the Latin word "tegula." Roof tiles are primarily intended to keep the rain out and are usually made from locally accessible materials such as terracotta or slats.

Most significantly, a roof tile provides excellent insulation and protection such as heavy rains, severe climate conditions burns, and natural disasters [1]. Since the concrete tiles are vulnerable to the outdoors on structure rooftops, it also vulnerable to repeated moisture as a result of rainwater, and weather differences every day as well as more severe weather conditions like hailstorms and heavy rains during storms. Concrete roof tiles are constructed of a sand, cement, and water combination that is molded underneath high heat and pressure. A colour substance can be used to finish the exterior surfaces of a tile. Concrete tiles contain extra water locks, or interlocking ribs, on the edges to prevent water penetration. Concrete tiles can imitate the look of conventional clay tiles, wood shake, slate, and stone [2].

There are various types of roof tiles and each type has its unique forms and styles, as well as its unique production technique. In comparison to ceramic tile, this roofing tile has the advantages which are high durability, the cheaper price of installation, and a wider variety of colours. However, the disadvantage of concrete roofing tiles is heavier than ceramic tiles and requiring a larger support framework. Other than that, zinc sheets are also applied to coat the roofs of buildings. Their benefits include cheap costs but are easily rusted. The poor thermal and acoustic insulation compared to other raw materials are disadvantages. As a result, it is mostly employed in low-cost buildings. As a sophisticated building material, plastic roofing tiles are also being used. Its technological benefits include improved thermal insulation, a wide range of colours, simplicity of fitting, and corrosion resistance. Since it has a shorter lifespan and is less resistant to UV radiation and rain than ceramic or concrete roofing tiles, it has not achieved an identical level of effectiveness [3].

Inadequate management procedures are the main difficulty of unsafe, waste management which has caused ecological difficulties. Solid garbage is mostly produced by outpost and residential, businesses building, and private companies [4]. Currently, the main difficulty in developing countries is a huge amount of waste materials from different types. Comprehensive waste recycling researches are being carried out to reduce environmental impact. Many sectors have done their respective responsibilities such as recycling and manufacturing businesses have also made progress in utilizing these waste items [5]. Among the examples of initiatives or innovations that can be done to reuse waste materials such as sea coat, ash rubber, crumb rubber, cider sand, copper slag, pistachio shell, sawdust, or stone dust to substitute fine aggregates partially in roof tile construction.

1.1. Sea shell

Sea shells have become a waste formed by the decomposition of deceased animals close to the seaside. It is deposited by waves on the shore of course. In a recent study in India, the use of sea shells has been widely used as a building material. The increased use of shells would not only save these building materials but also help with the problem of trash disposal. Therefore, the need to replace current materials such as concrete must be changed by making some innovations in the concrete mix to meet the structural demands. The availability of affordable raw resources can also reduce manufacturing costs coupled with recycled materials. Thus, the construction of housing can be saved and building materials affect whether the construction of the building is cheap or expensive [6].

1.2. Saw dust

Sawdust is generally removed through combustion and causes smoke that can be harmful to human health. The use of cement composites is more ecologically beneficial. A large amount of sawdust is used in producing furniture and is also discarded by the wood industry. As an innovation in the construction sector, wood dust can be used to replace fine aggregates in a mortar and horticultural goods to minimize the burden of scarce natural resources. Savings are characterized by composites with limited mechanical performance, low durability, and poor compatibility. The integration of sawdust into the concrete and mortar manufacturing process not only minimizes environmental damage but also preserves conventional components from concrete and mortar. It has numerous advantages over conventional concrete, including weight reduction (dead load reduction transmitted to the base), high economic efficiency over standard concrete weights, low damage, and longer life of coating because of lower exercising pressure, easy handling, mixing, and placing of other types of concrete. Although this is advantageous, the sawdust concrete is subject to irregular setting times and low flexibility [7].

1.3. Pistachio shell

The pistachio can be found throughout Asia Minor, from the Mediterranean islands in the west to India in the east. It is also widely found in Syria, Iraq, and Iran. It has probably been formed in wilderness areas since it requires lengthy warm summers for the development of organic products, is dry and salt-tolerant. Several studies discussed replacing sand with many different forms of waste materials such as the use of recycled glass bottles as fine aggregates in cement mixture was conducted by Ganiron [8]. He concluded that recycled glass bottles are a decrease in unit weight, the value of the elastic modulus and the cost of concrete as an alternative fine aggregate and that recycle bottles are not a useful alternative of fine aggregates for structural elements such as columns, beams, and suspended plates. Another researcher was investigated groundnut as a replacement to fine aggregates. They found that the use of groundnut shells in concrete decreases concrete workability because the groundnut shell is highly absorbed by water. The densities and compressive strength of concrete decreased with an increased percentage of groundnut shells [9].

1.4. Copper slag

Copper slag is an industrial by-product material formed by the copper process. About 2.2 tons of copper slag are created for every tone of copper output. The world copper sector has calculated that about 24.6 million tons of slag are produced. Although copper slag is commonly utilized in the sandblasting business and in the production of abrasive equipment, it is removed without further reuse or recycling. A group of researchers examined the influence on strength qualities of employing copper slag as a substitute for fine aggregate. M25 grade cement was employed and experiments were carried out with 0 % to 100 % of concrete for varying quantities of copper slag substitution with sand. The results show that the compressive strength of concrete increased by 55 % to 40 % when finer aggregates were replaced with copper lacquer [10].

1.5. Stone dust

There are many stones crushers in the country of Saurashtra producing enormous amounts of copper slag during stone cutting. The dust in sandstone causes a problem of disposal, environmental problems because of its non-fertile qualities, and health risks. In manufacturing development, the use of stone dust such as powder of marble, stone slurry, etc. as a substitute for the fine aggregate was implemented, and the substitution produced mixed results in the qualities of mortar [11]. In advanced industries, the use of stone dust such as stone slurry and marble powder have been well accepted as substitutes for fine aggregates. It can be concluded that replacement gives a batter capable of producing the properties of the mortar itself

1.6. Cinder sand

Cinder industrial waste is usually utilized in sunken areas of the slabs as a lightweight filler material. This is permeable in nature and irregularly available. It can be utilized as a coarse concrete aggregate in breakage to uniform sizes of 20mm or 12 mm, reducing the density of concrete and being used for structural purposes. By employing self-curing compounds instead of water treatment and M- sand in concrete rather than the river sand, natural resources like water and sand are conserved [12]. Cinder is a waste substance produced as coal leftover from the power plant blast furnace. The material has been taken from burnt store boilers in Jamshedpur and mostly contains silica and metal oxides. Slag is another waste substance formed during the production of molten iron as a byproduct. It is made through the fusing of limestone and also other streams with the coke ash and the ferrous ore's silicone and aluminum components. It comprises silica, aluminum, iron, and other oxides [13].

1.7. Rubber ash

The collection of waste tires, especially in Malaysia, is one of the biggest environmental challenges in the globe. Because scrap pneumatic materials are non-degradable, they can pose major environmental and public health hazards. One solution to solve the growing quantity of waste pipes is to include scrap rubber into concrete mixes to generate rubberized concrete. In a concrete mixture, rubber can replace fine particles. This strategy can help prevent sites from being landfilled. Pyrolysis is a word based on the Greek "pyr," meaning fire, and "lysis," meaning separation. Pyrolysis is a material disintegrating process and normally works at 430 °C (800 °F) at a higher temperature. This process is done if the organic material is burned otherwise without oxygen. In the absence of oxygen, pyrolysis is also characterized as a means of recycling discarded tires, by thermal decomposition [14]. In addition, discarded tires can lead to flames especially in the summer because it is difficult to extinguish. The potential options for reducing waste rubber are to integrate it as a partial substitute for natural aggregates into cement concrete. This approach could be environmentally benign (as it helps to reduce discarded material and minimize pollution) and more economic since some of the costlier natural aggregates can be preserved. The usage of rubber in concrete has attracted considerable interest. Previous research shows that replacing aggregates with waste pneumatics in the concrete mixture can significantly boost toughness and ductility [15].

1.8. Crumb rubber

Waste rubber can be used as a component of the fine aggregates, coarse aggregates, or both. The use of rubber in mixtures of concrete or mortar creates the prevention of waste and reduces the depletion of raw virgin substances. Crumb rubber is the name given to any substance generated by the reduction to uniform granules of waste tires and other rubber with intrinsic reinforcement components such as stain and fiber, removed together with inert impurities of any kind, such as dust, glass, and rock. Crumb rubber is produced from two basic feed stocks: tire retrieval, and tire scrap rubber [16].

Crumb rubber is used in concrete to generate rubbercrete as a partial replacement for fine aggregates. Compared to the fine aggregate it has a lower specific gravity of 0.51 to 1,2 kg/m3, low water absorption, resistance, and steepness of 524 kg/m3 to 1273 kg/m3. Crumb rubber is a hydrophobic and non-polar substance that repels the surface of the water and entails the air. It also has a different gradation than the finely aggregated aggregate which in particle size analysis falls below the lower curve limit. Therefore, it alters the grading to a not continuous aggregate gradation if the fine aggregate is partially replaced by rubbercrete. Partial substitution of fine aggregate with rubber cream is usually carried out by the material volume due to the lower specific gravity of the crumb rubber than the fine aggregate [17].

2. Mix proportion ratio of concrete with waste material as partial replacement

Table 1 shows selected mix proportions waste material as partial replacement in concrete. Researchers experimented with a variety of goal strengths, including varied waste material kinds, fine aggregate replacement levels, and water-cement ratio. The type of cement was using ordinary Portland cement (OPC) for all the experiments. Venkatesh [18] using cinder sand as a partial replacement for fine aggregate with 20% and 40%, and water-cement ratio 0.45 by using 1:3 (cement: fine aggregate).

Table 1. Mix proportion of waste material for partial replacement of fine aggregate in concrete.

Waste material used	Replacement level (%)	Water cement ratio	Mix proportion (cement: fine aggregate)	Size of aggregate	Reference
Rubber Ash	0 , 3, 5, 7	0.55	1:2	< 4.75 mm	[15]
Saw dust	5, 10, 15, 20, 30, 50	0.55	1:3	0.15mm to 2 mm	[7]
Sea shells and coconut fiber	0, 20, 30	-	1:4	-	[6]
Crumb rubber	5, 10, 15, 50	0.56	1:2	3mm to 5mm	[19]
Pistachio shells	0, 10, 20, 30, 40, 50, 60	0.48	1:3	< 4.75 mm	[9]
Cinder sand	20, 40	0.45	1:3	-	[18]
Stone dust	0, 25, 50	0.60	1:3	< 4.75 mm	[11]
Copper slag	20, 40, 50, 60, 80, 100	0.45	-	4.75mm to 20 mm	[10]

According to experiment conduct by Reddy et al. [10], the waste material that was used is copper slag to replace the fine aggregate by 20%, 40%, 50%, 60%, 80%, and 100% with a water-cement ratio of 0.45. A researcher conducts an experiment by replacing fine aggregate with crumb rubber by 5%, 10%, 15%, and 50% by using a ratio of 1:2 (cement: fine aggregate) and a water-cement ratio of 0.56 [19]. Experiment using pistachio shells as a partial replacement for fine aggregate has been conducted by Alsalami [9] with for 0%, 10%, 20%, 30%, 40%, 50%, and 60% with using a water-cement ratio of 0.48 and mix proportion ratio of 1:3 (cement: fine aggregate).

Senin [15] using rubber ash with 0%, 3%, 5%, and 7% as a partial replacement for fine aggregate with a mix proportion ratio of 1:2 (cement: fine aggregate) and water-cement ratio of 0.55. Jonathan and Charles [7] have been conducted partial replacement for fine aggregate using sawdust with 5%, 10%, 15%, 20%, 30%, and 50% with a mix proportion ratio of 1:3 (cement: fine aggregate) and a water-cement ratio of 0.55. Another waste material which is sea shells was used to replace the fine aggregate in the concrete mixture [6]. The percentage of sea shells to replace the fine aggregate is between 0% - 30% with a mix proportion ratio of 1:4 (cement: fine aggregate). Gamit and Patel [11] conducted stone dust as waste material by using 0%, 25%, and 50% for partial replacement of fine aggregate with a mix proportion ratio of 1:3 (cement: fine aggregate).

3. Compressive strength in concrete for each types of waste materials

This section discussed on strength of concrete mixed with waste materials from previous researchers. Table 2 was described the compressive strength of concrete mortar for combination from different types of waste materials used as partial replacement of fine aggregate. Among the types of waste materials are used in mix proportion of fine aggregate such as rubber ash, sawdust, sea shells, coconut fiber, rubber crumb, pistachio shells, cinder sand, stone dust and copper slag.

Table 2. Compressive strength of concrete for difference type of waste material.

Code	Waste material used	Replacement level (%)	Curing time	Control sample	Compressive strength	Reference
ASTM C109-02	Rubber Ash	0 , 3, 5, 7	28	30.97 MPa	3% = 28.90 MPa 5% = 31.50 MPa 7% = 25.60 MPa	[15]
(BS EN 1015-11)	Sawdust	5, 10, 15, 20, 30, 50	28	7.52 MPa	5%=6.99 MPa 10%=6.72 MPa 15%=4.71 MPa 20%=4.02 MPa 30% = 2.61 MPa	[7]
IS10262-2009	Sea shells and Coconut fiber	20 % seashell 30% seashell	28	7.95 MPa	20% = 8.7 MPa 30% = 9.1 MPa	[6]
-	Rubber crumb	5, 10, 15, 50	14	44 MPa	5% = 19 MPa 10% = 14.1 MPa 15% = 11.7 MPa	[19]
B.S 1881-part 4 - 1989	pistachio shells	0, 10, 20, 30, 40, 50, 60	28	71.47 MPa	10% = 51.54 MPa 20% = 14.10 MPa 30% = 6.20 MPa 40% = 4.35 MPa	[9]
-	Cinder sand	20, 40	28	36.31 MPa	20%= 47.04 MPa 40%= 37.90 MPa	[18]
IS 2250-1981	Stone dust	0, 25, 50	28	22.27 MPa	25% = 21.47 MPa 50% = 23 MPa	[11]
-	Copper slag	20, 40, 50, 60, 80, 100	28	42 MPa	20% = 43 MPa 40% = 45 MPa 50% = 44 MPa 60% = 38 MPa	[10]

A group of researchers has conducted the concrete using rubber ash as partial replacement to fine aggregate. They refer to the ASTM C109-02 standard as their procedure to conduct the test. The compressive strength test showed that 5% replacement of fine aggregate to rubber ash can increase the strength of concrete from 30.97 MPa to 31.50 MPa. However, concrete containing 3% and 7% of rubber ash showing a reduction in compressive strength which is 28.90 MPa and 25.60 MPa compare to the control sample [19].

An experiment was performed using sawdust as a partial replacement for fine aggregate in concrete by using a mix proportion ratio of 1:3 (cement: aggregate). The test was conducted according to BS EN 1015-11 standard and been test after 28 days. For the control sample which does not have any replacement of sawdust with the compressive strength is 7.52 N/mm^2. However, after replaced the fine aggregate with sawdust for 5%, 10%, 15%, 20%, 30%, and 50%, the compressive strength shows a reduction which is 6.99 N/mm^2, 6.72 N/mm^2, 4.71 N/mm^2, 4.02 N/mm^2, 2.61 N/mm^2, and 2.33 N/mm^2 follow by percentage of sawdust. The porosity of each component of a multiphase material such as mortar is the main reason for a source of strength loss. Due to their crystalline form (aggregate or sand) as traditional (natural) aggregates are often thick and powerful, while sawdust is woody and mushy when approaching water [7].

Many researchers have been conducted compression tests with sea shells and coconut fiber as a partial replacement for fine aggregate in concrete. The results showed that the addition of sea shells and coconut fiber increased the compressive strength of concrete. The test was done referring to IS10262-2009 as the standard for the compressive test of concrete by using a mix proportion ratio of 1:4 (cement: aggregate). A control sample was used as a guideline and the result shows the strength of concrete without any replacement is 7.95 N/mm^2. The compressive strength of concrete increased to 8.7 N/mm^2 when the combination of 20% of sea shells and 0.5% of coconut fiber are replaced in fine aggregate. It is also the same when the combination of 30% of sea shells and 0.6 % of coconut fiber has been replaced as fine aggregate and gives the strength incensement to 9.1 N/mm^2. The addition of coconut fiber and sea shells as an admixture will improve the compressive strength of the materials [6]. Natural fiber is one of the alternatives in composite parts that can be replaced or added into the original or conventional material of the composite itself. Among the natural fibers that have good tensile strength values are such as kenaf, banana leaves, coconut and pineapple leaves [20].

Crumb rubber was used as a partial replacement of fine aggregate in concrete to test its compressive strength. In the test, the partial replacement has been divided into 5%, 10%, 15%, and 50%. The result of compressive strength was taken after 14 days. A control sample was prepared and showed a compressive strength of 44 MPa. Based on the results, the compressive strength on partial replacement of fine aggregate with crumb rubber was decreased to 19 MPa, 14.1 MPa, 11.7 MPa, and 2 MPa. The researcher also highlighted that 50% of replacement is not suitable because the compressive strength is lowest [19].

The compressive strength of pistachio shell as a partial replacement for fine aggregate in concrete based on B.S 1881-part 4-1989 was tested by the researchers. They reported that pistachio shell substitution as a fine aggregate showed a drastic decrease in compressive strength after 28 days. The compressive strength for the control sample (purely cement and aggregate) is 71.47 MPa. However, after replacing fine aggregate with 10%, 20%, 30%, 40%, 50%, and 60% of pistachio shell, the compressive strength decreased to 51.54 MPa, 14.1 MPa, 6.2 MPa, 4.35 MPa, 2.5 MPa, and 1.75 MPa. The reduced workability of the mixture caused by the absorption of water by the pistachio shells during mixing could explain the decrease in compressive strength as replacement levels increased. Furthermore, the low density of pistachio shells relative to fine aggregate may lead to the compressive strength being lesser [9].

Venkatesh [18] has conducted the compressive strength of cinder sand to replace a fine aggregate in concrete. The mix ratio for the concrete is 1:3 (cement: aggregate) and a water-cement ratio of 0.45. The percentage of cinder sand that was used in the experiment to replaced fine aggregate is 20% and 40%. The compressive strength increased at 20% of replacement which is from 36.31 N/mm^2 (as control sample) to 47.04 N/mm^2. However, the compressive strength is decreased from 47.04 N/mm^2 to 37.90 N/mm2 when a partial replacement of cinder dust is 40%. The use of lightweight industrial waste aggregates like cinder with a combination of natural sand (20% from fine aggregate) and industrial waste admixture, has proven to be rather adequate in terms of various strengths evaluated [18].

All the procedures and testing such as casting, compaction and curing were accomplished according to IS 2250-1981. The compression testing has been conducted after 28 days of curing with 0%, 25%, and 50% of replacement percentage of fine aggregate. A control sample has been conducted first and after 28 days, the compressive strength showed 22.27 MPa. For the concrete mixture of a partial replacement of 25% stone dust, the compressive strength is decreased to 21.47 MPa. Besides that, a partial replacement for 50% stone dust, the compressive strength is increased to 23 MPa. It can be concluded that there is not much increase in compressive strength by the replacement percentage of sand stone dust [11]. However, the results show with added stone dust in the concrete mixture can be increased the compressive strength.

The researcher has cast the concrete using copper slag as a partial replacement of fine aggregate with 20%, 40%, 50%, 60%, 80%, and 100% replacement percentage. The compression testing after 28 days of the curing process has been conducted. A control sample has been tested first and showed a compressive strength is 42 MPa. For partial replacement results, the compressive strength of concrete containing 20%, 40%, and 50% of copper slag was higher than the control sample concrete. The strength is increased from 42 MPa to 43 MPa, 45 MPa, and 44 MPa according to the replacement percentage. The compressive strength starts to drop when the replacement percentage reaches 60% - 100% of copper slag. Copper slag replacement can provide compressive strength that is comparable to or higher than that of the control mixture in cement mortars [10].

4. Water Absorption

Water absorption and porosity are crucial markers of hardened concrete's longevity. The properties of concrete such as performance and life expectancy can be considerably improved by reducing water absorption and porosity. Reducing porosity also helps compressive and flexural strengths, as porosity and solid strength have a fundamental inverse relationship [21]. A water absorption test was performed on a concrete specimen to assess the amount of water that was getting through the concrete. Table 3 shows the water absorption percentage values of pistachio shells, sea shells with coconut fiber, rubber crumb, and sawdust with different replacement percentages. It shows that the percentage of replacement becomes increases, the water absorption also increases. The value of water absorption for pistachio shells is increased from 0.47% (control sample) to 6.04% (60% of fine aggregate replacement). The sample from the combination pistachio shell mixture is dried in a 105°C oven for 24 hours. The dried specimens were weighed after cooling until reach room temperature (25 °C). The dried specimens were submerged into water, and then the weight of the specimen was collected at predefined intervals [9]. For sea shells with coconut fiber combination, the water absorption value also increases to 6.8% (at 30% replacement) compare to a control sample with only 5.4%.

Most researchers followed standards to AS 4046.4-2002 (Methods of testing Roof tiles – Determination of Water Absorption). For rubber crumbs replacement, the amount of water absorption shows increased from 1.40% (control sample) to 4.83% (50% rubber crumb replacement) when it has been tested after 14 days of curing. Due to rubber particles are much softer than the surrounding particles, a mixture of concrete and rubber will make it easy to bend. Based on the findings obtained, table 3 shows that combination of rubber and concrete mixtures produce the lowest water absorption compared to other mixtures. Rubber is among the materials that are easily compacted aid to fill pores in concrete mixtures and will produce small porosity [19]. The same incensement also happens when replacing sawdust as partial replacement to fine aggregate. The amount of water absorption for a control sample is 11.97% and increased to 20.54% with the replacement of 50% sawdust.

Table 3. Water absorption of concrete for difference type of waste materials

Waste material used	Replacement level (%)	Testing day	Control sample	Absorption percentage	Reference
Pistachio shells	0, 10, 20, 30, 40, 50, 60	28	0.4695%	10 = 0.80% 20 = 2.04% 30 = 3.20% 40 = 4.01% 50 = 5.03%	[9]
Sea shells + coconut fiber	0, 20, 30	28	5.40%	20 = 6.09% 30 = 6.80%	[6]
Rubber crumb	5, 10, 15, 50	14	1.40%	5 = 2.56 % 10 = 3.19% 15 = 3.91% 50 = 4.83%	[19]
Saw dust	5, 10, 15, 20, 30, 50	28	11.97%	5 = 12.13 % 10 = 12.96% 15 = 14.55% 20 = 16.72% 30 = 18.92% 50 = 20.54%	[7]

5. Conclusion

Replace waste material as a partial replacement to fine aggregate in concrete is has a positive result in the properties of concrete. Compare to normal concrete, the strength of concrete can be seen increased in a few types of waste materials such as cinder sand, copper slag, and sea shells with the replacement percentage that has been studied. For materials such as rubber ash, cinder sand, and cooper slag the compressive strength of concrete was increased when the percentage of partial replacement was added in little and then decreased when replacing in a large percentage. For sea shells with coconut fiber, the compressive strength of concrete increased with the review percentage of replacement. Lastly, stone dust can be used until 50% replacement of fine aggregate to obtained high compressive strength. It has been observed by replacing fine aggregate with waste material in concrete has shown the amount of water absorption is increase. There is no waste material that is being used to show low water absorption. The physical form of waste materials that are sensitive to water such as pistachio shells and sawdust are unnecessarily used because moisture affects the weight and strength. For the other waste material, the incensement of water absorption level is still low if being replaces for a small amount of percentage especially rubber crumb. This paper focused on compressive strength and water absorption for the combination of waste material and fine aggregate as concrete mortar (roof tile). This is because these elements are very important to determine the effective combination of mixed proportion.

6. References

[1] Topacio A, Eng M, Arkuino R C, Romano K V and Jayvee C 2018 Development of hydrophobic composite roof tiles utilizing recycled plastic materials, *Materials Engineering and Development* **6** 1-13

[2] Rajeev P, Sanjayan J G and Seenuth S S 2016 Assessment of thermal cracking in concrete roof tiles, *Materials and Design* **107** 470–477

[3] Gaggino R, Kreiker J, Filippín C, Amono M P S, Laría J G and Peisino L E 2018 The comprehensive comparison of thermal and physical-mechanical properties of the recycled rubber and plastic roofing tiles versus roofing tiles made with different traditional materials, *Adv. Civ. Eng.* **2018** 1-11

[4] Alaloul W, Loganathan R, Mohammed B and Adamu M 2018 Deformation properties of concrete containing crumb rubber as partial replacement to fine aggregate, *Int. J. Civ. Eng. Technol.* **9** 317–326

[5] Rohit P, Varma P V S S P G and Sk A 2018 Partial replacement of fine aggregate with crumb rubber in concrete, *Int. J. Civ. Eng. Technol.* **9** 895–903

[6] Elwin S R, Abarna K, Aiswarya A, Anuruthra V and Atchaya A 2019 Experimental investigation on roofing tiles by partial replacement of seashell and using coconut fibre as an admixture, *Int. J. of Adv. Research in Basic Eng. Sci. and Technol.* **5** 102–105

[7] Jonathan S and Charles K K 2017 Study of brick mortar using sawdust as partial replacement for sand, *J. Civ. Eng. Constr. Technol.* **8** 59–66

[8] Ganiron T J 2013 Use of recycled glass bottles as fine aggregates in concrete mixture, *Int. J. Adv. Sci. Technol.* **61** 17–28

[9] Alsalami Z H A 2017 Study the effect of partially replacement sand by waste pistachio shells in cement mortar, *Appl. Adhes. Sci.* **5** 1–12

[10] Reddy K B P, Tanuja K and Naidu N V D 2016 Use of copper slag in concrete and cement mortar as replacement of sand, *Int. Research J. of Eng. and Technol.* **3** 254-260

[11] Gamit K C and Patel H S 2018 Evaluation of compressive strength of mortars containing stone dust as partially replacement with fine aggregate, *Int. Research J. of Eng. and Technol.* **5** 2373-2375

[12] Gunasekar S and Santhi M H 2020 Mix design of self- compacting self-curing concrete using m-sand and cinder aggregate, *J. of Xi'an University of Archit. & Technol.* **7** 1666–1673

[13] Sinha A K, Havanagi V G, Ranjan A, Kanaujia V K and Mathu S 2015 Stabilised cinder waste material for construction of pavement layers, *CSIR-Central Road Research Institute (New Delhi)* **7** 1–17.

[14] Senin M S, Shahidan S, Shamsuddin S M, Ariffin S F A, Othman N H, Rahman R, Khalid F S and Nazri F M 2017 The optimum content of rubber ash in concrete: flexural strength *IOP Conf. Ser. Mater. Sci. Eng.* **271** 1-8

[15] Senin M S, Shahidan S, Leman A S and Hannan N I R R 2016 Properties of cement mortar containing rubber ash as sand replacement *IOP Conf. Ser. Mater. Sci. Eng.* **160** 1-9

[16] Resmi G, Prasobh P, Krishnan R R C and Sarath J 2018 Experimental investigation on utilization of crumb rubber waste in tile manufacturing, *Int. Research J. of Eng. and Technol.* **5** 2670-2674

[17] Mohammed B S, Adamu M and Shafiq N 2017 A review on the effect of crumb rubber on the properties of rubbercrete, *Int. J. Civ. Eng. Technol.* 1 **8** 599–615

[18] Venkatesh D N 2016 Replacement of natural sand with cinder as a fine aggregate in cement mortar with compressive strength, *Int. J. for Research in Emerging Sci. and Technol.* **3** 33–36

[19] Zainun N Y, Mohammad H, Bahron N, Hamedon N N and Seman M 2021 Sustainable concrete roof tiles, *Sustain. of Construc. Engi.* **7** 69–80

[20] Ahmad H and Supar K 2016 Experimental Strength of woven fabric kenaf composite plates with different stacking sequences, *Applied Mechanics and Materials.* **833** 27-32

[21] Nassar R U D and Soroushian P 2012 Strength and durability of recycled aggregate concrete containing milled glass as partial replacement for cement, *Construc. Build. Material.* **29** 368–377

Acknowledgments

The author would like to thank Research Fund (TIER 1 Vot. H856), Research Management Centre, Universiti Tun Hussein Onn Malaysia (UTHM) in providing financial support for this project work.

Study on End-of-Life Tires (ELTs) Recycling Strategy and Applications

Habiba Afrin[1], **Nazmul Huda**[1] **and Rouzbeh Abbasi** [1]

[1] School of Engineering, Macquarie University, Sydney, Australia
Corresponding author email: habiba.afrin@students.mq.edu.au

Abstract. Due to modernization and urbanization, the number of vehicles on the road has been increased. Around 3 billion tires have been sold, and an equivalent number of tires have been discarded each year. Even though the lifetime of the tires has been increased but according to the Australian Bureau of Statistics, the number of end-of-life tires is going to rise approximately 5 billion in a year. Its complex composition, make it the most tricky and difficult waste in the world to handle. Because it creates significant health and environmental problem by emitting harmful chemicals in the environment, working as a birthplace for pests, and prone to fire hazards. Recycling waste tires can add economical value also creating a sustainable way to dispose of them. This paper presents different recycling strategies and civil engineering applications of end-of-life tires. Reduction, reuse, recovery, and recycling have been applied. Application of waste tire as reinforcing layers in landfill, road pavement, drainage system, fuel source in the kiln, playground surface makes it an ideal material for affordable, medium-density, low-rise buildings that are highly valued worldwide. Moreover, the sound insulation and absorption with enhanced seismic resilience properties of the end-of-life tire can provide novel and effective engineering solutions.

1. Introduction

The tire is a unique material that facilitates everyone's life. But despite its many valuable application to make our life faster, tire creates various pollution during different stages of its life cycle. It has durable, non-compostable, and resilient properties which make tires difficult to recycle. A large amount of non-biodegradable end-of-life tires causes critical environmental impacts, particularly by occupying a large landfill area. It is also dangerous for human health because of styrene, which is a strong toxic chemical [1]. In Australia, end-of-life tires are considered hazardous materials [2]. According to the data provided by the Australian Bureau of Statistics, for general use, the average lifespan of a passenger tire is approximately four years [2]. Because of the increasing use of vehicles and their short life span, a large number of tires accumulate [3][4]. Researchers have developed different ways for reducing environmental impacts and collecting valuable resources to contribute economically from the end-of-life tire [5]. Nowadays, sustainable awareness has raised pressure on sustainable waste management and alternative applications by adding value to the economy. One efficient alternative is to follow a circular economy by closing the materials loop by recycling, reusing, or reducing materials [6]. Feasible applications and markets for recycled end-of-life tires need to be developed and maintained worldwide. This research aims to find the feasibility of using ELTs reducing pressure on raw material use, energy consumption, water requirement, and environmental impact (reuse, recycling, recovery, regrooving, material recovery, and energy recovery) and study different recycling strategies and civil engineering applications of end-of-life tires. The study adopted a mixed-methods to observe ELT management practice worldwide. This included a range of data gathering techniques, combining both qualitative and quantitative methods. The qualitative and quantitative data were gathered from tire stewardship Australia (TSA) and official reports, legal documents, and scientific literature.

In 2016 the global market value of recycling tires was USD 0.95 billion, and it is growing by 2.1% per year [1]. According to tire stewardship Australia, it is estimated to use up to 30,000 tonnes of crumb rubber in the pavement as a modifier and over 12000 tonnes in asphalt by 2026, making road construction sustainable [1][7]. In figure 1, It shows the disposal of end-of-life tire in Australia and USA.

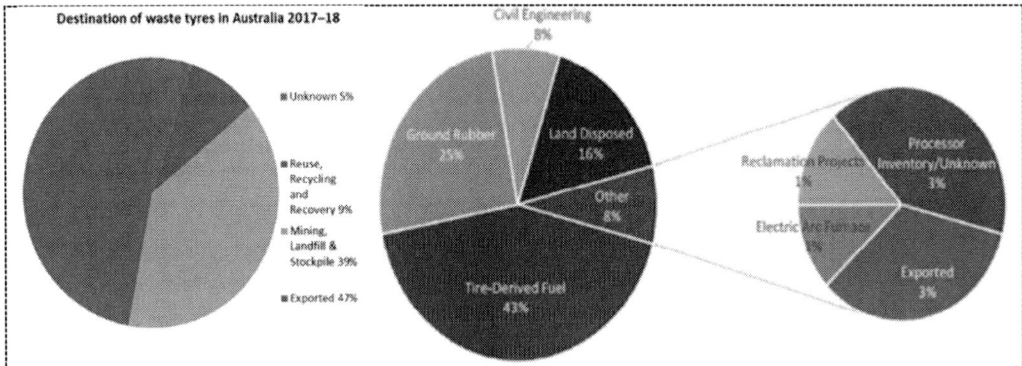

Figure 1. (a) Disposal of ELTs in Australia in 2017–18 [8] (b) Disposal of ELTs in USA [9]

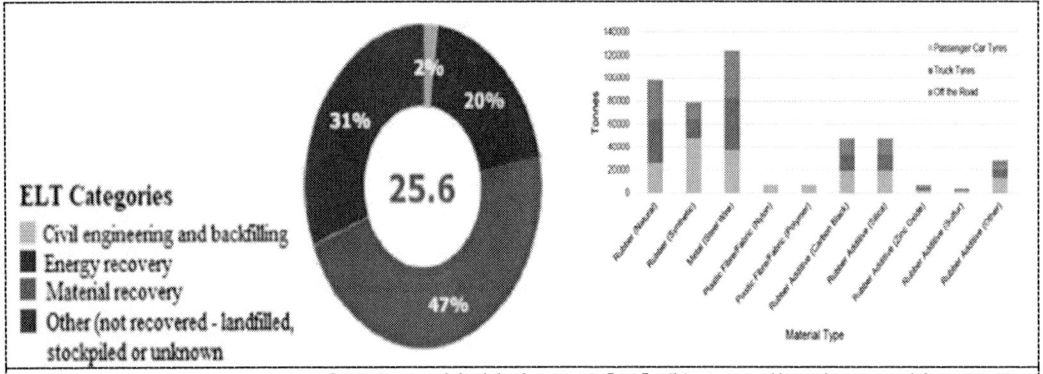

Figure 2. (a) Recovery rate of ELTs worldwide in 2019 [10], (b) Australian tires reaching ELTs by material in 2015–16 [11]

Around 16% of end-of-life tires are dumped in landfills and while others are being recycled in the USA. Around 6% to 8% of the waste tire are used as civil engineering materials in the USA and EU countries, but only around 0.4% of waste tires are being recycled in Australia [1]. In figure 2 shows the utilization of end-of-life tires worldwide and Australian tires reaching end-of-life in 2015-16. Figure 3 provides an inside view of the composition of a tire. Depending upon the manufacturers, their locations (i.e. locally or internationally), and type of vehicles (passengers and trucks), the composition of tires varies [12][13]. Manufacturing truck tires require a higher proportion of natural rubber, whereas passenger vehicle tires require a higher synthetic rubber. As natural rubber prices are gone up, the use of synthetic rubber has increased [7]. Table 1 shows the general components of passenger and truck tires.

IOP Conf. Series: Materials Science and Engineering 1200 (2021) 012009 doi:10.1088/1757-899X/1200/1/012009

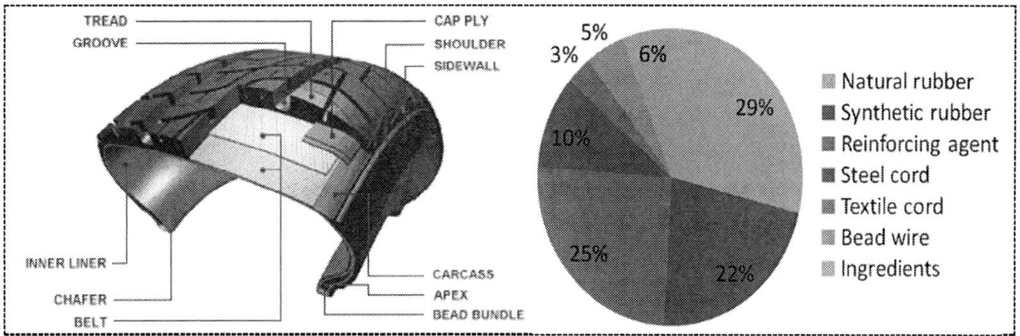

Figure 3. Different sections of a tire [8, 10].

Table 1. Components of truck and passenger vehicle tires

Material	Australia		United Kingdom		USA
	Passenger cars	Truck/buses	Passenger cars	Truck/buses	Components of passenger and truck tires in the USA is roughly the same
Rubber	16% Natural	29% Natural	17% Natural	28% Natural	14–27% Natural
	29% Synthetic	13% Synthetic	31% Synthetic	15% Synthetic	14–27% Synthetic
Carbon black	23%	24%	22%	21%	28%
Metal	16%	25%	15%	27%	14%-15%
Textile	6%	–	5%	–	
Additives	8% (Additives) 1% (Zinc oxide) 1% (Sulphur)	2% (Zinc oxide) 1% (Sulphur)	8% (Additives) 1% (Zinc oxide) 1% (Sulphur)	6% (Additives) 2% (Zinc oxide) 1% (Sulphur)	16–17% (Processing Oils)
References	[11]		[14][15]		[16]

2. Recycling strategy of end-of-life tire

Recycling end-of-life tire is an environmental concern as it takes around 80-90 years to degrade naturally. So it has a long life in the environment [17]. To alleviate the environmental impacts of end-of-life tires, reusing, retreading, regeneration, co-processing, pyrolysis, and landfills are the most common practice [10][11].

2.1. Reuse

Reusing is finding a new alternative to adopt usable tires for other applications. It can be used as a whole or cut into smaller pieces. This procedure creates an opportunity to produce other valuable materials. It presents a new strategy to cope with environmental issues associated with end-of-life tires [18]. The tire can be used in roadside barriers, parks and playgrounds, structures, channels, artificial reefs, and biogas drainage [19, 20]. Figure 4 shows the different stages of tire recovery.

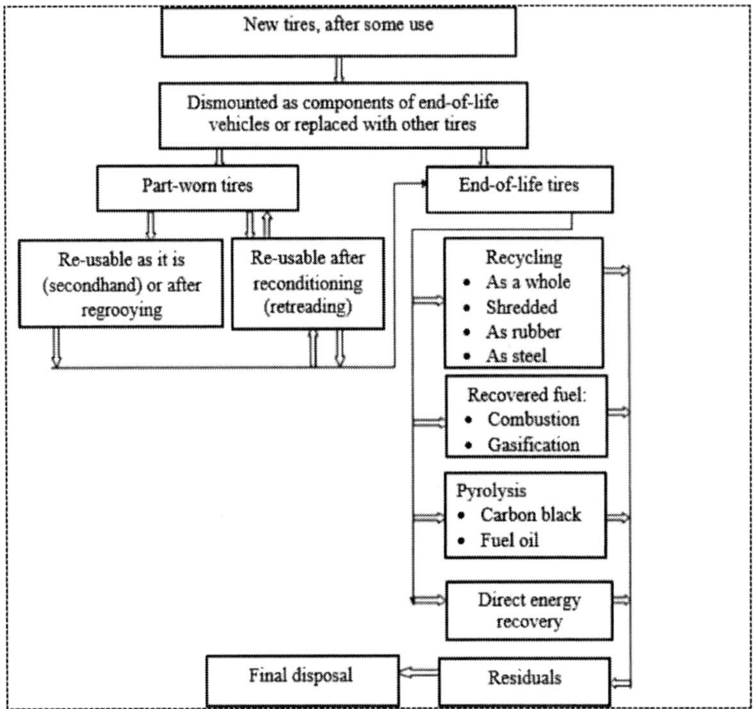

Figure 4. Different forms of tire recovery [21].

2.2. Reforming

Reforming is an essential process for end-of-life tire reclamation since it comes up with savings in raw resources and gets rid of the difficulties related to the discarding of end-of-life tires [22][23], and is also economical. Reforming is often done three to four times in the truck tire [24][25]. Manufacturing a new tire, energy can cost around 2.3 times higher than retreading [25][26]. As a thermosetting material, so it does not melt in temperature. But while heating up, it degrades, loses its elastic properties, and eventually burns and release energy [27]. In the reforming process, raw rubber is needed to mix and vulcanize them to develop the properties of a new tire. Reforming of end of life tires can be done by recapping, retreading, and remolding. Recapping is the process of replacing the tread, retreading replaces both the tread and its shoulder, and remolding, also known as bead-to-bead retreading, involves replacing the tread, shoulder, and entire sidewall surface. But reformed tires should be as safe as new tires [28].

2.3. Ground Tire Rubber (GTR)

As an elastomeric material, tires require special care. Because of steel's presence, it hampers mechanical grinding and creates a complex molding process [29]. For grinding, it requires an ambient temperature. This recycled ground tire rubber can be done by ultrasound, or cryogenically to make small pieces of rubber [30][]31]. Vulcanized rubber goes through a grinding process to become 7–10 cm particles. After that, grinder and processed at ambient temperature to make smaller particles, removing steel (by magnetism) and fibers (using vibratory) sieves and screens. Depending on the required size, additional grinding (tertiary) can be done to produce smaller particle sizes [31][32]. Ground tire rubber is mostly used in asphalt pavement as a modifier besides playgrounds, and other surfaces to landfill liners, molded rubber products [27]. Figure 5 shows the different stages of tire recovery.

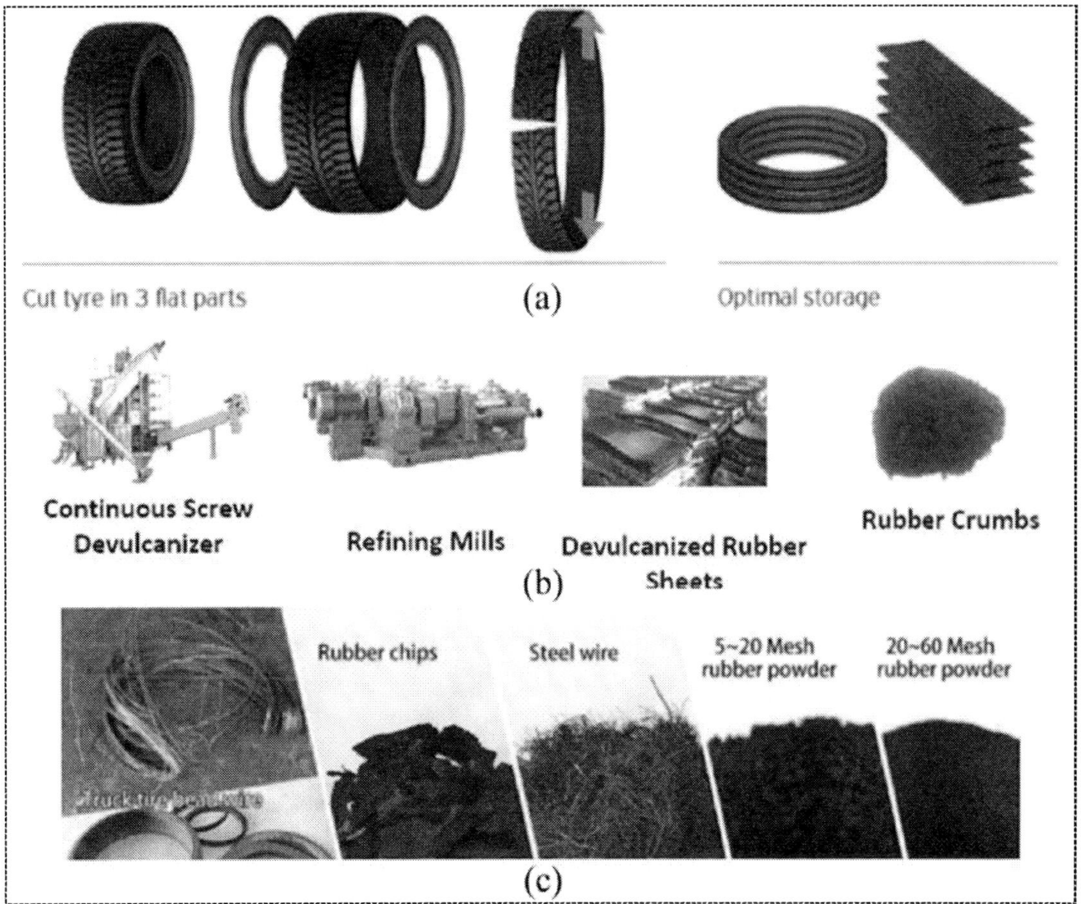

Figure 5. (a) Tire cut and storage [33], (b) process of powdering tires, (c) different forms of tire [34]

2.4. Regeneration of tire rubber

Regeneration is a new technology for recycling rubber. Rubber regeneration is a kind of renewal or restoration process of tire rubber. In this process, the rubber tire goes through a chemical reaction without going through permanent chemical change. This process requires heat, chemical products, and mechanical energy. Regeneration rubber is capable to replace virgin rubber with fewer technical requirements. It recovers its flow capacity as virgin rubber and the characteristics of a new one. It becomes more plastic, malleable, less viscous, and processable, with properties like new rubber. For achieving this property the chemical reaction need to be breaking of covalent carbon-carbon (C-C), carbon-sulfur (C-S), and sulfur-sulfur (S-S) bonds [23]. The quality of recovered regenerated rubber varies according to the source of the tire (different kinds of tire bus or truck and different companies) and the selectivity of the methods used in terms of the type and number of bonds to be broken. Regenerated rubber can be used in carpets, furniture, asphalt mixtures, glues, and adhesives [35]. In table 2 shows different sizes of material-derived products from end-of-life tires.

Table 2. Material Sizes of End-of-life Tire Derived Products (TDPs) [4].

Material	Size	Application
Cuts	>300 mm	Playground, Footpath, Animal farm as a slip resistance
Shred	50-300 mm	Rubber modified concrete
Chips	10-50 mm	Soil Moisturizer, fuel
Granulate (Crumbs)	1-10 mm	Size reduced rubber for insulation products: lumber and other construction product
Powder	<1 mm	As a replacement for cement
Fine powder	<500μm	As a replacement for cement or binding materials
Buffing's	0-40 mm	To assemble construction machines and transport technology
Reclaim	Dependent on input	In tires as inner tubes, tire lining, tire repair, rethreading, general molding, belting, adhesives, mastics, footwear, sheeting, matting, belting, cable bedding compound, and sound reduction
Regenerated rubber	Depend as require	Carpets, furniture, asphalt mixtures, glues, and adhesives
Devulcanisate	Depend on use	As a sheet
Pyrolytic char	<10 mm	Reprocess the tires into fuel gas, oils, solid residue, and low-grade carbon black
Carbon products	<500μm	Replacing coal or coke in steel manufacturing

2.5. Co-processing in cement production kilns

In this process, end-of-life tires at a cement production kiln go through material and thermal recovery. Due to the high temperature in cement kilns, often waste material is used as an alternative of fuel to dispose effectively and saving natural sources of fuel, and also cost-effective. Whole or ground tires can be burned in a cement kiln to produce clinker, an intermediate product in cement production [30]. The tire gives instantaneous, complete, and smokeless combustion burning in high gas temperatures (1000–1600 °C). The high temperatures (1500–1600°C) and oxidizing atmosphere in the cement production kiln provides total ignition of the tire and volatile material [19][22]. The iron content of steel-belted tires is beneficial for manufacturing cement [16].

2.6. Co-processing in thermoelectric power stations

In this process, ground tires are used with coal in the combustion reactor section to produce electrical energy and thermal energy. Tires generate 25–30% more energy than traditional power stations. Moreover, CO_2 emissions are also reduced by around 23% [30][36][37][38]. Traditionally, coal, oil, and gas provide 80% of the global energy demand [39][40]. So, alternative use of tires as a fuel can provide a good source of energy with less burden on disposal.

2.7. Pyrolysis of tire rubber

Pyrolysis is a process to recycle tire rubber in a high-temperature chemical process inside a chemical reactor to produce oil, gas, and carbon black by utilizing whole or shredded tires. The oxygen-free reactor is commonly produced around temperature (400–700°C) and pressure (0.01–0.04 MPa) to degrade the composition of tire rubber. The procedure consists of: gaseous (hydrogen, methane, and carbonic oxides), liquid (water and oils), and residual solids (metals and dust) [30][41]. The rubber polymers break down into smaller molecules in this process. The tire pyrolysis process is one of the most used clean procedures to produce negligible emissions or waste [20]. In table 3 shows the different applications of end-of-life tires in civil engineering and geotechnical engineering.

Table 3. Application of End-of-life Tire Products [4].

Industry	Application	Properties
Civil Engineering	Concrete Freeze/Thaw Protection	• To increase ductility • To achieve greater durability
	Flowable Concrete fill	• To reduce the density, • To minimize the overburden pressure of materials. • To replicate well-compacted soils.
	Replacement Material in Bitumen and Asphalt	• To replace sand at 20% • To limit permanent deformation
	Railway Maintenance	• To use as subbase materials • To reduce vibration and sound absorption
	High-Strength Concrete	• To utilize 0–12.5% as aggregate substitution can achieve up to 60 MPa after a cure time of 90 days
	Playgrounds and Sporting Surfaces	• Rubber chip products may be used as flexible coverage
	Sound Absorption Properties	• Sound insulation and absorption [42][43]
Geotechnical Engineering	Soil Stabilisation	• As a good reinforcement substitute in deep foundations and raft foundations [44]
	Unbound Pavements	• To use as a replacement material in sands used in unbound pavements [45]
	Use in Sub-Ballast Layers	• To replace conventional granular sub-ballast with recycled crumb rubber, • To reduce the demand on finite resources, transportation cost of raw materials, maintenance costs, and • To improve the bearing capacities and impermeability.
	Seismic Isolation Systems	• To provide effective seismic isolation • To decrease horizontal ground acceleration by 60–70 % and vertical ground acceleration by 80–90 % [46]
	Whole-Tire Embankments	• As a gravity retention system • As a replacement for energy-intensive brick and concrete structures. • To provide effective drainage properties • To reduce the embodied energy and costs associated with soil retention.

2.8. *Landfill disposal*

Even though an increasing number of wastes are getting reused or recycled, an average of 20 million tonnes of waste in Australia makes its way to hundreds of landfill sites each year. This is 40% of the total waste generated each year in Australia. Landfill disposal is simply the disposal of the tire without any treatment. Some nations completely suspended landfilled like Europe, according to directive 2000/53/EU [47]. Discarded tire takes a long time to degrade and emit toxic chemical in the environment. Due to their large volume and 75% void space, discarded tires are a convenient birthplace for pests [48, 49] and prone to fire hazards.

3. Conclusion

As a non-biodegradable material, extra care is required to eliminate the environmental impacts associated with the tires. Utilising tires in pavements presence of synthetic and natural rubber often makes it difficult to adopt. It is important to understand the application of reusing synthetic rubber and its performance as a road construction material and fuel. In Australia, the recovery rate is quite low compare to other developed countries. Public participation is also necessary to create societal acceptance of recycling, recovery, and reuse activities of end-of-life tires. End-of-life tires can be a good source of energy in cement power plants or cement kilns by lessening the demand for raw materials also sending fewer tires in the landfill.

4. References

[1] Siddika A, Al Mamun M A, Alyousef R, Amran Y M, Aslani F and Alabduljabbar H 2019 Properties and utilizations of waste tire rubber in concrete: A review, Construction and *Building Materials* **224** 711-731

[2] Australia,Y.B.(2019-2020). Australian bureau of statistics. *Canberra,Australia*.Retrived from "https://www.abs.gov.au/statistics/environment/environmental-management/waste-account-australia-experimental-estimates"

[3] Studdert D M, Gurrin L C, Jatkar U and Pirkis J 2010 Relationship between vehicle emissions laws and incidence of suicide by motor vehicle exhaust gas in Australia, 2001–06: an ecological analysis, *PLoS medicine* **7** e1000210.

[4] Oikonomou N and Mavridou S 2009 The use of waste tyre rubber in civil engineering works *In Sustainability of construction materials* (Sawston: Woodhead Publishing) pp. 213-238

[5] Siddique R and Naik T R 2004 Properties of concrete containing scrap-tire rubber–an overview, *Waste management* **24** 563-569

[6] Colangelo F, Petrillo A, Cioffi R, Borrelli C and Forcina A 2018 Life cycle assessment of recycled concretes: A case study in southern Italy, *Science of the Total Environment* **615** 1506-1517

[7] Tyre Product Stewardship Scheme, Tyre Stewardship Australia 2017 Retrieved on February 20, 2017 from "www.tyrestewardship.org.au/about-tsa/tyre-product-stewardship-scheme"

[8] Roychand R, Gravina R J, Zhuge Y, Ma X, Youssf O and Mills J E 2020 A comprehensive review on the mechanical properties of waste tire rubber concrete, *Construction and Building Materials* **237** 117651

[9] Kempf R C, Lyon W, Trueman W C, Boyce R and Brunot C 1995 American Automobile Manufacturers Association Heavy Truck Brake Tire Test *SAE Technical Paper* (No. 952663)

[10] Hamdi A, Abdelaziz G and Farhan K Z 2020 Scope of reusing waste shredded tires in concrete and cementitious composite materials: A review, *Journal of Building Engineering* **35** 102014

[11] Schmidt B, Quin D G and Steele W K 2018 Modelling wetland habitat preferences of Lewin's Rail'Lewinia pectoralis pectoralis' near Melbourne in southern Victoria, *Australian Field Ornithology* **35** 136-145

[12] Kashani A, Ngo T D, Hemachandra P and Hajimohammadi A 2018 Effects of surface treatments of recycled tyre crumb on cement-rubber bonding in concrete composite foam, *Construction and Building Materials* **171** 467-473

[13] Rao G V and Dutta R K 2006 Compressibility and strength behaviour of sand–tyre chip mixtures, *Geotechnical & Geological Engineering* **24** 711-724
[14] Presti D L 2013 Recycled tyre rubber modified bitumens for road asphalt mixtures: A literature review, *Construction and Building Materials* **49** 863-881
[15] Abdelrahman M 2006 Controlling Performance of Crumb Rubber–Modified Binders Through Addition of Polymer Modifiers, *Transportation research record* **1962(1)** 64-70
[16] March F, Ghabchi R, Zaman M and Arshadi P A 2016 Use of ground tire rubber (GTR) in asphalt pavements: literature review and dot survey
[17] Fazli A and Rodrigue D 2020 Recycling waste tires into ground tire rubber (GTR)/rubber compounds: A review, *Journal of Composites Science* **4** 103
[18] Chen T C, Shen Y H, Lee W J, Lin C C and Wan M W 2010 The study of ultrasound-assisted oxidative desulfurization process applied to the utilization of pyrolysis oil from waste tires, *Journal of Cleaner Production* **18** 1850-1858
[19] Clauzade C, Osset P, Hugrel C, Chappert A, Durande M and Palluau M 2010 Life cycle assessment of nine recovery methods for end-of-life tyres, *The International Journal of Life Cycle Assessment* **15** 883-892
[20] Ortíz-Rodríguez O O, Ocampo-Duque W and Duque-Salazar L I 2017 Environmental impact of end-of-life tires: Life cycle assessment comparison of three scenarios from a case study in Valle Del Cauca, Colombia, *Energies* **10** 2117
[21] Hilal A A 2011 Effect of crumb tyres rubber on some properties of foamed concrete, *Anbar Journal for Engineering Sciences* **4** 1-17
[22] Amari T, Themelis N J and Wernick I K 1999 Resource recovery from used rubber tires, *Resources Policy* **25** 179-188
[23] Cui X, Zhao S and Wang B 2016 Microbial desulfurization for ground tire rubber by mixed consortium-Sphingomonas sp. and Gordonia sp., *Polymer Degradation and Stability* **128** 165-171
[24] Van Beukering P J and Janssen M A 2000 A dynamic integrated analysis of truck tires in Western Europe, *Journal of Industrial Ecology* **4** 93-115
[25] Cooper D R and Gutowski T G 2017 The environmental impacts of reuse: a review, *Journal of Industrial Ecology* **21** 38-56
[26] Ferrão P, Ribeiro P and Silva P 2008 A management system for end-of-life tyres: A Portuguese case study, *Waste management* **28** 604-614
[27] Dierkes W and Saiwari S 2021 Regeneration and devulcanization *Tire Waste and Recycling* 97
[28] Mohajerani A, Burnett L, Smith J V, Markovski S, Rodwell G, Rahman M T and Maghool F 2020 Recycling waste rubber tyres in construction materials and associated environmental considerations: A review, *Resources, Conservation and Recycling* **155** 104679
[29] Karger-Kocsis J, Mészáros L and Bárány T 2013 Ground tyre rubber (GTR) in thermoplastics, thermosets, and rubbers, *Journal of Materials Science* **48** 1-38
[30] Fiksel J, Bakshi B R, Baral A, Guerra E and DeQuervain B 2011 Comparative life cycle assessment of beneficial applications for scrap tires, *Clean technologies and environmental policy* **13** 19-35
[31] Myhre M, Saiwari S, Dierkes W and Noordermeer J 2012 Rubber recycling: chemistry, processing, and applications, *Rubber chemistry and technology* **85** 408-449
[32] Thomas B S and Gupta R C 2016 A comprehensive review on the applications of waste tire rubber in cement concrete, *Renewable and Sustainable Energy Reviews* **54** 1323-1333.
[33] Valente M and Sibai A 2019 Rubber/crete: Mechanical properties of scrap to reuse tire-derived rubber in concrete; A review, *Journal of applied biomaterials & functional materials* **17** 2280800019835486
[34] Siringi G, Abolmaali A and Aswath P B 2015 Properties of concrete with tire derived aggregate partially replacing coarse aggregates, *The Scientific World Journal* **2015** 863706
[35] Pacheco E B A V, Visconte L L Y, Furtado C R G and Neto J R A 2012 Recycling of rubber: Mechano-chemical Regeneration *Advances in Materials Science Research* (New York: Nova Science Publishers)

[36] Liu Z, Guan D, Wei W, Davis S J, Ciais P, Bai J, ... and He K 2015 Reduced carbon emission estimates from fossil fuel combustion and cement production in China, *Nature* **524** 335-338
[37] Singh S, Nimmo W, Gibbs B M and Williams P T 2009 Waste tyre rubber as a secondary fuel for power plants *Fuel* **88** 2473-2480
[38] Webb A H and Hunter G C 1998 Power-station contributions to local concentrations of NO2 at ground level, *Environmental Pollution* **102** 283-288
[39] Muneer T, Asif M and Munawwar S 2005 Sustainable production of solar electricity with particular reference to the Indian economy, *Renewable and Sustainable Energy Reviews* **9** 444-473
[40] Feraldi R, Cashman S, Huff M and Raahauge L 2013 Comparative LCA of treatment options for US scrap tires: material recycling and tire-derived fuel combustion, *The International Journal of Life Cycle Assessment* **18** 613-625
[41] Li X, Xu H, Gao Y and Tao Y 2010 Comparison of end-of-life tire treatment technologies: A Chinese case study, *Waste management* **30** 2235-2246
[42] Zhao J, Wang X M, Chang J M, Yao Y and Cui Q 2010 Sound insulation property of wood–waste tire rubber composite, *Composites Science and Technology* **70** 2033-2038
[43] Harjana H, Sabino U, Yahya I and Kristiani R 2014 Sound insulation and absorption properties of re-claimed waste tire rubber *ICAMST2014* (Solo)
[44] Hambirao G S and Rakaraddi P G 2014 Soil stabilization using waste shredded rubber tyre chips, *IOSR Journal of Mechanical and Civil Engineering* **11** 20-27
[45] Speir R H and Witczak M W 1996 Use of shredded rubber in unbound granular flexible pavement layers, *Transportation research record* **1547** 96-106
[46] Tsang H H, Sheikh M N and Lam N 2007 *Rubber-soil cushion for earthquake protection* 1-8
[47] Gomes T S, Neto G R, de Salles A C N, Visconte L L Y and Pacheco E B A V 2019 End-of-life tire destination from a life cycle assessment perspective *In New Frontiers on Life Cycle Assessment-Theory and Application* IntechOpen
[48] Liu H S, Mead J L and Stacer R G 1998 Environmental impacts of recycled rubber in light fill applications: summary & evaluation of existing literature (Massachusetts: UMass Lowell Chelsea Center for Recycling and Economic Development)
[49] Chen R, Li Q, Zhang Y, Xu X and Zhang D 2019 Pyrolysis kinetics and mechanism of typical industrial non-tyre rubber wastes by peak-differentiating analysis and multi kinetics methods, *Fuel* **235** 1224-1237

The Study of Kenaf Powder Performance in Mortar

S S Ayop[1, 2], S Rasli[1], S N Moharam[1], M S Wahab[3] and N Sa'ude[3]

[1] Faculty of Civil Engineering and Built Environment, Universiti Tun Hussein Onn Malaysia, Batu Pahat, Johor, 86400, Malaysia
[2] Jamilus Research Centre, Faculty of Civil Engineering and Built Environment, Universiti Tun Hussein Onn Malaysia, Johor, 86400, Malaysia
[3] Faculty of Mechanical Engineering and Manufacturing, Universiti Tun Hussein Onn Malaysia, Batu Pahat, Johor, 86400, Malaysia
Corresponding author: sallehs@uthm.edu.my

Abstract. As cement is essential in producing mortar, reforms and innovations are needed to reduce the adverse effects of CO_2 in the future. The objectives of this study were to determine the optimal amount of kenaf powder in mortar strength and to examine the mechanical and flexural properties of kenaf powder in the mortar. Thirty-six mortar cubes of 50 mm x 50 mm x 50 mm and 12 number of mortar prism were cast using mix ratio of 1:3 with two different water-cement ratios. The replacement of kenaf powder ranging from 10% to 20% to the cement matrix were tested to investigate the mechanical parameters such as workability, density, compressive strength, and water absorption capabilities. Furthermore flexural strength was assessed using a similar mix proportion. The mortar containing 10% of kenaf powder and superplasticizer additives had demonstrated the best performance in workability and improved the compressive strength and flexural strength.

1. Introduction

Every tons of Portland cement produced, about one tons of CO_2 is released into the environment. The cement sector is, in fact, responsible for around 7% of worldwide CO_2 emissions. Research on this issue is starting to be carried out on potential materials for the use of cement in mortar, as not only society but also the environment are adversely affected by carbon dioxide emitted from cement. In order to overcome this problem and because of its easy growth and abundant availability in Malaysia, kenaf is a more promising green-based adsorbent utilized to replace commercial adsorbents [1]. Many studies on mechanical properties of kenaf powder are needed to gain environmental benefits without compromising mortar performance.

In Malaysia, mortar is widely used for wall plastering and brick binding. Mortar is a combination of cement, fine aggregates, and water used to improve the aesthetic value of a building and the sound and water-resistance of the wall construction [2]. However, mortar is not as effective as concrete and is usually not used as a single construction material [3]. Therefore, to improve the mechanical properties of the mortar, various forms of replacement and additional materials have been developed. Though the growth of the construction industry in this country, multiple studies that contribute to the efficiency and quality of construction have been done.

The objectives of this study were, to investigate the mechanical properties of mortar containing kenaf powder; to determine the optimum content of kenaf powder in the mortar strength and to study the effects of kenaf powder in flexural strength.

2. Sample Preparation and Testing

In this project, a total of 36 mortar cubes size 50 mm x 50 mm were used to determine the compressive strength on 7 and 28 days and 12 prisms size 40 mm x 40 mm x 160 mm were used to determine the flexural strength.

2.1. Materials

Preparation of sampling materials has been undertaken to ensure that the samples comply with the appropriate requirements. Several of the materials used were Ordinary Portland Cement (OPC), untreated kenaf powder, superplasticizer and fine aggregate. OPC used in this research was manufactured in compliance with the quality standards set out in the Malaysia Standard, MS 522:1:1989 OPC Specifications. Subsequently, the material passed the sieve analysis test to find the proportion of kenaf powder smaller than cement. The scale of the kenaf powder granules with the full size is physically identical to the cement's size. The fine aggregate gradation was finer than 5.0 mm. In this experiment, the superplasticizer (SP) used is Estop Admix AP. The water used in the mix was free from organic matter, chloride, acidic substances and silt and safe to drink in order to produce a good mortar mix. Table 1 shows the mix proportion used in this study. Kenaf powder was used as a cement replacement materials ranging from 10 % to 20 % by cement volume. Two w/c ratios were used were 0.4 and 0.5, while SP was used to check the workability effects on the mortar mix. From total 36 number of cube specimens, 27 specimens containing kenaf powder and 9 samples were used as a reference. The workability of the fresh mortar mix was examined according to BS EN 1015-3:1999.

Table 1. The mix proportion of 1:3 (cement: sand) for single mortar.

Mix Designation		Sand (kg)	Cement (kg)	Kenaf Powder (%)	Water (kg)	w/c
Control	M11	0.114	0.034	0%	0.014	0.4
	M12				0.014	0.4 + SP
	M13				0.017	0.5
KP 10%	M21	0.114	0.031	10%	0.012	0.4
	M22				0.012	0.4 + SP
	M23				0.015	0.5
KP 15%	M31	0.114	0.029	15%	0.011	0.4
	M32				0.011	0.4 + SP
	M33				0.014	0.5
KP 20%	M41	0.114	0.027	20%	0.011	0.4
	M42				0.011	0.4 + SP
	M43				0.014	0.5

For hardened mortar samples as in figure 1, the density test was conducted to study the effects of kenaf powder replacement on dry density of mortar. Water absorption test was conducted to examine the effects of kenaf particles when exposed with water. The strength characteristic of the mix was assessed based on the compressive strength on 28 days after cast. Furthermore, flexural strength test was conducted to assess the Modulus of Rupture of the prism.

Figure 1. Cube and prism samples.

3. Results and discussion

The results from the laboratory tests on the mortar specimen has been analysed and discussed accordingly.

3.1. Slump flow test

As shown in figure 2, for a similar mix proportion, the flow of the wet mortar was increased with an increasing of w/c ratio ranges from 22% on the control specimen up to 28% on the specimen with 20% kenaf powder. The additional of SP in the mix had improved the flow of the mortar. However the present of the kenaf powder had reduced the flow significantly when compared with the control samples. Kenaf powder is polar and it contains hydroxyl groups that have the capability to bind with the water molecules via hydrogen bonding during immersion [4]. Therefore it had reduced the free water content of the mortar mix thus reduce the ability of the materials to flow. The trend in figure 3 had proved this assumption. As can be seen, a linear increment of water absorption was observed on the sample with the increment of kenaf powder content.

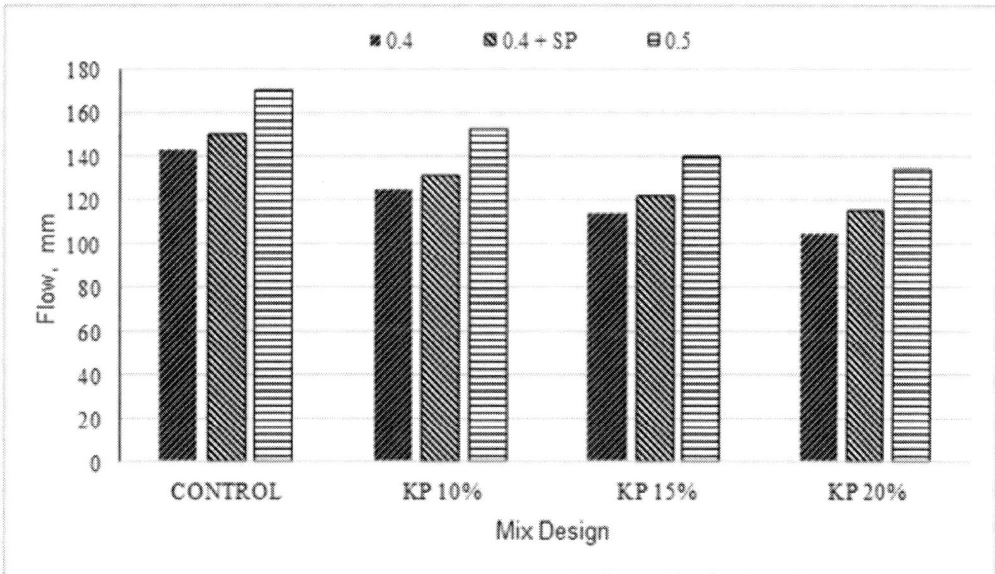

Figure 2. Flow test result for different w/c ratios.

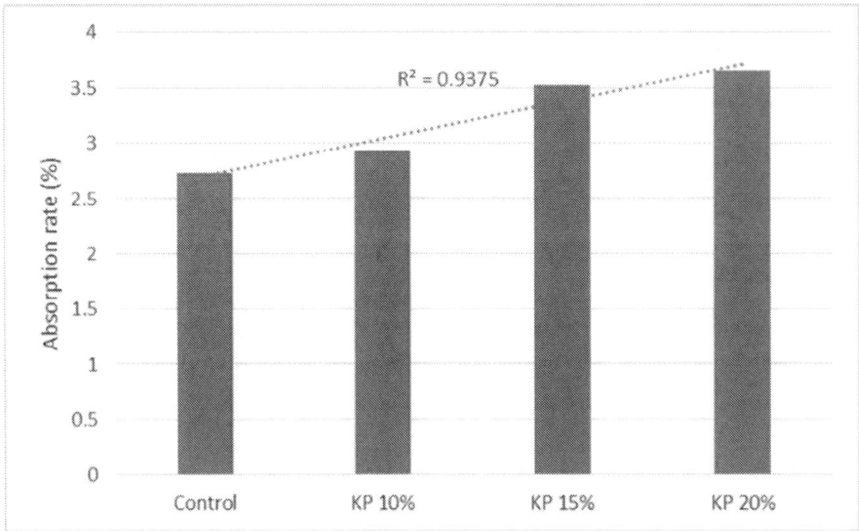

Figure 3. Absorption rate for different kenaf powder content.

3.2. Compressive strength

Figure 4 shows the result of compressive strength on 28 days after cast. The replacement of 10 % kenaf powder in the cement content had shown its effects in enhancing the compressive strength up to 5 % regardless its w/c ratio when compared with control specimen. However, this trend reduced steadily with the increment of kenaf powder replacement. Opposite trend was observed in Zhou et al. [5] study when they used kenaf fiber with 5 to 15 mm length in high strength cement composite study. The fiber only effective in enhancing the flexural strength but not in the compressive strength.

For 0.4 w/c ratio, the additive that used in the mix had significantly improved the compressive strength. The higher changes were observed on 10 % kenaf replacement with the increment of 16 %. Furthermore, higher compressive strength was found on the mix with SP regardless the kenaf powder content. Therefore, besides workability, SP had works significantly in increasing the compressive strength of the mortar.

Higher compressive strength was observed on 0.5 compared to 0.4 w/c ratio. The lower of water content in the 0.4 w/c ratio mix had affected the mixing process where the mix became dry and difficult to compact. It can be seen in the density test results of the samples in table 2 where the density of the 0.5 w/c was higher than 0.4 w/c.

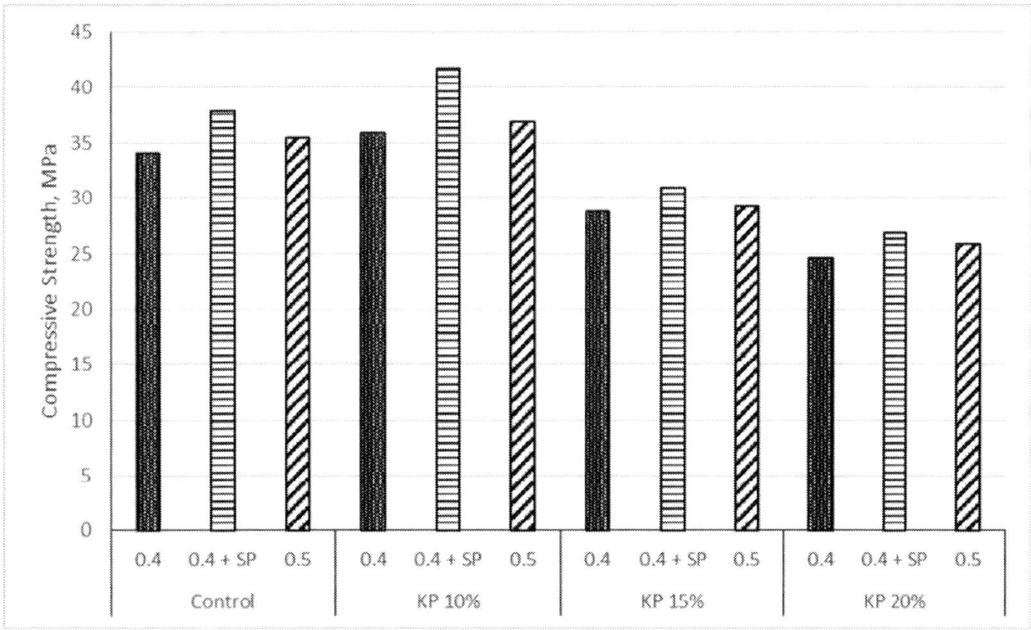

Figure 4. Mortar compressive strength at 28 days.

Table 2. Density test of mortar cube.

Mix Designation		Average Density for 28 Days (kg/m³)
Control	M11	2156
	M12	2175
	M13	2163
KP 10%	M21	2169
	M22	2184
	M23	2177
KP 15%	M31	2142
	M32	2158
	M33	2155
KP 20%	M41	2120
	M42	2137
	M43	2133

3.3. Flexural strength

3-point bending test was conducted on the prism sample and the modulus of rapture (σ) was calculated accordingly using the following equation (1):

$$\sigma = 3FL/2bd^2 \quad (1)$$

where, F is a force at crack point, L is a support span, b is width and d is depth of the prism. The span was set at 120 mm and the summary of the flexural test is depicted in figure 5. Flexural strength was higher on specimen with 10% kenaf powder replacement. By adding superplasticizer in the mix have improved the flexural strength of the mortar sample when compared with sample without superplasticizer.

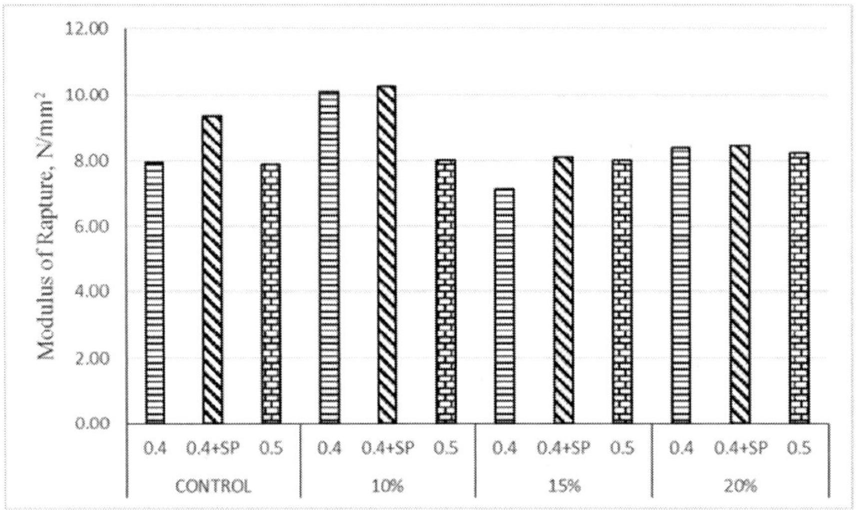

Figure 5. Modulus of Rapture of mortar samples.

4. Conclusion

Based on the result from the experimental works done, it shown that kenaf powder can be used as a cement replacement in the mortar. 10 % of kenaf powder replacement in the mix had improved the compressive strength and the flexural strength of the mortar prism. Furthermore, the absorption rate of the mix was higher when compared to the control specimens, and the trend of increment was observed when the percentage of replacements were increased. It is worth to mention that the superplasticizer used in this study plays an important role in enhancing the workability, compressive strength and flexural strength of the mortar mix.

5. References

[1] Bamigboye G O, Davies I, Nwanko C, Michaels T, Adeyemi G and Ozuor O 2019 Innovation in Construction Materials- A Review *IOP Conf. Ser. Mater. Sci. Eng.* **640** 012070

[2] Tam M T and Weng C C 1994 A study on acoustic emission characteristics of fly ash cement mortar under compression *Cem. Concr. Res.,* **24** 1335–1346

[3] Ahmadi Z, Esmaeili J, Kasaei J and Hajialioghli R 2018 Properties of sustainable cement mortars containing high volume of raw diatomite, *Sustain. Mater. Technol.* **16** 47–53

[4] Alias N F, Ismail H and Ku Ishak K M 2019 The effects of kenaf loading on water absorption and impact properties of polylactic acid/ natural rubber/ kenaf core powder biocomposite *Materials Today: Proceedings* **17** 584-589

[5] Zhou C, Cai L, Chen Z and Li J 2020 Effect of kenaf fiber on mechanical properties of high-strength cement composites, *Construction and Building Materials* **263** 121007

Acknowledgements

The authors would also like to thank the Material Engineering Laboratory and Advanced Materials Engineering Laboratory, Faculty Civil Engineering and Built Environment, Universiti Tun Hussein Onn Malaysia for their facilities support

A Review on the Performance of Waste Glass as Partial Replacement of Fine Aggregate

Norfaniza Mokhtar[1] and Nurul Atikah Jalani[1]

[1]Universiti Tun Hussein Onn Malaysia, Department of Civil Engineering, Faculty of Civil Engineering and Built Environment, University Tun Hussein Onn Malaysia
Corresponding author: norfaniza@uthm.edu.my

Abstract. Waste glass is one of the biggest contributors in waste production in Malaysia which can be used as partial sand replacement in concrete. Silica exists in waste glass components made it becomes a pozzolanic material and suitable to be used in a concrete mixture. The performance of the waste glass as fine aggregate was reviewed by considering the workability of fresh concrete, the strength, and the splitting tensile of hardened concrete. The range of the replacement waste are 0%, 10%, 20% and 30%. The influence of the waste glass on the microstructure of the concrete also have been evaluated. A total of seventeen previous research papers were collected and review based on the parameters selected. The results shows that replacement of waste glass in concrete give a positive impact for all selected parameters with 20% replacement represent as an optimal percentage replacement. The microstructure of the waste glass concrete resulting that more voids created after 20% replacement of waste glass which affects the compressive strength and split tensile strength of the concrete thus make the optimum replacement percentage is 20% replacement.

1. Introduction

Concrete is one of the main construction materials besides steel and timber. Typical concrete mix composed of 10% -15% cement, 60% - 75% aggregate, 15% - 20% of water, and lastly 5% - 8% of entrained air. From there, it is clear that more than half the volume of concrete was made from aggregate. Glass is one of the waste materials that were study as partial fine aggregate in concrete mixture. Some researcher has done laboratory experiment in order to evaluate the properties of fresh concrete and hardened concrete consists of waste glass as a partial replacement of fine aggregate [1][2]. The findings shows that the workability, compressive strength, and split tensile strength give a positive impact until certain replacement percentage.

Glass is a material that can be recycled several times because it can maintain its purity and quality, [3]. Recycling glass in daily live will drives into energy saving and lowering glass disposal cost [4]. Silica that exists in glass makes it pozzolanic cementitious and can improve the bond between cement-aggregate [5].

Waste glass is one of the best materials to replace natural sand because both waste glass and natural sand share lots of similarities in its physical properties [6]. Waste glass able to improve the fresh concrete properties as it has smooth surface and low water absorption that can influence the workability of concrete [7]. Replacement of sand with waste glass has been done in the past and all of this was done to evaluate the waste glass potential but a certain limitation has been applied since severe segregation with bleeding and inferior mechanical performance can be detected if the 100% replacement was done [8]. An optimum percentage of waste glass can increase the hardened concrete properties such compressive strength and split tensile strength. The replacement percentage of the waste glass in the concrete are 0%, 10%, 20%, 30% and this study aims on determining the optimum percentage of waste glass replacement. Environmental issue has increase due to large amount of waste glass disposal at the landfills. This study is important as it can help to decrease the percentage of overall glass waste in the landfill area thus led to the decrease of waste disposal [9].

2. Methodology

Planning is the first step of this project to make sure that the project was done systematically without missing an important part. All the important sub-topic such as the parameters that considered to be reviewed were list out in this process. The parameter chosen were workability of fresh concrete, compressive strength, and split tensile strength of hardened concrete and lastly, the microstructure of the concrete after replacing the sand with waste glass. The research papers that were related to this study were collected and reviewed. Most research papers that were collected is the research papers that were published in the past 5 years (including this year) but several papers are much older.

Seventeen research papers were taken to collect the data for workability, compressive strength, and split tensile strength while there are only three research papers were collected to get the microstructure results. Only the data for 0%, 10%, 20%, and 30% waste glass replacement were selected.

2.1. Workability

Table 1 represent the workability data that has been selected for concrete grade 20 and 25 from previous research.

Table 1. Workability result for concrete G20 and G25 from previous research.

References	Grade concrete	Workability of concrete (mm)			
		0%	10%	20%	30%
[1]	20	65	68	73	60
[2]		250	270	290	260
[3]		52	48	60	59
[4]		97	92	75	65
[7]		52	38	43	48
[10]		250	280	290	300
[11]	25	35	53	65	72
[12]		25	29	34	40

2.2. Compressive strength

Table 2 present the compressive strength data that has been selected for concrete grade 20 and 25 on 28 days.

Table 2. Compressive strength result for concrete G20 and G25 from previous research.

References	Grade concrete	Compressive strength (MPa)			
		0%	10%	20%	30%
[1]	20	30.33	31.33	31.00	29.66
[2]		22.77	25.88	29.84	24.66
[3]		26.00	27.10	27.50	28.80
[4]		23.57	26.88	27.11	27.76
[7]		25.00	36.80	25.80	24.00
[10]		33.80	35.60	38.40	34.25
[12]		37.00	35.90	34.20	28.88
[10]	25	29.66	29.33	30.00	29.66
[11]		29.68	33.03	27.14	25.11
[12]		27.30	29.25	30.25	29.90
[13]		28.07	33.69	35.11	30.82

2.3. Split tensile strength
Table 3 present the split tensile strength data that has been selected for concrete grade 20 and 25.

Table 3. Split tensile strength result for concrete G20 and G25 from previous research.

References	Grade concrete	Split tensile strength (MPa)			
		0%	10%	20%	30%
[1]	20	2.30	2.52	3.21	3.1
[12]		3.75	3.27	3.55	3.29
[7]		2.90	3.83	3.80	3.24
[3]		2.80	2.87	2.90	2.98
[4]		3.45	3.40	3.46	3.04
[10]	25	2.40	2.20	2.40	1.90
[13]		2.55	2.48	2.30	2.16

2.4. Microstructure
Scanning electron microscope (SEM) analysis is an analysis that has been done to analysed by referring to the changes in the microstructure of the waste glass concrete [14].

3. Results and discussion

3.1. Workability
Figure 1 and figure 2 shows the slump value result for waste glass concrete grade 20 and 25.

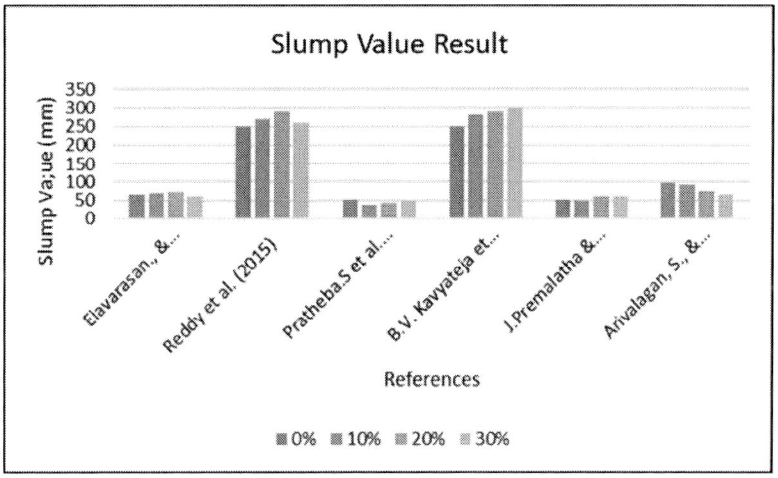

Figure 1. Value of slump test of waste glass concrete with different percentage of replacement of glass for G20.

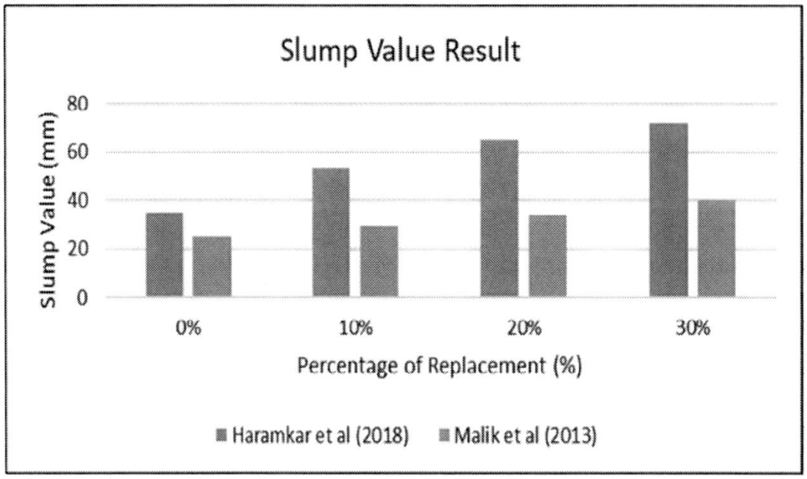

Figure 2. Value of slump test of waste glass concrete with different percentage of replacement of glass for G25.

From the graph in figure 1, 20% - 30% waste glass replacement shows highest result. The positive result of the workability is due to the smooth surface of the waste glass that absorb less water during mixing process. Figure 2 shows that slump value for [10] appears to be lower than [12] and the reason for this difference is the water-cement ratio used for the design mix proportion. Different water-cement ratio of 0.55 and 0.45 resulting this situation. The overall workability result from both figures shows positive achievement especially at 20% of waste glass replacement. It shows that 20% of waste glass replacement is the optimum replacement to increase the workability of the fresh concrete. The amount of admixture added in the waste glass concrete increase with the increase of waste glass percentage used in order to maintain the compaction factor value. This shows that the workability of waste glass concrete can be increased by and adding admixtures in the mix design proportion [13].

3.2. Compressive strength

Figure 3 and figure 4 shows the compressive strength result for waste glass concrete grade 20 at 28 days.

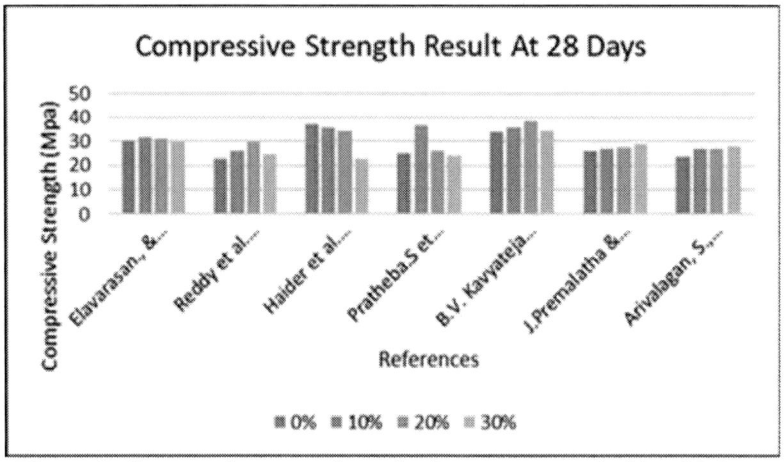

Figure 3. Compressive strength results for concrete grade 20 at 28 days.

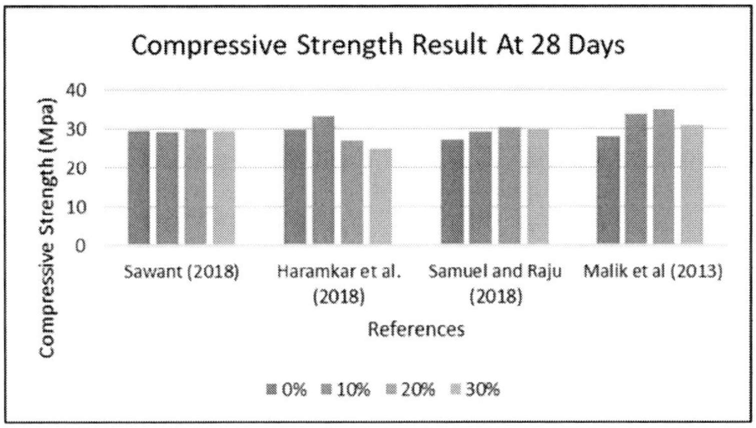

Figure 4. Compressive strength results for concrete grade 25 at 28 days.

Concrete was projected to achieve approximately 100% of its strength in 28 days. From the result in figure 3, all the results have obtained a minimum compressive strength of 28 days for concrete grade 20. However, there is only one sample continuously drop as the percentage replacement of waste glass increases which due to texture of glass aggregate has a smoother texture relative to sand, thereby minimizing the concrete mixture's paste-aggregate bond [12]. The poor resistance to aggregate fracture since the glass is easily broken and fragile. Besides that, most of the compressive strength results for waste glass concrete is higher than control concrete.

Figure 4 showed that all of the results have achieved the minimum compressive strength of concrete with grade 25 since all the results were higher than 25 MPa. This proves that waste glass can increase the compressive strength of concrete and produce concrete with higher strength compared to control concrete.

3.3. Compressive strength

Figure 5 and figure 6 shows the split tensile strength result for waste glass concrete grade 20 and 25 at 28.

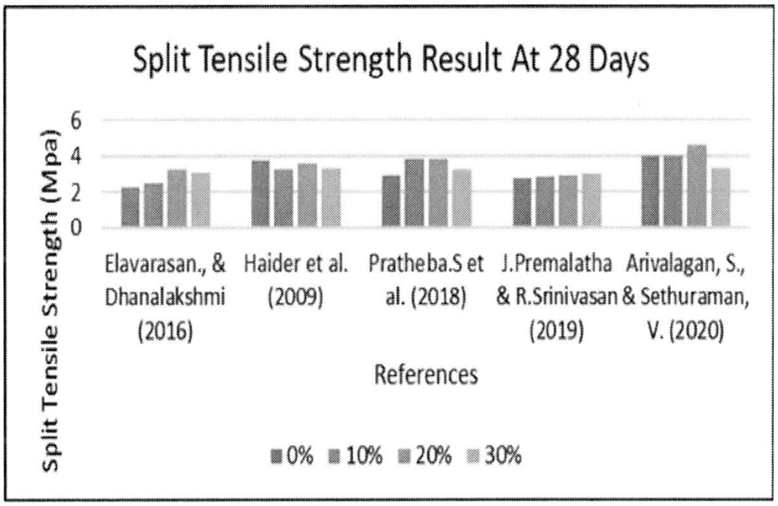

Figure 5. Split tensile strength results for concrete grade 20 at 28 days.

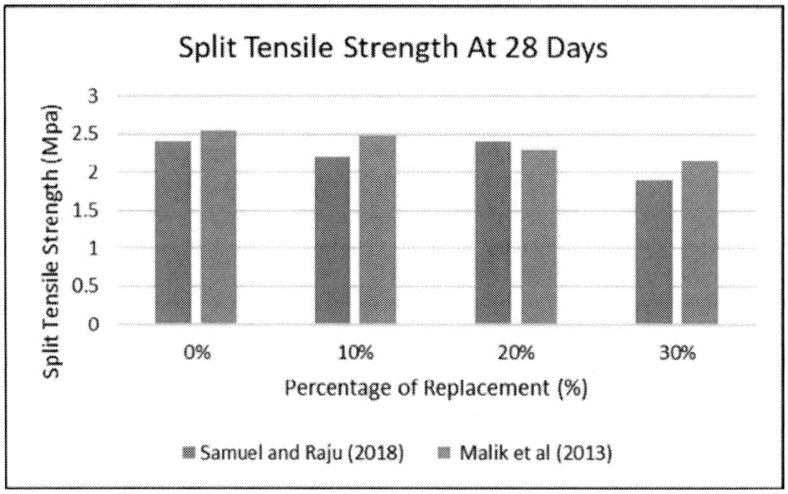

Figure 6. Split tensile strength results for concrete grade 25 at 28 days.

Figure 5 shows that Elavasran [1] has the highest result at 20% waste glass replacement while Pratheba [7] have the highest split tensile strength at 10% waste glass replacement. Premalatha [3] results increase marginally along with the increase in the replacement of waste glass. Most results in figure 5 shows the positive split tensile strength mostly at 10% - 20% waste glass replacement. The graph shown in figure 6 were based on the results of split tensile strength for concrete grade 25 on 28 days. Both results do not meet the minimum split tensile strength that should be reached with a grade of 25 for concrete. After 28 days, the split tensile strength at 20% waste glass replacement increases and have the same strength as control concrete [12]. This can be concluded that the optimum waste glass replacement percentage to increase the split tensile strength of grade 25 concrete is 20% [12].

3.4. Microstructure

Figure 7 shows the result of Scanning Electron Microscopy (SEM) analysis [15].

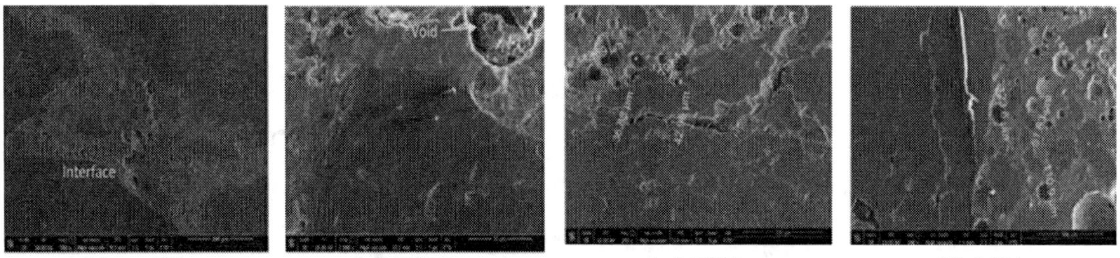

(a) 0% (b)10% (c) 20% (d) 30%

Figure 7. SEM analysis result [15].

From figure 7, the image for the concrete without waste glass shows a boundary between the cement and aggregates but the boundary become more invisible after 18% waste glass replacement. This is due to the waste glass that was dispersed uniformly in the concrete structure thus creating denser concrete and helps to improve the compressive strength and split tensile strength of the waste glass concrete. However, voids started to appear in the images of 18% waste glass replacement and 20% waste glass replacement. The number of voids increased as the waste glass replacement increased and cracks can be seen when the replacement percentage is higher than 20%. The existence of the cracks will be one of the reasons for the decrease of compressive strength and split tensile strength.

4. Conclusion

The optimum replacement percentage of waste glass in concrete mixture is 20% replacement. 20% replacement. The results for the workability, compressive strength, split tensile strength and microstructure shows positive result at 20% replacement. Voids start to appear in the concrete structure when the replacement percentage is more than 18% as in the microstructure result. Cracks start to form when the replacement percentage is higher than 20% thus effecting the compressive strength and the split tensile strength of the concrete.

5. References

[1] Elavarasan and Dhanalakshmi 2016 Experimental Sudy on Waste Glass as a Partial Replacing Material in Concrete for Fine Aggregate, *International Journal of Advanced Research in Biology Engineering Science and Technology* **2** 116-120
[2] Vijaya M, Reddy S, Sumalatha P, Madhuri M and Ashalatha 2015 Incorporation of Waste Glass Powder as Partial Replacement of Fine Aggregate in Cement Concrete, *International Journal of Scientific & Engineering Research* **6** 405-409
[3] Premalatha J and Srinivasan R 2019 Properties of Concrete with Waste Glass Powder (GP) as Fine Aggregate Replacement, *International Journal of Recent Technology and Engineering (IJRTE)* **8** 2277-3878
[4] Arivalagan S and Sethuraman V 2020 Experimental study on the mechanical properties of concrete by partial replacement of glass powder as fine aggregate: An environmentally friendly approach, *Materials Today: Proceedings* **45** 6035-6041
[5] Olofinnade O, Ade A, Jm N, Ngene B U, Ofuyatan O and Akinwumi I I 2018 Strength and Microstructure of Eco-concrete produced using Waste Glass as Partial and Complete Replacement for Sand, *Cogent Engineering* **5** 1-20
[6] Gerges N N, Issa C A, Fawaz S A, Jabbour J, Jreige J and Yacoub A 2018 Recycled Glass Concrete: Coarse and Fine Aggregates. *European Journal of Engineering Research and Science* **3** 1-10
[7] Pratheba S, Gowsalya M, Ishwariya A M, Jayasri A and Kalaiyarasi R 2018 An Experimental Study On Waste Glass As Partial Replacement For Fine Aggregate In Concrete, *International Research Journal of Engineering and Technology* **5** 1-5
[8] Lu J X, Zheng H, Yang S, He P and Poon C S 2019 Co-utilization of waste glass cullet and glass powder in precast concrete products, *Construction and Building Materials* **223** 210–220
[9] Yang S, Lu J X and Poon C S 2021 Recycling of waste glass in cement mortars:Mechanical properties under high temperature loading, *Resources, Conservation and Recycling* **174** 105831
[10] Malik M, Bashir M, Ahmad S, Tariq T, & Chowdhary U 2013 Study of Concrete Involving Use of Waste Glass as Partial Replacement of Fine Aggregates, *Journal of Engineering* **3** 8-13
[11] Haramkar P N, Maske S, Kushwaha R M, Sawalakhe R R, Mahajan J R and Kamdi V 2018 Partial Replacement of Fine Aggregate by Using Glass Powder, *International Conference on Emanations in Mordern Engineering Science & Management* (Maharashta) 1-5
[12] Ammash H and Muhammed S and Nahhab A 2009 Using of waste glass as fine aggregate in concrete, *Journal for Engineering Sciences* **6** 173–214
[13] Vinod T, Kunal B, Syed A K and Ramana P V 2021 Experimental investigation of mechanical properties and resistance to acid and sulphate attack of GGBS based concrete mixes with beverage glass waste as fine aggregate. *Journal of Building Engineering* **41** 102372
[14] Bisht K and Ramana P V 2018 Sustainable production of concrete containing discarded beverage glass as fine aggregate, *Construction and Building Materials* **177** 116–124
[15] Samuel J and Raju K L 2018 Experimental Study on Partial Replacement of Fine Aggregate by Waste Glass Powder and CNC Lathe Waste, *International Journal for Scientific Research & Development* **6** 557-56

IOP Conf. Series: Materials Science and Engineering 1200 (2021) 012012 doi:10.1088/1757-899X/1200/1/012012

Study of Life Cycle Assessment of Bricks and the Impacts to the Environment in Malaysia

H Adnan[1] and A T Balasbaneh[1]

[1] Faculty of Civil Engineering and Built Environment, Universiti Tun Hussein Onn Malaysia, Batu Pahat, 86400, Johor, Malaysia
Corresponding author: tighnavard@uthm.edu.my

Abstract. Life cycle assessment (LCA) is conducted in order to evaluate the environmental impacts of products chosen from the manufacturing phase and the end-of life cycle of the material and in clay brick and concrete were chose as the observed products. Brick is one of the important construction materials that can be seen at the surrounding. Main objective for this study is to investigate the impact of production of different types of brick to the level of emissions of carbon dioxide to the environment. Four stages of life cycle assessment were conducted before the result for the study analysis can be obtained and that stages including goal and scope definition, life cycle inventory (LCI), life cycle impact assessment (LCIA) and the interpretation part. The results obtained from the simulation of the Simapro shown that the concrete contributes more negative impact compared production of clay brick in terms of global warming, ozone depletion, formation of fine particulate matter and ozone formation. Manufacture of clay brick contributes more negative impact to the ionizing radiation, freshwater eutrophication and mineral resource scarcity.

1. Introduction

The building division utilize between 30% and 40% of society's total energy usage and accountable for almost 1/3 from the total carbon dioxide emissions into the atmosphere. As the construction in Malaysia is growth rapidly thus, increasing numbers of construction material produced every day. Brick is one of the traditional construction materials used which estimated millions of bricks are manufactured to support the construction required numbers of bricks. This is one of the reasons studies should be conducted in order to investigate the impact produced from the production of clay brick and concrete bricks [1]. Brick is most common construction material that can be seen in our daily life even at our wall without knowing what the environmental impact was coming from this material. Various types of bricks can be found these days for example sand lime bricks and engineering bricks. Clay and concrete brick are types of brick that usually being used for construction in Malaysia.

CO_2 emissions has big impacts towards the environment which around 60% of the greenhouse gases are comes from the carbon dioxide emission and a study to identify way to lower the CO_2 gasses released is a major concern as to protect control the effect on climate change [2]. This responsibility is not only on the authority's shoulder, but the society need to play their own role. Globally, 1.6 billion tons of CO_2 produced every year which around 8% from the total emissions of Carbon Dioxide from human activities. In the manufacturing of the clay brick, combustion phase contributed to high quantity of emission of SO_2 and NO_2 after the emissions of CO_2 which takes the first place as the highest percentage of gasses released to the atmosphere [3]. The main concern is actually the condition of the air that we breath in since the production of both products is actually contaminating our clean air. IEA data on 2018 stated that 35 Mt of CO_2 has been released from the industrial activities. Action needs to be taken by the engineers to investigate the process of manufacturing the construction material as for the outcome good decision in choosing a sustainable material can be done.

The objective of doing this research is to analyse the environmental issues occurred from the manufacturing of the clay brick and concrete brick. Next, to identify types of brick that has less impact in terms of carbon dioxide emissions and more sustainable types of brick to use.

2. Application of life cycle assessment

A tool used to evaluate the environmental impact over the life cycle of a product, process or the operation involved [2]. There are four phases of LCA in conducting the LCA study, which are the Goal and Scope Definition, Life Cycle Inventory Analysis (LCI), Life Cycle Impact Assessment (LCIA) and Interpretation.

Goal and scope phase is where the reasons of conducting the research were identified and the purpose goal and scope definition being identified and expressed in terms of the functional units act as the prove of the functions that the system works. The outcome from the LCI phase is a collection of the inputs and the outputs of the product of the life cycle which related to the functional unit. Then, the objective of doing the LCIA is to enhance the understanding of the magnitude and significance of the possible environmental impacts from the analysis [4]. Impact Assessment phase is known as the Valuation or Value-choices which the most arguable part in LCA because it suggests the subjective value understanding in deciding on the importance of different impacts. For the interpretation part, the results observed from the earlier phases are analysed as for the goal and scope phase to achieve the conclusions and provide suggestion for future use. Life cycle analysis is a beneficial method used for the observation and documentation to reduce the environmental effect of the materials or productions.

2.1. Methodology

Four phases of the life cycle assessment being applied in the methods towards the investigation of the environmental impacts of both products clay brick and concrete brick.

2.1.1. Goal and scope definition. 1 Kg of brick is the functional unit applied for this study and every single input of raw materials, energy utilization, transportation, release of waste and wastewater must stick to it as benchmark for both types of bricks. Cradle to gate approach is applied for the boundary system which only considering the manufacturing phase and the acquisition phase of the unprocessed material in the observation. Raw materials used in the clay brick production is only clay and the process of manufacturing the brick are mixing, molding, drying and then firing phase. While for the production of concrete brick, materials used are cement with aggregates, sand and then water. The process occurred are mixing, shaping and curing process.

2.1.2. Life cycle inventory. Site investigation required as to obtain required data such as the size of the brick, raw materials used, fuels consumption and the process of manufacturing of the brick. All of these required data is gain by having phone call interview with the person in-charge from the selected factory.

2.1.3. Life cycle impact assessment. ReCiPe 2016 method is used for the impact assessment as the calculation setups through the Simapro simulation. ReCiPe 2016 method is applied in the impact assessment as an achievable implementation with the combination of midpoint approach. All types of the result of life cycle inventory are linked through 16 midpoint categories to three damage classification. The main highlight in assessing this ReCiPe 2016 assessment is for the global warming which the results of the CO_2 gas emissions [4].

2.1.4. Life cycle interpretation. The results from the Simapro simulation which the outcomes of the inventory analysis were analyzed and evaluated as for to come out with the conclusions of the assessment and hand the recommendations. This phase is where the objectives to be achieved at the end of the interpretation phase. The comparison of the data is made by plotting the results on bar chart so that easier to evaluate and choose which types of brick is more sustainable to the environment and better for the usage at the construction.

3. Results and discussion

Result and discussion part are the presentation of the data and analysis of the study. This section is where the analysis of the impact of the clay brick and concrete brick to the environment being discussed based on the results from the simulation of Simapro software.

3.1. Functional units

ReCiPe 2016 method has been chosen as the medium of the calculation setups in the analysis and. The functional unit or benchmark of the analysis is 1 kg for both types of brick and considering the raw materials used in the production and the other process happened in the making of the bricks. Table 1 is the data required for the simulation of the Simapro gathered from the interview.

Table 1. Details used for both types of bricks for the simulation of Simapro

Characteristic	Concrete brick	Clay brick
Brick Weight (Kg)	1	1
Dimension (m^3)	250x125x65	250x125x65
Distance travelled of material (Km)	150	100
Load of the Lorry travelled (Tonne)	10	10
Types of raw materials used	Sand, Aggregate, Cement	Clay

3.2. Life cycle impact assessment by applying end-point of ReCiPe 2016 method – Characterization analysis

Sixteen (16) categories of environmental impact were applied in the ReCiPe 2016 analysis method through the simulation of the Simapro software shown in the table 2 which proves that this analysis is run in a very detail way. Three main units used in the analyzation are DALY (Disability adjusted life years), Species.yr (species lost in overtime) and USD2013 (U.S dollar reference year in 2013) [5]. Based on the figure 1, the first two categories are focusing on the global warming with two different view which first one is human health and then is the terrestrial ecosystem. Both aspects were more affected by the production of the concrete brick with difference about 14.7%. While the other categories that has been affected more by the concrete brick production are stratospheric ozone depletion, formation of the fine particulate matter, impact of ozone formation and acidification occurred to the terrestrial ecosystem, ecotoxicity on the terrestrial area, impact to the land use and the last category is the scarcity of the fossil resources.

Emissions of carbon dioxide is the main key where all of these categories contributed to the environmental impacts due to the production of the concrete brick. The highest data difference that the concrete brick led is with 67.4% for the ozone depletion of the stratospheric where concrete brick with 1.2E-10 DALY while for the clay brick is only with 3.92E-11 DALY. Ozone depletion can be defined as the event of the ozone layer is getting thinner which allow more UV radiation enter the surface of the earth which endanger the human being. Past studies shown that production of the concrete has contributed to emissions of greenhouse gas which one of the main reasons of the reduction of the thick of the ozone layer and through this analysis is again proves that production of concrete bricks has become one of the reasons why the depletion of the ozone layer happened [6].

Land use for the concrete production also become crucial part because different types of raw materials required to produce one block of concrete brick so that more natural sources from the land need to be collected like the aggregate and sand [7]. More harmful activities towards the land also conducted in order to have the resources. Only 9.02E-11 species.yr for clay brick while 2.75E-10 species.yr for the concrete brick with 67.2% of difference for both results. Strict monitoring is required for the sources collection of the concrete brick's raw materials as to reduce the number of the species loss times year and this is important to protect the ecosystem and provide safer environment for all creatures.

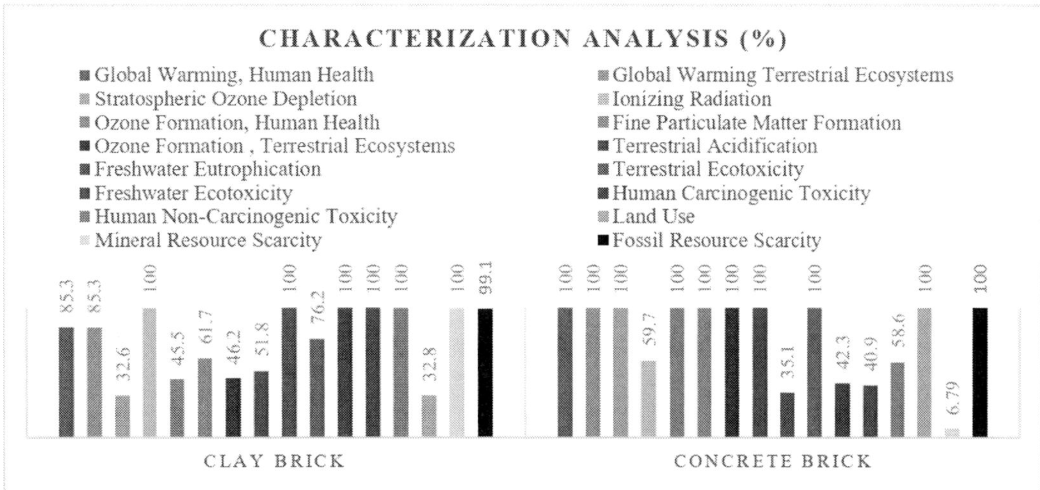

Figure 1. Results of the characterization through ReCiPe 2016 of End-Point Analysis based on the results in table 2.

Table 2. Results of the characterization from the Recipe 2016 of end-point analysis.

Impact category	Unit	Clay brick	Concrete brick
Global Warming, Human Health	DALY	2.94E-7	3.45E-7
Global Warming, Terrestrial Ecosystem	Species.yr	8.87E-10	1.04E-9
Stratospheric Ozone Depletion	DALY	3.92E-11	1.2E-10
Ionizing Radiation	DALY	5.81E-11	3.47E-11
Ozone Formation, Human Health	DALY	1.98E-10	1.75E-9
Fine Particulate Matter Formation	DALY	2.01E-7	3.26E-7
Ozone Formation, Terrestrial Ecosystem	Species.yr	1.16E-10	2.52E-10
Terrestrial Acidification	Species.yr	1.54E-10	2.97E-10
Freshwater Eutrophication	Species.yr	2.73-E11	9.57E-12
Terrestrial Ecotoxicity	Species.yr	1.51E-11	1.98E-11
Freshwater Ecotoxicity	Species.yr	6.44E-12	2.72E-12
Human Carcinogenic Toxicity	DALY	3.15E-8	1.29E-8
Human Non-Carcinogenic Toxicity	DALY	4.11E-8	2.41E-8
Land Use	Species.yr	9.02E-11	2.75E-10
Mineral Resource Scarcity	USD2013	0.00354	0.000241
Fossil Resource Scarcity	USD2013	0.0272	0.0275

Acidification of the terrestrial also happened more because of the production of the concrete brick with 2.97E-10 species.yr compared to only 1.54E-10 species.yr for the clay brick production. This is a serious matter where acidification of terrestrial means the changes of the soil properties where the nutrient inside the soil is no longer the same and even turn to bad condition of soil [6]. Acidifying compounds such as nitrogen and sulfur has deposited the nutrients contained inside the soil and even plant cannot stay green for long time once this incident happened. Most of the acidifying compound can be found through the production of concrete brick rather than clay brick.

Main problem happened to production of clay brick is shortage of sources of the clay which proves by the result that shown at the mineral resource scarcity data where almost 94% difference of comparison for the result of clay brick and concrete brick. Only required clay as the raw material for the clay brick formation has led to problem in controlling the extraction of the clay from the sources where this issues irregularly in concrete brick production [6]. Other than that, human carcinogenic also influenced more by the production of the clay brick with 3.15E-8 DALY compared to concrete brick production with only 1.29E-8 DALY. In aspects of exposing human being to cancer, clay brick production has contributed more. This is a very serious matter because cancer is one of the serious diseases which hard to cure so that enforcement to control and avoid this issue from happening without proper precaution plan.

3.3. Damage assessment analysis

In damage assessment analysis, there are three factors being considered as to observe the impacts of both products to environment which are the human health, ecosystems and the resources [5]. The major point here is the impacts toward the ecosystems which including the quality of the air, climate change and even the greenhouse effect. As we can see from the figure 2, concrete brick has more influence in the damage assessment where the human health and ecosystems factors has been affected more due to productions of the concrete brick. While clay brick only has more impact to the resources factor but with not so big difference compared to the impact given by the concrete brick.

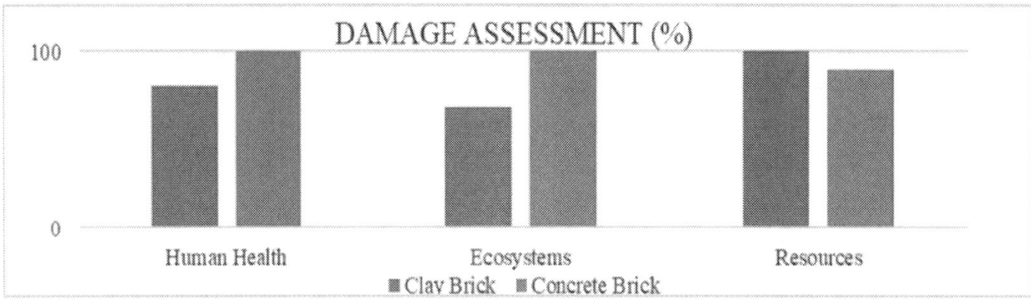

Figure 2. Results of the characterization through ReCiPe 2016 of End-Point Analysis based on the results in table 3.

Table 3. Results of the characterization from the Recipe 2016 of end-point analysis.

Damage Category	Unit	Clay Brick	Concrete Brick
Human Health	DALY	5.69E-7	7.1E-7
Ecosystems	Species.yr	1.3E-9	1.9E-9
Resources	USD2013	0.0308	0.0277

Human health and the ecosystems in more danger condition due to concrete brick production compared to clay brick production shown in table 3. 7.1E-7 DALY for damage assessment from human health category for concrete brick while only 5.69E-7 for clay brick which almost 20% more dangerous and this is not small amount not to be worried about and more precaution need to be prepared for the concrete brick production. For the ecosystem category, concrete brick also won this one where the result shown that 1.9E-9 species.yr for the concrete brick and 1.3E-9 species.yr for the clay brick production and this result is based on the evaluation of species loss times year. This result shown that the ecosystem we live in nowadays is threaten by the production of the concrete brick. For ecosystems category, concrete brick has contributed to 31.6% damage more than clay brick to the ecosystems.

Only for damage to the resources is more to clay brick production and this happened because one main raw material required for the clay brick production which is clay and amount of clay required for every tonne of production is very high so that the resources has become limited [8].

3.4. Life cycle impact assessment by applying mid-point of Recipe 2016 method – Characterization analysis

Mid-point analysis is as same as the end-point analysis based on table 4 but with different unit used for the results of the impact categories which more focus on the type of gas emitted for each impact category for example the amount of carbon dioxide is being calculated at for the global warming category and amount of sulphur dioxide released in terrestrial acidification is being compared for both types of brick. Our main focus in figure 3 is the global warming category where concrete brick production released 0.371 Kg CO2 eq while 0.317 Kg CO2 eq for clay brick production. The higher the carbon dioxide emissions, the lower the quality of the air we breathed in [9]. This is where more strict enforcement required for the concrete brick production as to provide better quality of the air [10].

Next, also from the concrete production, high amount of greenhouse gases released have been analyzed from the mid-point analysis where more than 50% difference of amount of Chlorofluorocarbon (CFC) released detected from the concrete brick compared to clay brick production. This is a very serious matter to be highlight and actions required from the authorities so that constant monitor of this type brick production can be provided as to ensure constant check of the amount of CFC gases released so instant action can be taken if the abnormal amount of CFC released is reported. Our ozone layer is part of the earth that we must protect because it is the layer protection to avoid us from exposed to high amount of UV radiation that can affect the skin and more negatives impact might happened.

The other serious category is the ozone formation where both the human health and terrestrial ecosystem are in danger due to production of the concrete brick. 54.5% for the human health factor while 53.8% for terrestrial ecosystem factor of difference for amount of Nitrogen Oxides for both types of brick. Nitrogen Oxide is one of the air pollutants which endanger the human being and the other living things. Different percentage with more than 50% is something to be worried especially who lives nearby the production area. Respiratory disease is the main concern based on the results obtained because most of the unit used is a measurement of the amount of gas released for each impact categories and most of it are dangerous pollutant which already inside the air particle that we breath in daily [11].

For the clay brick production based on the mid-point analysis, the most-risky category is the human carcinogenic and non-carcinogenic toxicity results where 59.1% and 41.3% of differences for the comparison of the results from the analysis for both categories of impact. Carcinogenic is substances that can cause cancer which very dangerous for human being and clay brick production has produced this hazardous substance more than the amount produce from concrete brick production [7]. From the midpoint analysis, the carcinogenic substance used for the evaluation is 1,4-dichlorobenzene as the human toxicity measurement unit.

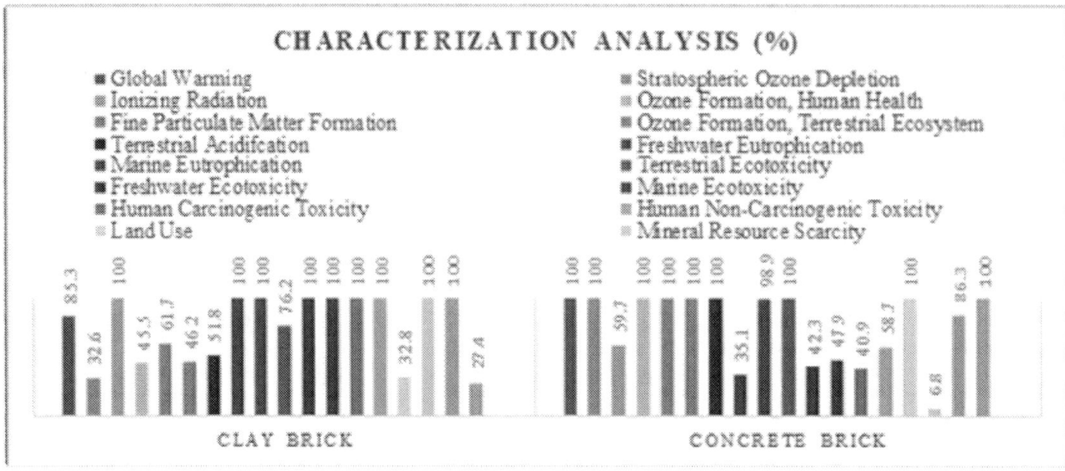

Figure 3. Results of the characterization through ReCiPe 2016 of mid-point analysis based on the results in table 4.

Table 4. Results of the characterization from the Recipe 2016 of mid-point analysis.

Impact category	Unit	Clay brick	Concrete brick
Global Warming	Kg CO2 eq	0.317	0.371
Stratospheric Ozone Depletion	Kg CFC11 eq	7.38E-8	2.26E-7
Ionizing Radiation	kBq Co-60 eq	0.00684	0.00409
Ozone Formation Human Health	Kg NOx eq	0.000877	0.00193
Fine Particulate Matter Formation	Kg PM2.5 eq	0.00032	0.000518
Ozone Formation, Terrestrial Ecosystem	Kg NOx eq	0.000903	0.00195
Terrestrial Acidification	Kg SO2 eq	0.000725	0.0014
Freshwater Eutrophication	Kg P eq	4.08E-5	1.43E-5
Marine Eutrophication	Kg N eq	3.04E-6	3.0E-6
Terrestrial Ecotoxicity	Kg 1,4-DCB	1.32	1.74
Freshwater Ecotoxicity	Kg 1,4-DCB	0.00931	0.00394
Marine Ecotoxicity	Kg 1,4-DCB	0.0126	0.00603
Human Carcinogenic Toxicity	Kg 1,4-DCB	0.00949	0.00388
Human Non-Carcinogenic Toxicity	Kg 1,4-DCB	0.18	0.106
Land Use	m2a crop eq	0.0102	0.031
Mineral Resource Scarcity	Kg Cu eq	0.0153	0.00104
Fossil Resource Scarcity	Kg oil eq	0.075	0.0647
Water Consumption	m^3	0.000639	0.00233
Global Warming	Kg CO2 eq	0.317	0.371

4. Conclusion

Construction industry nowadays must prioritize the material that has low negative impact to the environment because the product used might be one of the sources of the air pollution, degradation of soil or even the depletion of the thickness of ozone layer. From the damage analysis, we can see that concrete brick has been proved as the product where the production contributes more negative impact to the environment compared to production of clay brick which referring to the three damage categories which are human health, ecosystems and resources.

There are a few suggestions and recommendations that can be made regarding the discussion above in order to provide better management of the concrete brick and clay brick production:

- Application of high quality filtration equipment at the factory
- Usage of sun-dried clay brick in construction
- Enforce more strict law by the authorities to control the consumption of the natural resources used
- Use higher loads carrying capacity lorry as to reduce number of transportations involved in the production

5. References

[1] Dabaieh M, Heinonen J, El-Mahdy D and Hassan D M 2020 A comparative study of life cycle carbon emissions and embodied energy between sun-dried bricks and fired clay bricks, *Journal of Cleaner Production* **275** 1-14

[2] Begum R A, Sohag K, Abdullah S M S and Jaafar M 2015 CO2 emissions, energy consumption, economic and population growth in Malaysia, *Renewable and Sustainable Energy Reviews* **41** 594–601

[3] Koroneos C and Dompros A 2007 Environmental assessment of brick production in Greece, *Building and Environment* **42** 2114–2123

[4] Yau Y H and Hasbi S 2017 A Comprehensive Case Study of Climate Change Impacts on the Cooling Load in an Air-Conditioned Office Building in Malaysia, *Energy Procedia* **143** 295–300

[5] Huijbregts M, Steinmann Z J N, Elshout P M F M, Stam G, Verones F, Vieira M D M, Zijp M and Van Zelm R 2016 *ReCiPe 2016: A harmonized life cycle impact assessment method at midpoint and endpoint level Report I: Characterization* RIVM Report 2016–0104 (Bilthove: National Institute for Public Health and the Environment)

[6] Muñoz I, Cifrian E, Andrés A, Miguel G S, Ruiz D and Viguri J R 2018 Analysis of environmental benefits associated with the incorporation of Waelz slag into fired bricks using LCA, *Construction and Building Materials* **168** 178–186

[7] Mah C M, Fujiwara T and Ho C S 2018 Life cycle assessment and life cycle costing toward eco-efficiency concrete waste management in Malaysia, *Journal of Cleaner Production* **172** 3415–3427

[8] Christoforou E, Kylili A, Fokaides P A and Ioannou I 2016 Cradle to site Life Cycle Assessment (LCA) of adobe bricks, *Journal of Cleaner Production* **112** 443–452

[9] Kumanayake R and Luo H 2018 A tool for assessing life cycle CO2 emissions of buildings in Sri Lanka, *Building and Environment* **128** 272–286

[10] Tang K H D 2019 Climate change in Malaysia: Trends, contributors, impacts, mitigation and adaptations, *Science of the Total Environment* **650** 1858–1871

[11] Maia de Souza D, Lafontaine M, Charron-Doucet F, Chappert B, Kicak K, Duarte F and Lima L 2016 Comparative life cycle assessment of ceramic brick, concrete brick and cast-in-place reinforced concrete exterior walls, *Journal of Cleaner Production* **137** 70–82

Acknowledgments
Completion of this research was possible to be achieved by the help from my supervisor, Dr. Ali Tighnavard Balasbaneh who always shared his knowledge in this topic and guideance in using the Simapro software. I would also like to thank the Faculty of Civil Engineering and Built Environment, Universiti Tun Hussein Onn Malaysia for its support and convenient time given for me to finish this final year project on time. Not to forget, my beloved family and all of friends who always be there through my up and down in finishing this project.

IOP Conf. Series: Materials Science and Engineering 1200 (2021) 012013 doi:10.1088/1757-899X/1200/1/012013

Optimization of Bolted Connection in Steel Corbel Attach to RC Column of MRT Viaduct

Yazmin Sahol Hamid[1], Hazrina Mansor[1], Amir Atif Abdul Razak[1], Haikal Ajmal Bin Bukhory[1] and Nursafarina Ahmad[1]

[1] School of Civil Engineering, College of Engineering, Universiti Teknologi MARA (UiTM), Shah Alam 40450, Malaysia
Corresponding author: yazminsahol@uitm.edu.my

Abstract. MRT is one of the backbones of a city's public transportation system, capable of carrying large crowds. The use of a portal frame in the design of an MRT train station has raised significant concerns about how the portal frame's load will be supported by the extended structure element known as a corbel. A corbel is a protruding structural element that supports weights like primary beams and girders. Engineers must then decide how to properly bolt the steel corbel structure to the concrete pier segment or columns. Generally, most corbel structure designs were constructed of concrete; however, in this study corbel design was made of steel so that the steel portal frame could rest on the corbel structure, allowing for more usable area on the platform, such as kiosks and other amenities. Optimization of end plate thickness and beam web thickness is carried out. Manual calculations are used in addition to FEA modeling to examine the bolt's deflection, shear, bearing, tension, slip, and block tearing resistance. When using Eurocode, all three loadings, transverse force, vertical force, and transverse moment, produce values that are 10% lower than the British Standard. As a result, designers can optimize their designs using Eurocode.

1. Introduction

MRT is the country's largest infrastructure project, requiring massive tunnelling, underground construction, and viaduct work. If everyone drove, the city would be congested therefore this type of rail transportation system in Kuala Lumpur is expected to help alleviate traffic congestion as more people use public transportation in the Klang Valley.

Corbels structure usually consist of a Segment Box Girders, crossheads, bridge decks, piers, etc. It also serves as a support for important beams and girders and is a component of the MRT structure. However, it has been found out that there is less research on corbel particularly on bolted connection. Aside from the technical difficulties of designing and building a complex rail structure, the design of the corbel structure for the Mass Rapid Transit 2 (MRT 2) train station is fraught with problems and difficulties. Because of the sophisticated design of the architects, problems arose regarding how to support the load of the portal frame itself. To support the portal frame, they needed an extended structure element known as a corbel. Corbel is a protruding structure from the column face that is used substantially in supporting primary beams and girder [1]. Figure 1 shows a SketchUp drawing of the steel corbel structure supporting the extended portal frame. Figures 2(a) and (b) illustrate the actual steel corbel structure installation at the Damansara Damai construction site, as well as a close-up view of the steel corbel structure. It is designed to be installed outside of the existing platform to accommodate other platform elements like information kiosks and shop spaces. Unfortunately, engineers must decide the optimum bolt-type connection to be employed with sufficient strength to link the steel corbel structure to the concrete pier column, which is a concern and a challenge. The goal of this research is to better understand how steel corbels behave and how they are connected. This research focuses on existing MRT projects that use steel corbels attached to concrete pier segments rather than concrete corbels.

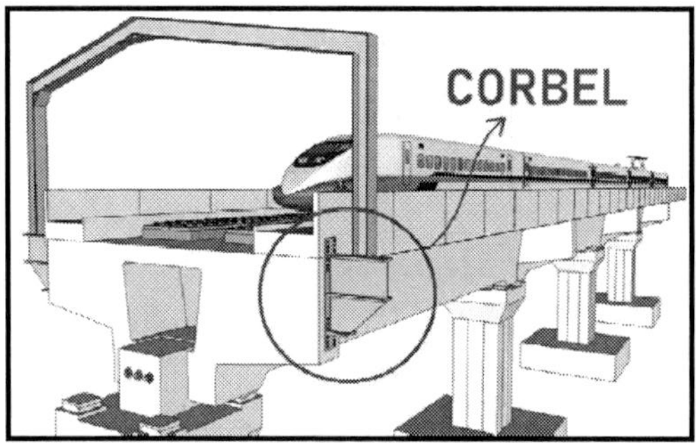

Figure 1. Illustration of corbel structure supporting load from portal frame.

(a) (b)

Figure 2. Steel corbel structure at Damansara Damai: (a) Installation of steel corbel, (b) close up view of steel corbel.

In this study the steel corbel connections are designed and analyzed using StaadPro V8i in line with Eurocode 3: Design of Steel Structures. The aims of this study are to optimize steel corbel bolted connections in terms of bolt numbers and size, end plate and beam web thickness. To determine the adequacy of the connection calculation of the connection's shear, bearing, tension, slip, and block tearing resistance is conducted and checked in accordance with Eurocode 3: Design of steel structures Part 1-8. Joint Design. Lastly, to analyze the stress distribution from the modelling outcome and recommend the optimum design is based on Joints in Steel Construction: Moment-Resisting Joints to Eurocode 3, Tata Steel UK 2011.Generally, these safety checks are crucial during initial construction, including corbel installation, temporary bolting, bracket and seating alignments, and supported element construction. The corbel structure design can be optimized by providing an economical design while maintaining connection strength. In conclusion, if the structural engineer underestimates the design force that the connection must withstand, a failure can occur.

1.1. Corbel

Corbels are often constructed to withstand vertical loads at the beam's end, but concrete corbel also support horizontal forces caused by creep, temperature changes, and shrinkage, which are as important [2]. The vertical load carrying capability of concrete corbel structures was found to be significantly reduced due to horizontal forces operating on the corbel, yet the corbel was nevertheless harmed due to horizontal limitations, according to a test conducted by the Portland Cement Association [1].

Steel corbels are manufactured off-site, away from the impact of site employees and weather [3]. Therefore, the environmental factor is reduced. Steel, unlike concrete, does not creep, shrink, or degrade when properly protected. Speed, site safety, lifespan, fire prevention, built-in quality, aesthetics, and sustainability are just a few of the benefits of steel corbels. Reinforced concrete corbels, on the other hand, need up to 28 days to cure before being used, whereas steel corbels can be used almost immediately once bolts are installed [4]. According to [1], the aesthetic and architectural manifestations of the steel corbel were important considerations in the selection as the use of steel corbels can reduce waste from formwork and abortive activities on the construction site.

Generally, most corbels structure designs were constructed of concrete; however, in this study corbel design was made of steel so that the steel portal frame could rest on the corbel structure, allowing for more usable area on the platform, such as kiosks and other amenities. The worry enginners had was how to build the most optimized steel corbel structure based on the number of bolts, moreover the majority of the design was done using a design software and no 2nd order analysis was done, plus validation was done using manual calculations. Besides that, it is a fact that failure due to a connection is the most serious as it can cause failure to the entire structure.

1.2. Steel Corbel Connections

Steel corbels are made up of several elements, as shown in figure 3, including an I-beam as the main structure, stiffeners, end plates, and bolts for connection to piers, portals, or columns [5]. I-beam that carry loads will experience bending, shear and deflection. The shear resistance of a beam depends on the shear capacity or shear buckling resistance. If the shear forces do not exceed 60% of the shear capacity, the shear effect on the bending capacity is considered low and can be ignored. However, if the shear loads exceed 60% of the shear capacity, the shear effect must be considered when calculating the moment capacity [6].

Structural failure can take various forms, but failure due to a connection is the most serious, and this is the type of failure that can occur in steel corbel constructions. The number of bolts is one of the parameters that affect the quality and durability of a bolted joint connection when it comes to maximizing its efficiency under load [7–10]. One element in bolt application is the insert bolted connection, which is utilized in grip joints as an additional friction purpose and is subjected to compressive force due to bolt preload. When estimating the strip mass per bolt, the thickness of the connecting strip and the distance between bolts should be assumed [7]. This can be done by dividing the external loads with the mass of the resulting joint. Thus, the joint efficiency, or load carrying capability per unit mass, may be calculated.

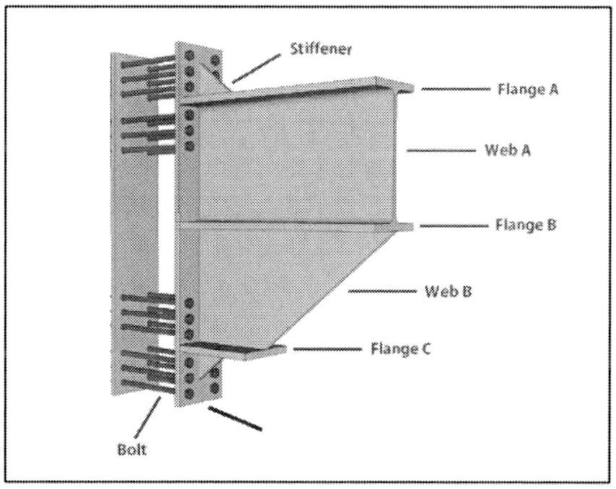

Figure 3. Steel corbel.

2. Finite element modelling process

In this study, the I-Beam section property of Advance UKB 686x154x170 has been proposed to be used. All dimensions were according to the actual dimension of the corbel structure as listed in Table 1. The proposed type of bolt that was used in the connection of the corbel structure was grade 8.8 M24, which shows the diameter of the bolt to be 0.024m. The corbel structural elements, i.e., web section, flange section, stiffeners, end plate section, and bolt structure are all steel. Support reactions were assigned only to the bolt structure's beam elements. To provide fixed support, it was necessary to restrain axial loading in the directions of the X, Y, and Z axes. It is assumed that the strength of concrete is equal to 4400 kN/m^2. The portal frame model shown in Figure 4 utilized in this study was produced by a group of students that competed in the MSSA-CIDB Open Ideas Competition (OIC) for the year 2018 (MSSA-CIDB OIC 2018) with the theme "REIMAGINE: MRT Stations." and has been remodeled using StaadPro software (see figure 5) to determine the maximum loading of the portal frame structure.

Figure 4. Portal frame structure.

Table 1: Details dimension of corbel structure.

Element	Section (see Figure 2)	Thickness / Diameter (m)	Length (m)	Width (m)	Shape
Plate	Flange A and B	0.0237	1.346	0.356	Rectangle
	Flange C	0.0237	0.800	0.356	Rectangle
	Web A	0.0145	1.362	0.696	Rectangle
	Web B	0.0145	0.800	0.602	Rectangle
	Stiffener	0.0150	0.275	0.275	Triangle
	End plate	0.042	1.820	0.430	Rectangle
Beam	Bolt	0.024	0.800	-	Circle

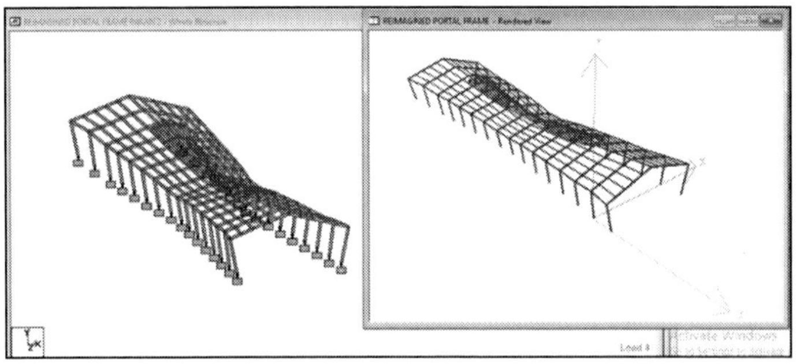

Figure 5. Structural design model of the MRT 2 train station comprises of portal frame section.

2.1. Proposed variation parameters

Four parameters were considered in this study to determine the best design of the steel corbel. The four parameters that have been proposed are bolts design arrangement configuration, diameters of bolts, dimensions of the endplates and web thickness of the beam.

2.1.1. Configuration of bolts design arrangement. The configuration of 24 numbers of bolts represents the original design proposed by the engineer. It is critical to optimize to ensure that the most effective or functional optimize strength is used for the bolted connection of the steel corbel structure. That is why the numbers of bolts, the thickness of the plate, and the beam web are also modelled with various sizes and thicknesses to achieve the most optimal size. In this study there are four (4) configuration sets of bolt number analysed in this work. The four configuration sets were described in the following paragraphs.

i) Configuration set 1: 24 number of bolts (original design by the design engineer)
The bolts in the top part of the corbel are created first followed by the bolts at the bottom parts. The equal amounts of bolts (12 number of bolts) were placed at both of the top and bottom part of the plate which gives a total of 24 number of bolts as per original design by the design engineers at the consultant firm.
ii) Configuration set 2: 18 number of bolts
The second configurations were proposed whereby the bolts were reduced to 10 number of bolts (upper) and 8 number of bolts (lower).
iii) Configuration set 3: 14 number of bolts
Next, for the third configuration of bolt arrangements, the bolts were reduced to 8 number of bolts (upper) and 6 number of bolts (lower).
iv) Configuration set 4: 10 number of bolts
The 4th configuration were arranged so that the bolts were reduced to 6 number of bolts (upper) and 4 number of bolts (lower).

The bolts are modelled with four different diameter sizes (16mm, 20mm, and 24mm), the end plates with three different thicknesses (15mm, 20mm, and 25mm), and the beam webs with three different thicknesses (11mm, 13mm, and 15.4mm).

2.1.2. Boundary conditions and loads applied. Corbel was supported on a concrete pier via steel bolt. Therefore, it is crucial to get an accurate value of bolt-concrete stiffness behavior. This is done by considering having fixed support at bolt end. Meanwhile, pinned supports were assigned at the middle nodes of each bolt which is in X-axis and Y-axis only.

Ulrike Kuhlmann, Frantisek Wald, and Jan Hofmann [11] conducted an experiment to determine how different concrete strengths affect the stiffness of the concrete under varying displacements. The point of zero displacement is regarded to be the point of failure. As illustrated in Figure 6, the lesser the displacement, the greater the stiffness of the concrete. Ulrike et al. [11] discovered that the optimal stiffness of concrete occurs at 1.5mm displacement, and the MRT project used a 35MPa concrete grade for the pier. Thus, we may deduce the stiffness value from the graph, which is 4300 N/mm. As a result, this value will be utilized to determine the stiffness of the X- and Y-axes.

This study used a linear static analysis to establish the maximum value of the support reactions studied using EC, and it was discovered that the transverse force, vertical force, and transverse moment were 128 kN, 523 kN, and 459 kNm, respectively.

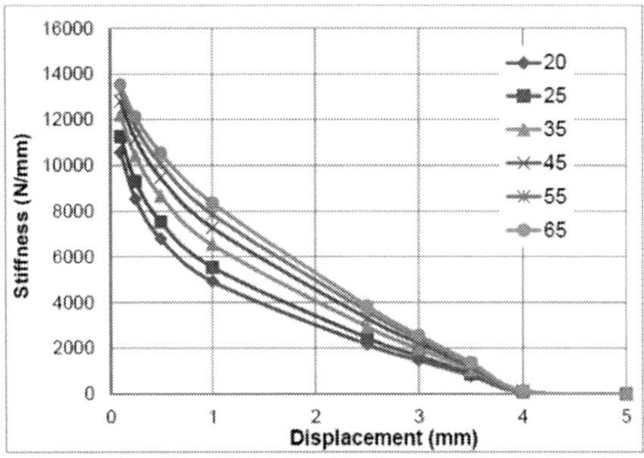

Figure 6. How concrete strength fck influence the concrete stiffness [11].

3. Validation of FEM Modelling

The vertical deflection of the steel corbel generated by StaadPro was compared against manual calculation. For the FEM, the taken point was at Node 56 and the vertical deflection values in Y-axis was considered. The analysis results comparison between StaadPro and theoretical calculations were presented in table 2.

Table 2. Comparison between StaadPro results and theoretical calculations.

Load (kN)	StaadPro Results (mm)	Theoretical Calculations (mm)	% Difference
100	1.793	1.49	14.81
200	3.009	2.98	0.96
300	4.226	4.47	5.46
400	5.442	5.95	8.54
523	6.938	7.79	10.94
700	9.090	10.42	12.76

It can be concluded that the difference between theoretical calculation and StaadPro results was very close and thus the modelling is validated and accepted.

Sensitivity analysis was also conducted to discover the ideal mesh sizes and computation effort. The study used mesh sizes of 5mm to 30mm. Beyond a mesh size of 10, the deflection values become nearly constant. The study used a 10-mesh size to reduce computing load.

4. Results and Analysis

Table 3 summarizes the outcomes of the bolt analysis results. From the study it was found that both 18 bolts and 24 bolts design are satisfactory, but for optimization purpose, 18 bolts design can be selected.

4.1. Verification of StaadPro analysis results with EC3

Based on Eurocode 3: Design of Steel Structures - Part 1-8: Design of Joints Detail, a hand calculation on bolt design was performed. Firstly, the clearance and bolt positioning are checked in terms of bolt clearance, bolt positioning, maximum end and edge distance, and minimum and maximum spacing. Second, the bolts are tested against several aspects, which are shear resistance, bearing resistance, tension resistance, and slip resistance. Third, in order to ensure the end plate is adequate to resist shearing force on it, its capacity is tested in terms of shear resistance (gross and net section), shear resistance of web and block tearing. From The results it was found that all bolt numbers passed the clearance and bolt positioning check.

Table 3. Summary of bolt analysis results.

Bolt Number	10		14		18		24	
	6 upper	4 lower	8 upper	6 lower	10 upper	8 lower	12 upper	12 lower
Highest Stress Value (N/mm^2)	559	386	517	358	348	221	334	212
Corbel Yield Stress (N/mm^2)	355	355	355	355	355	355	355	355
Comparison with Yield Stress	Over	Over	Over	Over	Under	Under	Under	Under
Design Acceptance	Unsatisfactory		Unsatisfactory		Satisfactory		Satisfactory	

Table 4 displays the shear resistance of bolts of various sizes. The values shaded in grey boxes represented insufficient shear resistance. For 16mm bolt sizes, all 10 nos., 14 nos., and 18 nos. of bolts are insufficient to resist the design shear of 651kN, whereas only 10 nos. of bolts are insufficient for 20mm bolt sizes. Aside from that, the checking is adequate.

Table 4. Shear resistance of bolts: Shear resistance (Required Value: 651kN).

Bolt sizes / Bolt numbers	16mm	20mm	24mm
10 nos.	301.4kN	470.4kN	677.8kN
14 nos.	422.0kN	658.6kN	948.9kN
18 nos.	542.5kN	846.7kN	1220.0kN
24 nos.	723.4kN	1129.0kN	1626.7kN

The results of bearing resistance design for various bolt sizes are shown in Table 5. According to the table, all bolt configurations were adequate to resist the required value of 651kN, and the checking was satisfactory.

Table 5. Bearing resistance of bolts: Bearing resistance (Required Value: 651kN).

Bolt sizes / Bolt numbers	16mm	20mm	24mm
10 nos.	1328kN	1660kN	1992.0kN
14 nos.	1859.2kN	2324kN	2788.8kN
18 nos.	2390.4kN	2988kN	3585.6kN
24 nos.	3187.2kN	3984kN	4780.8kN

Table 6 displays the tension resistance results for bolts with diameters of 16mm, 20mm, and 24mm. The configurations of 10 and 14 bolts for the 16mm diameter bolt are insufficient to resist the required value of 651kN. Aside from the aforementioned, the checking is adequate.

Table 7 shows the slip resistance design results, which show that nearly half of the bolt configurations were insufficient to meet the required slip resistance value of 651kN. According to the table, the configurations with 18 and 24 numbers of 20 mm and 24mm diameter bolts were the only ones that could withstand the slip resistance capacity of 651kN. As a result, such configurations are the fewest that can be used to optimize the corbel design.

Table 8 shows a summary of the shear (gross and net section) and block resistance design for various end plate thicknesses. It can be concluded that all plate thickness configurations passed the testing, and thus a 15mm plate can be chosen to optimize the design.

Table 6. Bearing resistance of bolts: Tension resistance (Required Value: 651kN).

IOP Conf. Series: Materials Science and Engineering 1200 (2021) 012013 doi:10.1088/1757-899X/1200/1/012013

Bolt sizes / Bolt numbers	16mm	20mm	24mm
10 nos.	452.2kN	705.6kN	1016.7kN
14 nos.	633.0kN	987.8kN	1423.4kN
18 nos.	813.9kN	1270.1kN	1830.1kN
24 nos.	1085.2kN	1693.4kN	2440.1kN

Table 7. Slip resistance of bolts: Slip resistance (Required Value: 651kN).

Bolt sizes / Bolt numbers	16mm	20mm	24mm
10 nos.	211.0kN	329.3kN	474.5kN
14 nos.	295.4kN	461.0kN	664.3kN
18 nos.	379.8kN	673.6kN	854.1kN
24 nos.	506.4kN	790.3kN	1138.8kN

Table 8. Checking of end plates: Checking of End Plates (Required Value: 651kN).

Plate Thickness	Shear Resistance (Gross Section)	Shear Resistance (Net Section)	Block Tearing
15mm	1706.5kN	2202.41kN	2644.4kN
20mm	2275.3kN	2936.5kN	3525.8kN
25mm	2844.1kN	3670.7kN	4407.3kN

Table 9 summarizes the local shear resistance for different web thicknesses. The web must carry a value of 651kN. All web thickness configurations passed the design check, and thus 11.0mm plate can be chosen to optimize the design. Table 10 shows the optimum parameter values that can be used to optimize the design of the corbel structure.

Table 9. Checking of beam web: Checking of Beam Web (Required Value: 651kN).

Web Thickness	Local Shear Resistance
11.0mm	1430.4kN
13.0mm	1690.5kN
15.4mm	2002.5kN

Table 10. Optimized corbel structure design.

Parameters	Selected value
Bolt numbers	18 nos.
Bolt sizes	20mm
End plated thickness	15.0mm
Beam web thickness	11.0mm

4.2. Optimized corbel structure design

Figure 7 displays the original corbel bolt location with 24 bolt numbers. To make a bolt with 18 bolt numbers, six bolts that are designated with a red circle must be removed.

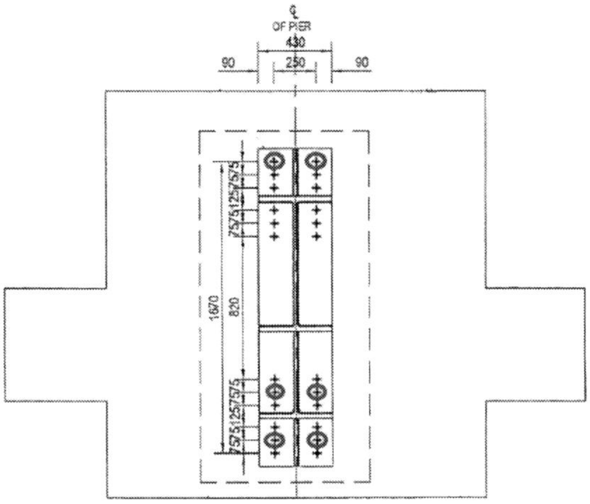

Figure 7. Bolts to be removed to form corbel with 18 numbers of bolts.

5. Summary and conclusions

Through this research, all of the research objectives have been completed. Using the four specified parameters, optimization of the bolted connection was achieved. The four parameters considered are 1) Configuration of the numbers of bolts 2) Diameter of the bolts 3) Plate thickness and 4) Web thickness. The numerical analysis of the optimization of bolted connections in steel corbels attached to RC columns of the MRT viaduct revealed several suggestions for optimizing the connections. The suggestions are as follows:

- Comparing Eurocode with British Standard for loading analysis. This was determined using portal frame modeling and Eurocode requirements. When utilizing Eurocode as a reference, all three loadings, transverse force, vertical force, and transverse moment, generate 10% lower values than the British Standard. As a result, designers can use Eurocode to optimize their designs.
- By optimizing the steel corbel's bolted connection, the total number of bolts required can be reduced from 24 to 18. The diameter of the bolts can be adjusted from 24mm to 18mm, and the end plates' thickness from 25mm to 15mm. The beam's web thickness can also be reduced from 15.4mm to 11mm. There were no problems resisting the design loads mentioned.
- The connection's sufficiency against shear, bearing, tension, slide, and block tearing was checked manually. The stress distribution produced by StaadPro finite element modeling was also analyzed to establish the ideal steel corbel design. The contour of the stress distribution was determined using FEA modeling results. In summary, FEA modeling and manual calculations in accordance with applicable rules of practice show that steel corbel structures can be adjusted to be more cost effective while still resisting design loads.

6. References

[1] Kriz L B and Raths C H 1965 Connections in Precast Concrete Structures-Strength of Corbels, *PCI J.* **10** 1 16–61

[2] Boon C W, Ooi L H, Tan J G and Goh C Y 2020 Deep excavation of an underground metro station in karstic limestone: a case history in the Klang Valley SSP Line, *Geotech. Sustain. Infrastruct. Dev.* 423–430

[3] Bjorhovde R, Brozzetti J and Colson A 1987 *Connections in Steel Structures: Behaviour, Strength & Design* (Amsterdam: Elsevier Applied Science) pp 1-394

[4] Dai X, Lam D, Sheehan T, Yang J and Zhou K 2018 Use of bolted shear connectors in composite construction *12th International Conference on Advances in Steel-Concrete Composite Structures (ASCCS 2018)* (Valencia)
[5] Joints in Steel Construction 2013 *Moment-Resisting Joints to Eurocode 3 (P398)* (Berkshire: The Steel Construction Institute and The British Constructional Steelwork Association)
[6] Eurocode 3 2005 *Design of steel structures - Part 1-8: Design of joints* (Brussels: European Committee for Standardization)
[7] Bianchi G, Aglietti G, and Richardson G 2007 Optimization of bolted joints connecting honeycomb panels *1st CEAS, 10th European Conference on Spacecraft Structures, Materials and Mechanical Testing* (Berlin)
[8] Thomsen O 1997 Sandwich Plates with 'Through-the-Thickness' and 'Fully-Potted' inserts: Evaluation of Differences in Structural Performance, *Compos. Struct.* **40** 159–174
[9] Thomsen O and Rits W 1998 Analysis and design of sandwich plates with inserts : a high-order sandwich plate theory approach, *Compos. Part B-engineering* **29** 795–807
[10] Ericksen W S 1960 *The bending of a circular sandwich plate under normal load* (Washington: Air Force-Navy-Civil Subcommittee)
[11] Kuhlmann U Wald F and Hofmann J 2014 *Design of Steel-to-Concrete Joints. Design Manual II* (Prague: Publishing House of CTU)

Acknowledgments
The authors would like to acknowledge The Institute of Research Management and Innovation (IRMI) UiTM, Shah Alam, Selangor, Malaysia and Ministry of Higher Education (MOHE) for the financial support of this research. This research is supported by MOHE under the Fundamental Research Grant Scheme (FRGS-RACER) with project code: 600-IRMI/FRGS-RACER 5/3 (060/2019).

Industrial Building System (IBS): A Unique Intra-Module Connection on Modular Steel Building (MSB)

M A Gapar[1], M F Razali[1], H Mansor[1], Y S Hamid[1] and N E A Subki[1]

[1] Civil Engineering, Universiti Teknologi MARA (UiTM) Shah Alam, Malaysia
Corresponding author: hazrina4476@uitm.edu.my

Abstract. Modular Steel Building (MSB) provide benefits towards green building technology such as minimum wastage, faster build time and cost-efficiency. The intra-module connection is the most important aspect of MSB construction since it has a significant impact on overall structural stability and robustness. A novel intra-module connection was proposed for the MSB. The proposal was designed to suit the illustrative five-storey hexagon shape modular steel building that possibly imagines by Architect. Two analyses phases are being presented, namely the Macro and Microanalysis model. The former is the stage for global analysis design of the proposed five-storey hexagon shape modular steel building via SAP2000. The latter is the local intra-module connection behaviour analysis using ABAQUS software. Linear and nonlinear static analyses were carried out on the proposed intra-module connection under the vertical applied load. In this work, the failure of the connection under the given load was governed by the hexagon diaphragm, while the fin plate demonstrates the least affected constitutive component. It anticipates that the suggested unique intra-module connection will encourage architects to employ modular steel construction designs with greater flexibility. Future research will concentrate on the parametric study to improve the performance of the diaphragm and the connection's limitations.

1. Introduction

One of the strategic thrusts to achieve the Sustainable Development Goals (SDG) is by pursuing green growth for sustainability and resilience. Sustainability is quantified in terms of various measures of environmental, social, and economic performance [1]. The sustainable benefit of modular steel construction in environmental, economic, and social has been widely discussed by Lawson et al. [2]. The construction of modular steel structures is actively being used over a decade ago. Yet, very limited studies, either experimentally or numerically, have discussed theoretical aspects, particularly inter and intra module connection. The standard intra-module connection used bolted connection to connect parallel flange channel known as PFC to mild steel square hollow section (SHS). Researcher developed double angle cleats connection to connect universal beam (UB) and rectangular hollow sections (RHS) [3]. While another researcher, in 2019, developed an innovative connection for intra-module and inter-module connection called Vector Bloc Modular Connector [4]. The rectangular hollow steel (RHS) section is welded to the connecter and bolted to the column-to-beam connection.

According to research performed by [5][6][7], the current key limits of inter and intra-module connection applications include the lack of an established guideline to assist structural designers in building economic and safe connection designs to exploit modular structural fully. A review carried out by [5] summaries many experimental works [6]-[13] being conducted that contributed to various new innovative inter-module connection designs to suits any architectural demand. For instance, studies conducted by Annan et al. [10], [11] give insightful knowledge for practices, revealing that consideration of the true behaviour of direct welding leads to a distribution of forces and moments that is different from those found in traditional steel buildings. As a result, the regression function was utilised to generate a simplified analytical model to capture such behaviour in actual practice, leading and assisting the structural engineer in making an accurate forecast on the design of modular steel flooring. Apart from [5], several other scholars have attempted to improve the inter-module connections. Park et al. [5] presented a bolted connection for modular steel construction with the adjacent module unit is bolted directly to the cruciform gusset plate. Chen et al. [6] proposed and tested a beam-to-beam bolted connection. Deng et al. [9] proposed a new connection with socket-shaped tendons and cruciform gusset plate to connect the module unit. Despite all the rigorous design [4]–[13], the proposed connections perhaps well suited for the structural component of the modules but may not work well with the internal finishes and the use of the floor and walls, which highlight the manufacturing efficiency.

Moreover, connections [4-13] connect the upper module and lower module by bolting the ceiling beam and floor beam without connecting the columns, which is presumed to lead to indirect uplift load transfer when subjected to lateral forces, such as wind and earthquake. Additionally, limited experience and technical knowledge among architects & engineers, plus no component standardisation causing the upfront cost increase for component mould. The issues have caused insecurity on the modular product quality and certification in terms of safety, affecting the economic growth of the modular construction. Research conducted by [14] highlighted that the current modular system is less adaptable, with a conventional rectangular shape that appears to constrain modern architectural design innovation. Research in [15] reported that the modular steel building is more suited for a lower-income family. These problems must be solved at the fundamental level, whereby there is an urgent need to formulate a robust universal design concept of intra and inter-module connection that suits the multi-purpose designs of the building. However, all of the connections listed are only relevant for 90° column-to-beam connections, hence the connections are only suitable for the modular's rectangular design. Some researchers stated that the conventional modular steel building existing in the industry is too plain and inflexible [16]. Usually, the architects tend to propose a complex yet aesthetically presentable building. This work proposed a novel intra-module connection for a modular steel construction building. The proposed novel intra-module connection was designed to suit the illustrative five-storey hexagon shape modular steel building that possibly imagine by Architect. The failure mechanism of intra-module connection under shear and the lateral load was analysed using linear and nonlinear static analyses via the ABAQUS Finite Element Analysis (FEA) software.

2. Proposed Modular Steel Building

The proposed modular steel building is shown in figure 1. All dimensions presented in figure 1 are in unit mm. The height of the proposed modular steel building is 3 meters with a side width of 2.5m each. The three-dimensional view and the plan views of the proposed modular steel building are illustrated in Sketch up and shown in figure 2. The proposed modular steel building is then modelled and designed in SAP2000. The dead load was directly taken from the default self-weight of the section material used in the SAP2000, while the other actions were the ceiling (1kN/m^2), floor (3.3kN/m^2), wall (3.8kN/m), and wind(0.65kN/m^2). The modular structure was designed to the ultimate limit state (ULS), in which the values of the partial and combination factors were taken as given in the UK National Annex. These values are subject to modification in the National Annex, which must be consulted. In this context, structural connections' components are steel channel section, fin plate, bolt, nut, and diaphragm. Table 1 provides the detailed design section that has been proposed. Later, all the listed structural section in table 3 was used for modelling in ABAQUS FEA .

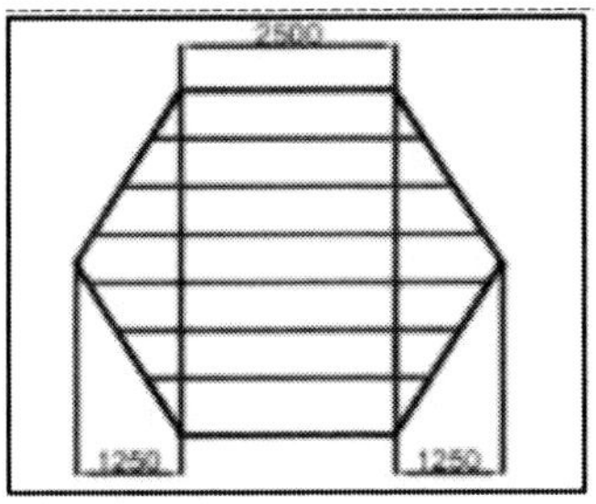

(a) Plan view 1

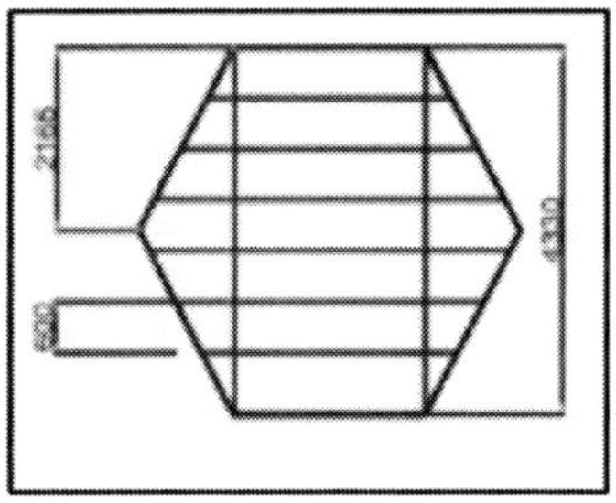

(b) Plan view 2

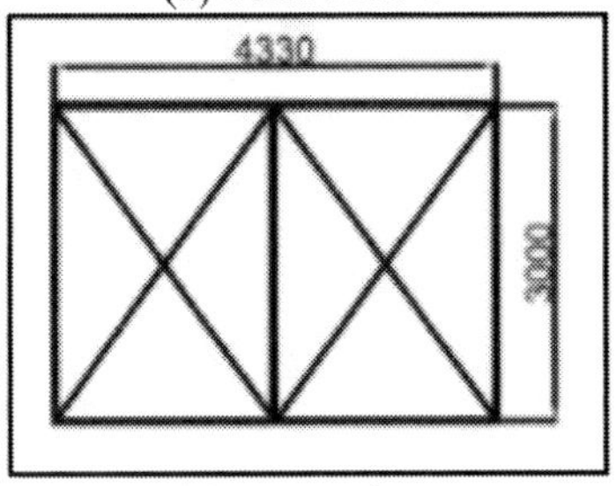

(c) Front view

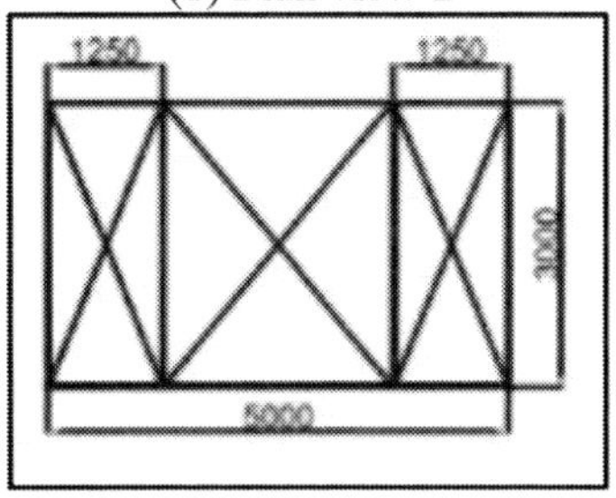

(d) Side view

Figure 1. Proposed modular steel building.

(a) Side view of the building

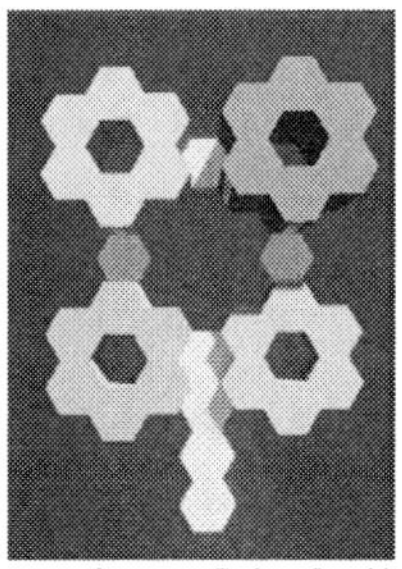

(b) Top view of the building

Figure 2. Illustration of Modular Steel Building

Table 1. Design sizes of component of the intra-module connection.

Component of the connection	Sizes of the component
Main Frame	Primary steel frame = C10 x 30 Secondary steel frame = C6 x 8.2
Circular Hollow Structural Section	Circular HSS 6.88 x 0.25mm
Diaphragm	Hexagon shape with a thickness of 20mm
Bolt and Nuts	Size 16mm with grade 8.8
Fin Plate	200 x 200 x 20mm

3. Verification of the finite element modelling method

The single-lap joint shown in figure 3 was used to examine the accuracy of the finite element modelling technique to model the connection. The top and bottom plates are grade S355 with 200mm x 200mm with a thickness of 25mm. The bolt size (d) is 16mm for the single lap joint, and the bolt hole size (d_0) is 18mm. It has a high tensile bolt strength grade of 8.8. A 10N/mm^2 pressure load, P, was applied parallel into the longitudinal direction x-axis. Next, the load capacity of the connection was calculated by using the formula as in equation (1). The resistance capacity for this connection is determined to be 60.288kN as in equation (2).

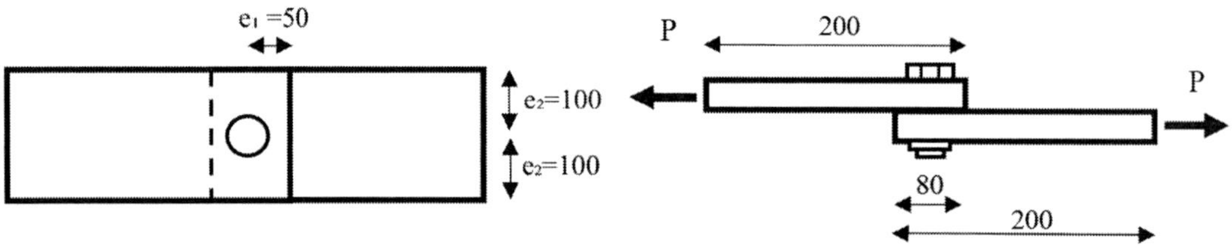

Figure 3. Single lap joint connection with dimension.

$$F_{v,Rd} = a_v\, f_{ub}\, \frac{A_s}{\gamma_{m2}} \tag{1}$$

$$F_{v,Rd} = 0.6\ (800)\, \frac{157}{1.25} = 60\ 288\ \text{N} = 60.288\text{kN} \tag{2}$$

The single-lap joint was then modelled in ABAQUS version 6.14. A continuum element that was used to model the two plates is naming as C3D6 whereby first letter is represent continuum stress/ displacement, second letter is its assymetric deformation (optional), third is the no of nodes, forth is reduce integration and lastly is small-strain formulation in ABAQUS/ Explicit. Meanwhile, the CCL24 element was used to model the bolt and nuts. The characteristic of the material assigned in the ABAQUS was presented in table 1. The deformable solid element with extrusion is used to model the steel channel section, circular hollow structural section, diaphragm, bolts, nuts, and fin plate in ABAQUS. For bolts, it was initially created with two different components: the head and the thread. The head and thread were then kept in intersecting boundaries to form a component. All the connection components created were assembled in the assembly section, as presented in figure 4. Linear static analysis was carried out to compare the shear resistance capacity design between the manual calculation and ABAQUS analysis. From the ABAQUS analysis conducted, the shear force reactions were then measured at the interface between the two plates and compared with the hand calculation designed according to calculation specified in [17]. The designed shear resistance per bolt comparison between ABAQUS and applied shear is shown in table 2 and indicates an excellent agreement with a difference of only 0.28 percent. The shear load obtained in ABAQUS is lesser than the ultimate design shear calculated using the calculation specified in [18], indicating that the single lap joint connection is able to resist the 10N/mm^2 (50kN) applied load. The deformed shape obtained from ABAQUS FEA software is shown in figure 5. The plates and bolt are allowed to move in X-direction. Thus, the moving of the plates will pull out the bolt from its original condition. Hence, the pulling out results show that the bolt experience high stress compared to other components. Figure 4 shows the undeformed shape before pulling out opposite the X-direction. Meanwhile, figure 5 shows the deformed shape of the single lap joint connection model after being pulled out by 10N/mm^2 pressure load (equal to 50kN force), respectively. Figure 6 shows the resultant force of the bolt that was pulled out.

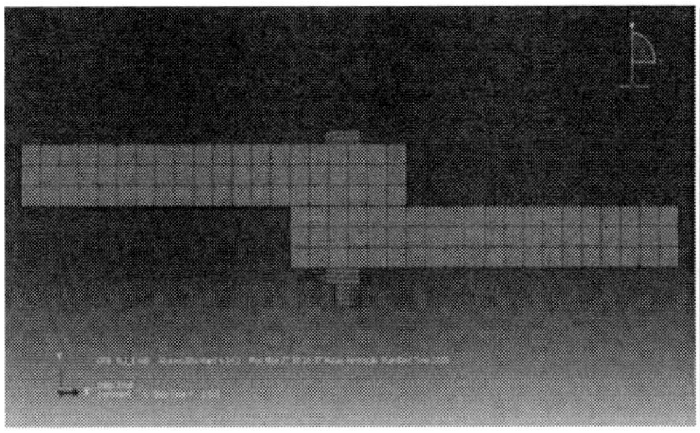

Figure 4. Undeformed shape of the SLJ connection.

Figure 5. Deformed shape of connection after being pull out by 10N/mm^2 pressure load.

Table 2. Comparison between manual calculation and ABAQUS FEA Software.

	Applied shear load	BS EN1993-1-1 (2015) Ultimate shear resistance	ABAQUS FEA Software Shear load resistance
Shear Force (kN)	50.00	60.29	49.86

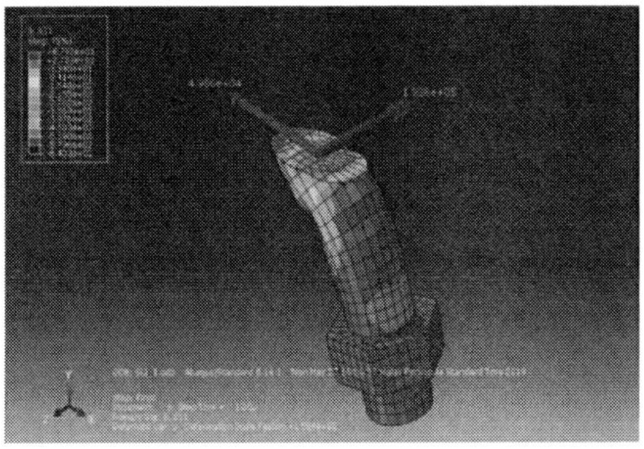

Figure 6. The resultant force of the bolt.

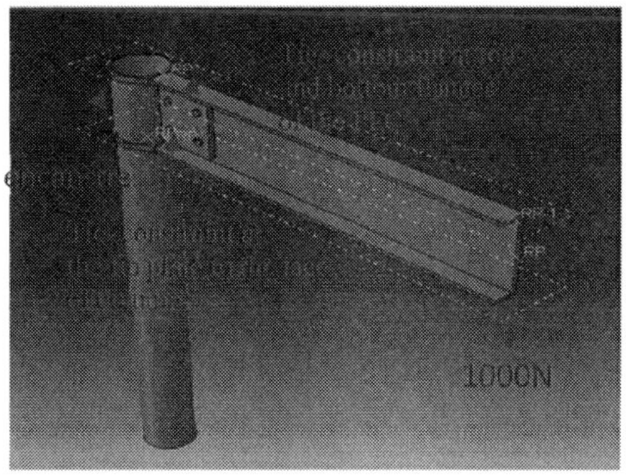

Figure 7. Assemble the proposed model in ABAQUS software.

4. Finite Element Modelling and Analysis of the proposed Intra-module connection.

The proposed intra-module connection shown in figure 7 was modelled and analysed in ABAQUS to investigate the governing failure mechanism. The intra-module connection consists of a steel channel, circular hollow, diaphragm, bolt, nut, and fin plate. The dimensions and materials for the steel C-channel, circular hollow, and fin plate are stated in table 1 and table 3. A continuum element C3D6 was used to model the circular hollow section, fin plate and diaphragm. Meanwhile, the CCL2 element was used to model the bolt and nut. The bolt and nut are fixed by using tie constraint. Nonlinear analyses via the ABAQUS FEA software is used to analyse the innovated novel intra-module connection. A general nonlinear Static, Riks method is used to simulate the intra-module connection's behaviour up to failure. The initial axial load is exerted 1000N downward at the end of the beam (refer figure 7), and it is incremented by time until failure.

Elastic-plastic materials are assigned in accordance with table 3. All steel elements for the diaphragm, circular column and c-channel are assigned as steel grade S275. The fin plate is assigned as steel grade S355, bolts and nuts are designated as high tensile steel Grade 8.8, respectively. For elastic characteristics, Young's Modulus is 210000 N/mm^{2,} and its Poisson's Ratio is 0.3. The density of steel is 7850 kg/m^3. The circular column is tied with a diaphragm using tie constraint. The fin plate is joined together by ties constraints perpendicular to the circular hollow column between the diaphragms. The c-channel is connected via tie constraint at the top and bottom flanged to the diaphragm. Contact interactions are used to connect the bolts and fin plate to the c-channel. All the contact interaction modelling used the technique master and slave whereby the slave element (loose part) will be tied to master elements that are rigid to form a body of the structure. The type of "encanstre" boundary condition is assigned at two places; at the bottom of the column and the side of the diaphragm (with an assumption that other c-channel members hold the other perimeter of the diaphragm). The modelling results such as stresses, strains, displacements, force/reaction and contact are requested and recorded in field and history output.

5. Result and Discussion

5.1 Failure mechanism of intra-module connection

The incremental vertical load was applied to the proposed novel intra-module connection until it failed. Every component of the connection was observed via step frame analysis to identify the critical failure mechanisms. The critical failure mechanism is the first component of the connection that has passed their Von Mises Stress. The first component that passed its original yield stress is the most severe, while the last component that passed its actual yield stress is the least severe. At the same time, the type of failure mechanism or failure mode acted on components of the connection have been identified and shown in table 4. The ABAQUS simulation analysis reveals that the diaphragm has achieved its yield capacity when it's reached 23440N (23.44kN) axial load, which is estimated about 23 times its initial applied load. In the proposed novel intra-module connection, the diaphragm is experiencing the most severe failure compared to other connected components. In comparison, the bolt and nut are the least severe among the members of the connection. The diaphragm was the first to experience local buckling at the top of the column facing the XY-axis for the finite element model under consideration. The increase of load and deformation of the diaphragm has indirectly pulled the circular hollow column and caused it to fail under flexural failure. The applied force on the c-channel has pushed the column and the diaphragm down, causing the torsional-flexural failure seen in figure 8.

Table 3. Stress and strain value for S355 Steel Plate and Bolt and Nut.

S355 Steel		Bolt and Nut		S275 Steel	
Stress (kN/m^2)	Strain (mm/mm)	Stress (kN/m^2)	Strain (mm/mm)	Stress (kN/m^2)	Strain (mm/mm)
0	0	0	0	0.0	0
355	0.002	640	0.003	275.0	0.001
355	0.015	750	0.008	275.0	0.015
397	0.026	800	0.028	315.4	0.037
475	0.057	640	0.118	347.7	0.058
510	0.110	66	0.139	373.0	0.080
504	0.140			392.4	0.101
489	0.170			406.9	0.123
				417.3	0.144
				424.1	0.166
				428.2	0.187
				429.9	0.209
				429.7	0.230

Meanwhile, the axial load acted on the structure cause the c-channel to experience local buckling at the top and bottom flanges. Flexural failure was observed occurred on both of the circular hollow sections and along with the diaphragm. The plate with the highest grade only accounted for the stress of 357.50 N/mm^2, which is the lowest of the connecting members. The mises stress that act on the bolt and nut section is the highest among others. The plastic deformation that occurs at the c-channel, column, diaphragm and plate connection are shown as in figure 9 and figure 10, respectively. The highest strain that happened was in the diaphragm, which probably indicated that the diaphragm governed the failure mechanism.

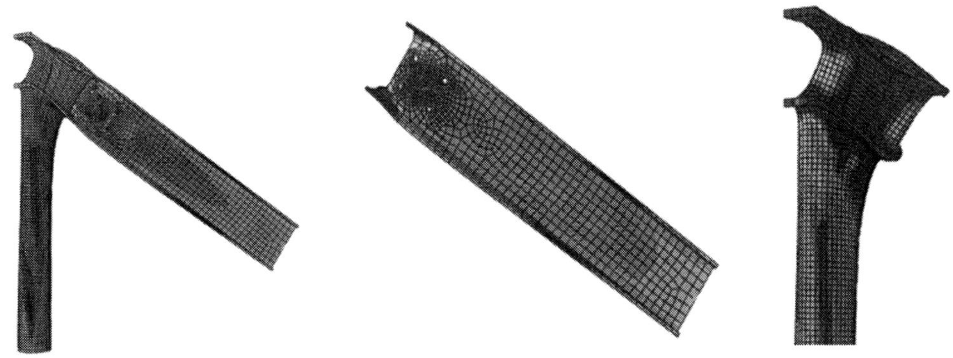

(a) Plate connection & bolts and nuts (b) Deformed Parallel Flange Channel (PFC) (c) Deformed Column and Diaphragm

Figure 8. The failure mechanism of the connection.

a) Side view b) Front view

Figure 9. View of the c-channel section under plastic strain deformation.

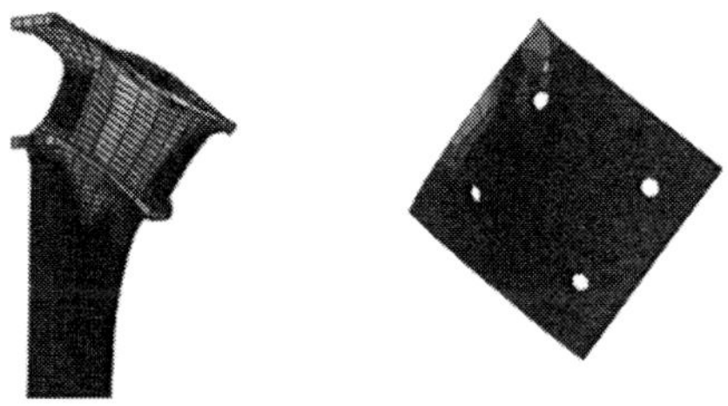

a) Column and diaphragm b) Plate connection

Figure 10. Isometric view of column, diaphragm and plate connection under plastic strain deformation.

Table 4. Summary of original yield stress, mises stress, plastic strain and failure mode of every part of the proposed novel intra-module connection.

Structure Component	Original Yield Stress (N/mm^2)	Mises Stress (N/mm^2)	Plastic Strain	Failure mode
C-Channel Section	275	429.70	0.73	Local buckling of the top and bottom flanges
Circular Hollow Column	275	429.80	1.57	Flexural-torsional buckling

Diaphragm	275	429.90	2.02	Flexural-torsional buckling
Plate Connection	355	357.50	0.02	Flexural failure concentrated
Bolt and nut	640	742.70	0.02	Flexural failure concentrated

5.2 The load-displacement behaviour

Figure 11 depicts the force vs displacement relationship measured at the point where the load is applied on the PFC. As can be seen, the ultimate load imposed on the PFC prior to the structural system becomes unstable is 249.63kN, with a vertical displacement of 435.71mm. The diaphragm was the first to yield at a load capacity of 23.4kN, followed by the top and bottom flanges of the PFC (refer to figure 8b). As the PFC deformed at large displacement, it caused the diaphragm and column to be pulled downward and strained to 1.57 to 2.02, respectively. In addition, the diaphragm and the column were observed to experience slight torsional-flexural deformation under the imposed load. One of the possibilities was due to the geometrical shape of the diaphragm. The intra-module structure member was joint at one of the adjacent diaphragm parameters and was rigidly restrained on the other adjacent diaphragm parameter, as shown in figure 7. As previously said, the diaphragm was welded to one side of the column's circumference; therefore, as the PFC moving downward, the adjacent side of the diaphragm parameter has also been pulled and twisted downward together with the column. The exclusion of the secondary beams in restraining the main PFC laterally was also the additional reason for such deformation behaviour. The behaviour of the moment-chord rotation characteristic was also observed under the incremental applied load. It can be seen that the highest moment measured at the PFC-diaphragm-column connections was about 280kNm, with chord rotation is at 0.35 mm/mm. The moment capacity obtained has exceeded the plastic yield moment capacity of the PFC, column and diaphragm. Such observation can be seen in figure 9 and figure 10, in which the equivalent plastic strain has occurred at the top and bottom of the PFC flanges, the diaphram and one spot at the fin plate.

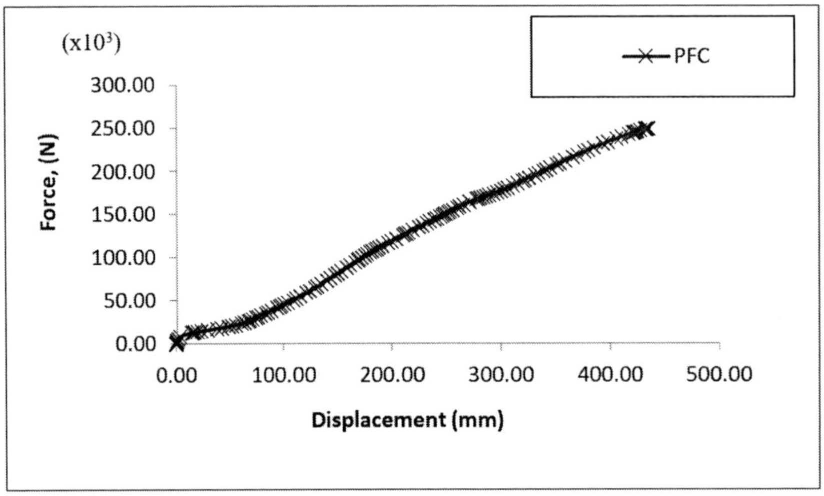

Figure 11. Force vs displacement curves of the intra-module structures.

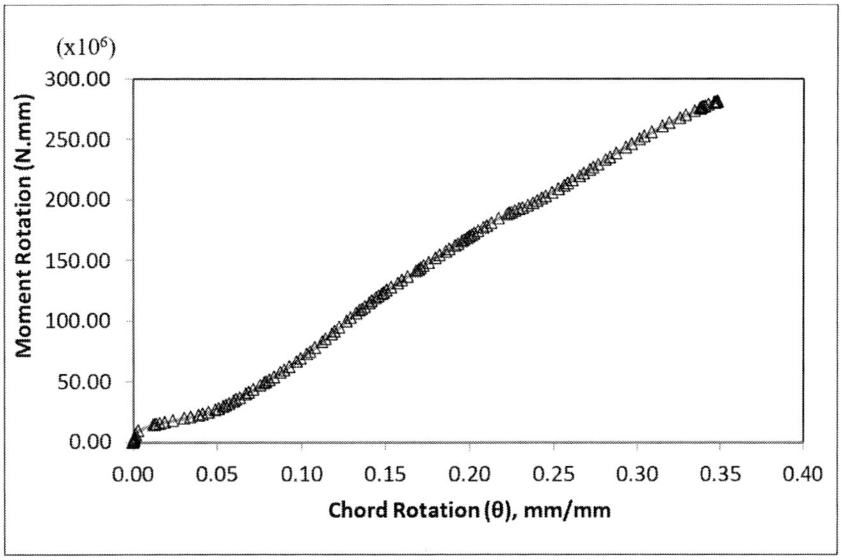

Figure 12. Moment vs Chord Rotation.

6. Conclusion

A new design scheme of the intra-module connection for the innovative hexagon modular steel building has been proposed. The novel intra-module connection has been successfully evaluated. The failure mechanism of the novel intra-module connection under the applied incremental vertical load using the ABAQUS FEA software has been determined. For the model under consideration, the failure of the proposed connection is probably governed by the diaphragm and depends on how the load is applied to the connection. The analysis shows that the proposed hexagon diaphragm, which was welded to the circular hollow column, tends to experience flexural torsional buckling. One of the main reasons is the hexagon shape diaphragm used to connect the intra-module structural members and the secondary beam's exclusion to restrain the main PFC laterally. In this work, the failure of the connection under the given load was governed by the hexagon diaphragm, while the fin plate demonstrates the least affected constitutive component. It anticipates that the suggested unique intra-module connection will encourage architects to employ modular steel construction designs with greater flexibility. Future research will concentrate on the in-depth parametric study to improve the performance of the diaphragm and the connection's limitations.

7. Recommendation

The ABAQUS FEA software modelling in this work does not restrain the steel channel section with any beams or slabs. As a result, the steel channel section should be restrained with a secondary beam of steel channel section spaced at 0.6 m intervals to prevent flexural-torsional buckling. It is also significant to physically test the unique intra-module connection that has been proposed. The performance of the suggested innovative intra-module connection under dynamic stresses such as fatigue and earthquake would be a fascinating future field to research.

8. References

[1] Aye L, Ngo T, Crawford R H, Gammampila R and Mendis P 2012 Life cycle greenhouse gas emissions and energy analysis of prefabricated reusable building modules, *Energy and Buildings* **47** 159-168

[2] Lawson M, Ogden R and Goodier C I 2014 *Design in modular construction* (Boca Raton, FL: CRC Press) Vol. 476 p. 280

[3] Styles A J, Luo F J, Bai Y and Murray-Parkes J B 2016 Effects of joint rotational stiffness on structural responses of multi-story modular buildings *Proceedings of the International Conference on Smart Infrastructure and Construction Construction* (Cambridge)
[4] Dhanapal J, Ghaednia H, Das S and Velocci J 2019 Structural performance of state-of-the-art VectorBloc modular connector under axial loads, *Engineering Structures* **183** 496-509
[5] Hong S G, Cho B H, Chung K S and Moon J H 2011 Behavior of framed modular building system with double skin steel panels, *Journal of Constructional Steel Research* **67** 936-946
[6] Jaganathan S, Nesan L J, Ibrahim R and Mohammad A H 2013 Integrated design approach for improving architectural forms in industrialized building systems, *Frontiers of Architectural Research* **2** 377-386
[7] Jaillon L, Poon C S and Chiang Y H 2009 Quantifying the waste reduction potential of using prefabrication in building construction in Hong Kong, *Waste management* **29** 309-320
[8] Kyjakova L and Baskova R 2016 Advantages and Disadvantages of Modern Methods of Construction Used for Modular Schools in Slovakia *5th International Conference on Zero Energy Mass Customised Housing - ZEMCH 2016* (Kuala Lumpur)
[9] Lacey A W, Chen W, Hao H and Bi K 2019 Experimental and numerical study of the slip factor for G350-steel bolted connections, *Journal of Constructional Steel Research* **158** 576-590
[10] Lawson R M, Ogden R G, Pedreschi R, Grubb P J and Popo-Ola S O 2005 Developments in pre-fabricated systems in light steel and modular construction, *Struct. Eng.* **83** 28–35
[11] Lawson R M, Ogden R G and Bergin R 2012 Application of modular construction in high-rise buildings, *Journal of Architectural Engineering* **18** 148-154
[12] Quale J Design in Modular Construction, *J. Archit. Educ.* **71** 119-120
[13] Rahman M M 2014 Barriers of implementing modern methods of construction, *Journal of Management in Engineering* **30** 69-77
[14] Kamali M and Hewage K 2016 Life cycle performance of modular buildings: A critical review, *Renewable and Sustainable Energy Reviews* **62** 1171-1183
[15] Generalova E M, Generalov V P and Kuznetsova A A 2016 Modular buildings in modern construction, *Procedia engineering* **153** 167-172
[16] Aziz S and Abdullah M R 2015 Modular Construction System in Malaysia : Issues for Research in Sustaining an Affordable Home Project *Proceedings of Postgraduate Conference on Global Green Issues (Go Green)* (Seri Iskandar)
[17] Eurocode 3 2005 *Design of steel structures — Part 1-8 : Design of joints* (Brussels: European Committee for Standardization)
[18] Eurocode 3 2005 *Design of steel structures — Part 1-1 : General rules and rules for buildings* (Brussels: European Committee for Standardization)

Acknowledgements

The Authors would like to show gratitude to Ministry of Higher Education Malaysia and Research Management and Innovation (IRMI) UiTM, Shah Alam, Selangor, Malaysia for providing financial support through Fundamental Research Grant Scheme (FRGS). (File no: FRGS/1/2018/TK01/UiTM/02/08)

Shear Strengthening of Reinforced Concrete Beams Using Fibre Reinforced Polymer: A Critical Review

S J S Hakim[1], M A H Mohd Rodzi[2], S S Ayop[2], S Shahidan[2], S N Mokhatar[1], N Salleh[2]

[1] Structural Dynamics and Computational Engineering Focus Group (StrucDyCE), Faculty of Civil Engineering and Built Environment, University Tun Hussein Onn Malaysia (UTHM), 86400 Parit Raja, Johor, Malaysia
[2] Department of Civil Engineering, Faculty of Civil Engineering and Built Environment, University Tun Hussein Onn Malaysia (UTHM), 86400 Parit Raja, Johor, Malaysia
Corresponding author: seyedhakim@uthm.edu.my

Abstract: The primary purpose of reinforcing bar stirrups in a reinforced concrete beam is to improve shear strength. The FRP system may significantly improve a concrete beam's ultimate shear strength, serviceability, and ductility. The application of FRP for the repair and reinforcement of the structures has become very popular due to its low weight, high tensile strength, and simplicity of installation on uneven surfaces. FRP material outperforms other traditional materials in strengthening applications due to its high strength-to-weight and stiffness-to-weight ratios, resistance to corrosion, and ease of handling. The overall objective of this research is to investigate and improve the understanding of the recent research in the area of shear FRP strengthening of reinforced concrete beams. In this paper, recent publications were reviewed to see how different anchoring procedures, different factors that affect FRP performance and different failure scenarios affect the shear strengthening of concrete beams. The benefits and limits of FRP systems, as well as some current research trends are discussed in this project. From the research, it can be stated that type of anchorage technique and different parameter give a different impact to failure mode of the beam.

1. Introduction

Failure in structures often leads to damage and can be defined as a weakening of the structure which may cause undesirable displacements, stresses or vibrations to the structure that negatively affects the functionality of the structure [1][2][3]. Concrete is one of the most well-known materials to be reinforced in terms of strength since it is unable to sustain a considerable amount of weight alone. Using internal steel reinforcing bars in concrete buildings is a common technique of improving structural strength. Reinforcement bars, on the other hand, will lose their strength as time goes on due to their construction. As a consequence, an external fiber-reinforced polymer must be used to aid reinforced concrete in performance. FRP is a composite material consisting of a polymer reinforced with carbon, glass and aramid fibres.

Because of its excellent strength-to-weight ratio, FRP is applied in many different sectors, including construction, marine, aerospace and automotive. FRP may be used to enhance material with high corrosion resistance and can also be utilized to maintain damaged structures. Carbon was the most commonly used composite material in construction, exceeding all other composite materials in terms of strength and weight. This is because carbon is stronger and lighter than other materials.

Comparing carbon fibre to glass fibre, carbon has a fibre content and density that is appropriate for materials with normal weight percentages. Moreover, compared to the other composite materials, carbon has the largest longitudinal tensile modulus and tensile strength. Consequently, carbon is a very suitable and extensively utilized material for increasing structural strength.

Concrete is a key building material in the construction industry. To make the concrete strong, the presence of steel reinforcement is needed, however this is not enough to support a greater weight. As a consequence, the presence of internal steel reinforcement coupled with external FRP was needed to create concrete that could resist a higher weight. Reinforcing steel bars have been utilized to construct most existing structures. This gives the concrete structure strength, but the problem is that the internal reinforcement state will decrease based on the time. There are numerous historically important structures should be conserved for future generations. However, if preventive maintenance is overemphasized, it may adversely affect the uniqueness of the component structure as well as increase the component cost owing to the use of new materials.

Additionally, another way to cope with this problem is to demolish the old structure and replace it with a new one. This action will be costly owing to the new design, the huge quantity of materials needed, the high construction cost, and the time required to create the new structure. Another solution to this question is the usage of FRP. The correct technique for applying this external reinforcement as well as the kind of anchoring strategy to employ is important. According to the literature, different studies have been done to improve the flexural and axial strengthening of RC structures using FRP. But still little research has been done on utilizing FRP for shear strengthening. Moreover, the findings are restricted and sometimes disputed even with typical reinforced concrete components without FRP. Due to the lack of experimental data, several analytical models were developed to estimate the shear contribution of externally bonded FRP. However, a research was needed based on various anchoring methods and several variables affecting FRP's shear strengthening. The aim of this research is to review the effect of different anchorage techniques, different parameters on the shear strength of RC beams using FRP and the effect of RC beams strengthened with FRP in failure modes.

2. Effect of different anchorage techniques on shear strengthening

Kim and Smith [4] recognized the need for anchoring the ends of FRP strips or sheets in their experimental to prevent premature debonding failure and to efficiently manage shear stress without debonding, Uji [5] stated that sufficient fibre sheet anchoring is required, comparable to steel stirrups. However, this was observed in all situations at the time, except for columns. When adequate anchoring is utilized, failure is avoided since the strength development in the FRP depends completely on the anchor's strength and not on the strength of the connection between that FRP and the concrete layer. Using a practical anchoring technique, the debonding failure mode is successfully prevented, resulting in a more desired FRP rupture mode than elsewhere [6][7].

This is worth to mentioning that installing an anchoring system does not avoid debonding completely. Along the length of the FRP strips or sheets, a certain degree of debonding must occur in order to be successful. In the absence of an anchoring mechanism, the strength of the whole reinforcing system depends solely on the connection between the FRP material and the concrete substrate. This section discusses the various anchoring methods and split into three categories: (a) FRP anchors using FRP products for anchoring, (b) metal anchors using steel or aluminium for anchoring, and (c) mixed anchors using anchoring metals and FRP. Anchorage systems for externally bonded FRP methods are extremely beneficial in preventing or delaying interfacial crack opening, increasing the total possible transmission of interfacial shear stress and providing a stress transfer mechanism where there is no bond length beyond the critical section. Different methods of anchoring FRP utilized in the literature, together with their advantages and disadvantages, have been studied. The details of different anchorage techniques are described in the following sections.

2.1. Near surface mounted anchorage technique

Lorenzis and Teng [8] indicated that the enhanced effectiveness of the near surface mounted (NSM) anchorage method is significantly influenced by the bond behaviour of NSM FRP reinforcement and the components to be reinforced. Khalifa and Nanni [9] initially proposed NSM anchoring method.

Figure 1 details a typical aspect of the NSM anchoring technique. A groove is formed across the whole length of the shear span as illustrated in Figure 1, situated either at the web-flange or on the flange's soffit face. FRP sheets are then connected to the concrete surface and groove walls applying glass FRP (GFRP) bars [6], carbon fibre (CFRP) strips, CFRP coat [10], or hardwood strips [11]. The FRP bar diameter utilized in these experiments varied from 10 mm to 13 mm. Micelli et al. [12] used NSM method to connect two CFRP sheets. Researchers utilized the NSM anchoring technique to connect continuous fiber-reinforced polymer sheets and FRP sheet strips.

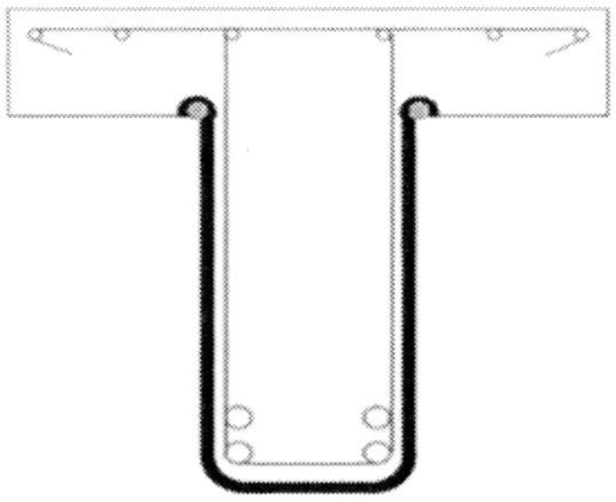

Figure 1. An illustration of the NSM anchoring method in action.

2.2. FRP extension anchorage technique
According to figure 2, this method consists of extending U-wrap legs under the flange. Application of FRP U-wraps is very attractive since it is a simple technique that substantially delays the debonding of longitudinal FRP layers, thereby improving flexural capacity. Owing to load concentration, FRP U-wraps may fracture at the beam soffit corners. The proposed method aims to improve effectiveness by preventing premature debonding and reducing stress concentration. The use of this technique was shown in the anchoring of CFRP sheet strips [13], GFRP sheets [13], CFRP sheets with 45 degree fibre direction [13], GFRP continuous sheets [13], CFRP continuous sheets [14], and triaxial GFRP continuous sheets and a total of GFRP continuous sheets. Despite the fact that this anchoring technique is easy and cheap.

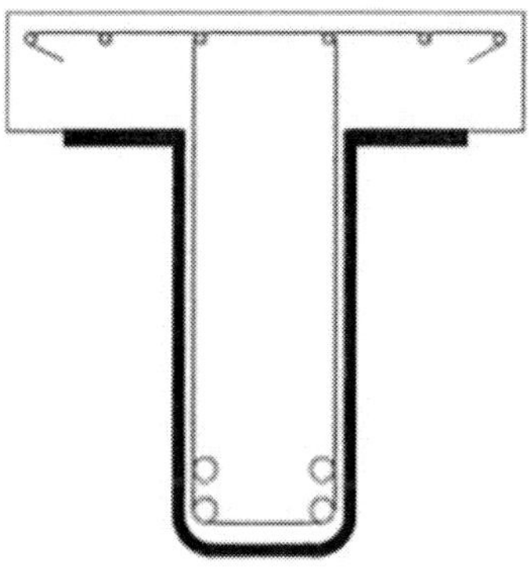

Figure 2. An example of the FRP extension anchoring method in action.

2.3. Horizontal anchorage system
As shown in Figure 3, this method includes attaching horizontal laminate sheets of FRP to U-wrap scheme's free ends. It was chosen to use a horizontal strip constructed of CFRP sheet with an internal cross-section for the horizontal strip [15]. Although the FRP horizontal anchoring method considerably increased the FRP shear involvement, the methodology did not completely prevent debonding failure.

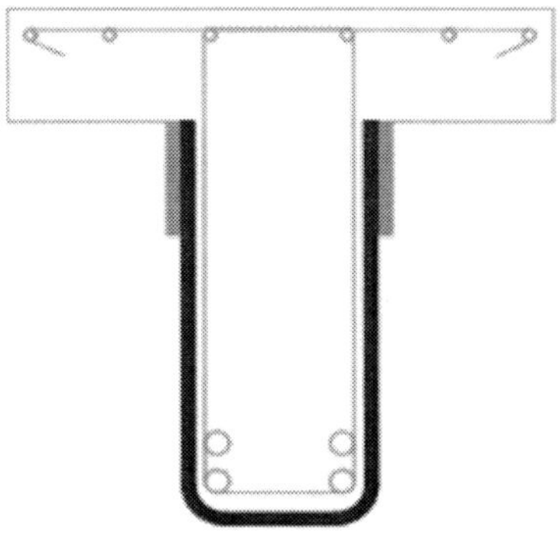

Figure 3. An illustration of the horizontal FRP anchoring method in action

3. Effect of different parameters on shear strength using FRP

The findings of the published experimental research were provided to recognize the roles performed by the various factors affecting the performance of FRP shear-strengthened anchored RC beams. The following parameters must be considered when designing a FRP shear reinforced beam. Moreover, curves are used to connect the dots showing the apparent outcomes for the various anchoring techniques to provide a more accurate representation of the data set. Specimens with 200 mm or less beam height were utilized to avoid interference with the results. Shear span to effective depth ratios below 2.5 are regarded excellent results. Finally, the research eliminated specimens with a flexure failure mechanism.

3.1. Beam depth

According to Collins and Kuchma [16], RC beams lose their shear strength as the beam expands. Other than those reported by some researchers [17][18][19][20][21], most specimens for anchored FRP shear reinforced beams were below 350 mm. Godat et al. [22] analyzed the average of the FRP shear strength as a percentage of the overall FRP shear strength. The FRP rope anchoring which was followed by FRP extension and sandwich plates methods offered the greatest shear strength. The lowest shear strength for FRP was attained by combining the extension anchoring and horizontal FRP techniques. This is essential to highlight that the number of research examining the efficacy of these methods is low, requiring further investigation. Shear contribution of FRP varies with the depth of beam is shown in figure 4.

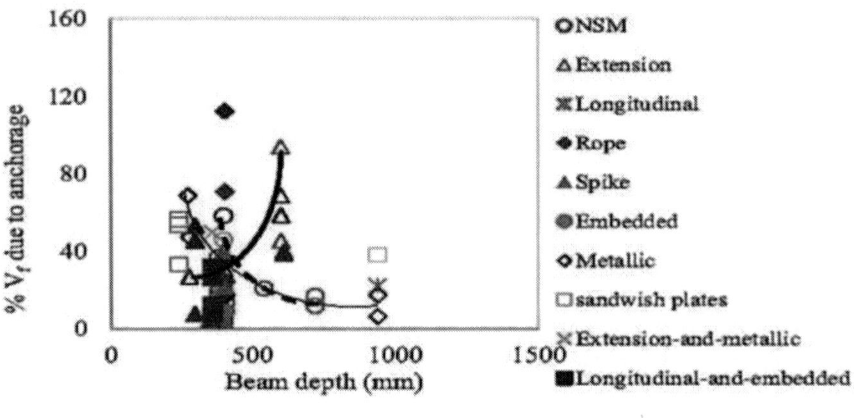

Figure 4. Shear contribution of FRP varies with the depth of beam [22].

According to figure 4, only metallic, FRP extension, near surface mounted, and spikes anchoring methods have a substantial impact on depth while the other approaches have no effect. On deeper beams, increases in shear strength and strain are obtained via the FRP extension and spike anchoring techniques. There was a connection between shear contribution and beam depth and the size of the bonding area in

FRP. The higher the FRP shear contribution, the higher the beam and the greater the bond surface must be in the beam. FRP debonding was the failure mode for all anchored specimens, save those reported by Kim et al. [23], where the beam depth surpassed 450 mm.

3.2. Ratio of shear span (a) to the effective depth (d)

The gap between load and closest support is called shear span. To differentiate between shallow and deep beams, two a/d ratios were computed. If shallow beams reveal shear or flexural cracks, the ratio of shear span to the effective depth (a/d) is required. In the research by [24], the effect on FRP shear influence is removed by omitting a/d 2.5 values. The shear strength of FRP seems to drop as the ratio of shear span to the effective depth rises for the different methods evaluated in this study. Both NSM and spike anchoring methods demonstrated this trend. This implies utilizing FRP for shear enhancement becomes less attractive as this ratio rises. Conversely, with the FRP spike anchoring method, the FRP strain decreased as the a/d ratio rose. When the a/d ratio exceeded 3.5, no FRP debonding failures occurred. The importance of the ratio of shear span to the effective depth in calculating FRP shear strength influence is not addressed in the current FRP enhancing design guidelines, despite many research acknowledging the usefulness of this ration in design of RC beams.

3.3. Compressive Strength of Concrete

The compressive strength of concrete (fc) is one of the main factors to influence the FRP shear contribution and affects the FRP/concrete interface behaviour and failure mechanism. FRP shear strength rises with the value of compressive strength of concrete. For spike anchoring method, the strain has a tendency to drop. Except for a sample in Baggio et al. [25], there was FRP debonding with f_c greater than 35 MPa in the database. There is no scientific explanation for this phenomenon since it is intended to prevent FRP debonding failure and avoid concrete crushing in the compression zone.

4. Failure modes of FRP

Failure processes in reinforced concrete beams include flexural failure and shear failure, the two most common kinds of failure. These failure modes are further split into many kinds of failure, including tension, compression, debonding, crushing, rupture, flexure, and compression shear failure for the former. In design, some are favored above others. However, others must be avoided if one is to prevent catastrophic structural failures. For example, in concrete, crushing and shear failure modes are unwanted since they occur failure quickly and without notice. Due to its brittleness and complexity of processes, a typical RC beam's shear behavior is difficult to predict [26][27].

Because of the impact of different bonded FRP failure modes such as debonding failure and tensile rupture failure, calculating the shear capacity of a reinforced concrete structure externally reinforced with fibre-reinforced polymer (FRP) sheets is even more challenging [28][29]. Therefore, the methods for estimating shear capacity and identifying the failure mode recommended in design codes are based on empirical or semi-empirical equations derived from design equations for typical RC structures rather than analytical solutions [30]. This section will discuss the failure modes identified via a review of research on shear strengthening of RC beams employing FRP to provide a comprehensive overview of the failure modes observed in experimental testing.

4.1. Externally applied FRP debonding failure modes

When CFRP materials are placed in a "U"-wrap or side-bonded fashion, debonding failures are a significant issue because if the CFRP starts to separate from the concrete substrate, the beam may collapse extremely rapidly, reducing member ductility. Recent design guidelines for externally applied FRP materials restrict the effective tensile strain of the material to 0.004 in/in, or 40% of its ultimate value to avoid this kind of failure [31]. This implies that 60% of the potential of the CFRP system is never utilized in practice. When tested in debonding mode, nearly all side-bonded beams and the overwhelming majority of "U"-wrapped beams failed. Although debonding is regarded as a failure mechanism in CFRP systems, the carbon fibre sheets must operate correctly. Large stresses in CFRP near cracks are produced by strain incompatibility with concrete substrate. Concrete fracture results in localized CFRP material debonding at the crack location is illustrated in figure 5.

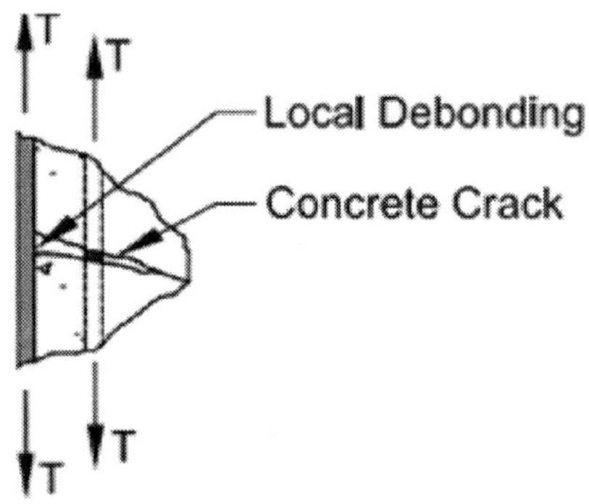

Figure 5. The concrete surface is covered with CFRP after cracking [32].

To ensure debonding does not result in a structural collapse, many measures are required. The quality of the surface preparation performed before the CFRP is installed, as well as the effective bond length established between the CFRP and the concrete surface, as well as the compressive strength of the concrete substructure and the axial rigidity of the applied system, all have an impact on CFRP debonding. In relationship applications, a strong connection between the CFRP and the concrete surface is needed since the CFRP material cannot be fully wrapped around a concrete component. Any laitance, dust, oil, existing coatings or other elements that may interfere with the CFRP system should be removed from the surface as well as any other components that may interfere with the CFRP system. A sufficient bond length is required for CFRP sheets to withstand shear pressures. However, the amount of shear stress the CFRP can take does not grow linearly with the bonding length. After shear crack development, only the bonded part of CFRP material that extends beyond the fracture can resist shear forces.

When it comes to bond strength between concrete and CFRP, one of the most important variables affected by this issue is the concrete's compressive strength. Debonding occurs most frequently in concrete a short distance from the concrete CFRP contact, and is usually always evident in concrete. A piece of concrete remains attached to the CFRP even after debonding occurs. It is evident from the fact that the concrete failure plays an essential part in determining the total structural strength of the entire system [33]. The debonded strip pushed a piece of the concrete substrate away from the beam, causing the concrete substrate to fracture. Last but not least, the system's axial stiffness affects its surface debond propensity. Triantafillou [34] stated the effective bond length required to achieve the ultimate tensile force carried by the CFRP strips depends nearly proportionately on the axial rigidity of the CFRP utilized in the application.

4.2. Failure modes of anchor

The FRP debonding, FRP rupture, flexural failure and concrete crushing have been considered as failure modes in this study. Godat et al. [36] tried with a variety of anchoring techniques, however none were successful in decreasing the FRP debonding failure mechanism as shown in figure 6. Graph illustrating the ratio of failure modes produced by each anchoring method FRP debonding is a characteristic of fiber-reinforced polymer extension, horizontal and patch anchoring techniques.

The CSA/S606 [35] standard recommends using the horizontal FRP anchoring technique to secure shear-reinforced polymer beams. During concrete crushing, FRP debonding failure occurred in all specimens connected to FRP rope, showing the FRP was debonded. Data indicated that debonding failure is the most common failure mechanism for NSM and FRP spike anchoring techniques, the second most common being debonding failure. The FRP rupture failure mechanism characterizes horizontal and insertion anchoring techniques. This anchoring technique can only be used for FRP sheets and needs prior slab reinforcement locations information.

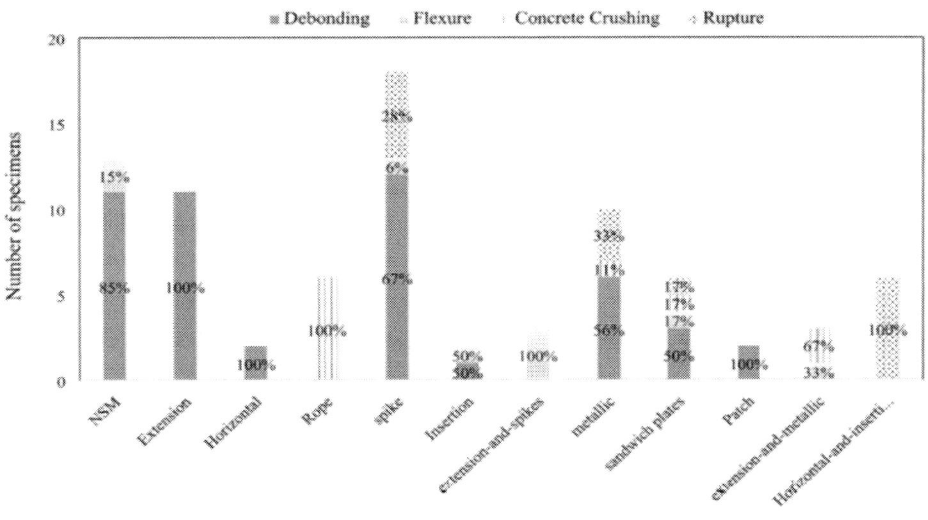

Figure 6. Failure mechanisms of shear-strengthened FRP beams [36].

Longitudinal reinforcement succumbed before failure when the anchoring method was employed to secure the structure. Debonding was found to cause failure in 56 percent of the 11 specimens secured utilizing the metallic anchoring technique. The average shear strength for each of the various anchoring methods investigated in this research is approximately 80%. The total shear strength of the anchored and unanchored beams is deducted from the overall shear strength (V_{total}) computed as ($V_f = V_{total} - V_{control}$) to determine the shear strength (V_f) owing to anchoring. It is important to highlight that in certain instances; some experiments did not utilize a reinforced specimen or anchorage. In this situation, $V_{control}$ was the weakest. This assumption is unlikely to have any major influence on findings.

The FRP shear strength of anchored specimens utilizing FRP extension and extension-and-metallic methods was lower than the shear strength of the control specimen. In certain instances, FRP sheet expansion beneath the flange has been found to reduce FRP sheet shear strength and increase the probability of FRP debonding failure. It is produced by the existence of residual stresses at the web-flange corner, resulting in high stress concentrations and reduced anchoring technique performance. In this study, the relationship between failure mode and FRP shear strength can be shown. In certain instances, while the FRP is brittle, the FRP rupture and flexure failure modes do not result in significant FRP shear strength.

For example, FRP extension anchoring method, failure to bind together is characterized as FRP debonding. However, after compared to other anchoring techniques, average shear strength showed the greatest. In addition, all horizontal specimens failed according to FRP break, while their shear strength showed half that of the FRP rope anchoring technique. Anchoring methods may prevent or postpone debonding, which improves the shear strength of FRP. When proposing an anchoring technique for inclusion in existing design standards, these two problems should be considered.

5. Conclusions

In this research, different anchoring methods for FRP shear strengthening were emphasized and analyzed the efficiency of each approach in strengthening shear involvement and preventing FRP debonding failure. The influence of variables that significantly affect the shear contribution of FRP may negate the benefits of anchoring techniques. This will also result in failure modes in anchoring technique. The NSM technique may improve bonding efficiency while minimizing exposure to unfavorable conditions. This technique has the potential to increase the shear contribution, but due to the requirement of specialized equipment, this strategy is extremely difficult to implement. From the literature, sandwich plate systems have a good strength-to-weight ratio, making them increasingly desirable for building form use. With increasing beam depth for NSM and metallic anchoring methods, shear strength begins to drop. There was a connection between shear contribution and beam depth and the size of the bonded area in FRP.

Using FRP rope anchoring method, the maximum shear strength was obtained, followed by FRP sandwich plate and FRP extension anchorage methods. Only a few anchoring methods were effective in producing FRP rupture and flexural failure modes. The FRP extension anchorage approach exhibited yielding in longitudinal reinforcement. The finding of this paper showed that, more research is needed to enhance the contribution between the parameters and the shear strength and FRP strain, as well as the failure mode utilizing anchoring method.

6. References

[1] Hakim S J S and Abdul Razak H 2014 Frequency response function-based structural damage identification using ANNs-a review, *Research Journal of Applied Sciences, Engineering and Technology* **7** 1750-1764

[2] Noorzaei J, Hakim S J S, Jaafar M S, Abang Ali A A and Thanoon W A M 2007 An optimal architecture of artificial neural network for predicting compressive strength of concrete, *Indian Concrete Journal* **81** 17–24

[3] Hakim S J S and Razak H A 2011 Application of artificial neural network on vibration test data for damage identification in bridge girder, *International Journal of the Physical Sciences (IJPS)* **6** 7991- 8001

[4] Kim S J and Smith S T 2009 Strengthening of RC slabs with large penetrations using anchored FRP composites *Proceedings of the 2nd Official International Conference of International Institute for FRP in Construction for Asia-Pacific Region (APFIS2009)* (Seoul, South Korea)

[5] Uji K 1992 Improving shear capacity of existing reinforced concrete members by applying carbon fibre sheets, *Transactions of the Japan Concrete Institute* **14** 253-266

[6] Triantafillou T C and Antonopoulos C P 2000 Design of concrete flexural members strengthened in shear with FRP, *Journal of composites for construction* **4** 198-205

[7] Teng J G, Lam L and Chen J F 2004 Shear strengthening of RC beams with FRP composites, *Progress in Structural Engineering and Materials* **6** 173-184

[8] De Lorenzis L and Teng J G 2007 Near-surface mounted FRP reinforcement: An emerging technique for strengthening structures, *Composites Part B: Engineering* **38** 119-143

[9] Khalifa A and Nanni A 2000 Improving shear capacity of existing RC T-section beams using CFRP composites, *Cement and Concrete Composites* **22** 165-174

[10] Mofidi A, Chaallal O, Benmokrane B and Neale K 2012 Performance of end-anchorage systems for RC beams strengthened in shear with epoxy-bonded FRP. *Journal of Composites for Construction* **16** 322-331

[11] Chen G M, Zhang Z, Li Y L, Li X Q and Zhou C Y 2016 T-section RC beams shear-strengthened with anchored CFRP U-strips, *Composite Structures* **144** 57-79

[12] Micelli F, Annaiah R H and Nanni A 2002 Strengthening of short shear span reinforced concrete T joists with fiber-reinforced plasic composites, *Journal of Composites for Construction* **6** 264-271

[13] Sato Y, Katsumata H and Kobatake Y 1997 Shear strengthening of existing reinforced concrete beams CFRP sheet. *Proceeding of the Third International Symposium on Non-Metallic (FRP) Reinforcement for Concrete Structures (FRPRCS-3)* (Sapporo, Japan)

[14] Tanarslan H M, Ertutar Y and Altin S 2008 The effects of CFRP strips for improving shear capacity of RC beams, *Journal of Reinforced Plastics and Composites* **27**1287-1308

[15] Belarbi A, Bae S W and Brancaccio A 2012 Behaviour of full-scale RC T-beams strengthened in shear with externally bonded FRP sheets, *Construction and Building Materials* **32** 27-40

[16] Collins M P and Kuchma D 1999 How safe are our large, lightly reinforced concrete beams, slabs, and footings, *Structural Journal* **96(4)** 482-490

[17] Galal K and Mofidi A 2010 Shear strengthening of RC T-beams using mechanically anchored unbonded dry carbon fiber sheets, *Journal of Performance of Constructed Facilities* **24** 31-39

[18] El-Maaddawy T, Chekfeh Y 2013 Shear strengthening of T-beams with corroded stirrups using composites.,*ACI Structural Journal* **10** 779-790
[19] Panigrahi A K, Biswal K C and Barik M R 2014 Strengthening of shear deficient RC T-beams with externally bonded GFRP sheets, *Construction and Building Materials* **57** 81-91
[20] Frederick F F, Sharma U K and Gupta V K 2015 Effect of end anchorage in external CFRP confinement on shear damaged RC beams, *Procedia Eng.* **125** 953-958
[21] Routray S 2015 *Shear Behaviour of BFRP Strengthened RC T-Beams* PhD Thesis (Rourkela, India)
[22] Godat A, Hammad F and Chaallal O 2020 State-of-the-art review of anchored FRP shear-strengthened RC beams: A study of influencing factors, *Composite Structures.* **254** 112767
[23] Kim Y, Quinn K T, Satrom C N, Ghannoum W M and Jirsa J O 2011 Shear strengthening RC T-beams using CFRP laminates and anchors, *ACI Structural Journal.* **275** 1-18
[24] American Concrete Institute (ACI) 2019 *Building code requirements for structural concrete and commentary* Report No. 318-19 (Michigan: American Concrete Institute)
[25] Baggio D, Soudki K and Noel M 2014 Strengthening of shear critical RC beams with various FRP systems, *Construction and Building Materials* **66** 634-644
[26] Mohammadhassani M, Jumaat M Z B, Chemrouk M, Ghasemi A, Hakim S J S and Najmeh R 2011 An experimental investigation of the stress-strain distribution in high strength concrete deep beams, *Procedia Engineering* **14** 2141–2150
[27] Ravanfar, S A, Razak H A, Ismail Z and Hakim S J S 2016 A two-step damage identification approach for beam structures based on wavelet transform and genetic algorithm, *Meccanica.*, **51** 635–653
[28] Kalfat R, Al-Mahaidi R, Smith S T 2013 Anchorage devices used to improve the performance of reinforced concrete beams retrofitted with FRP composites: State-of-the-art review, *Journal of Composites for Construction* **17** 14-33
[29] Liu S, Yuan H, Wu J 2019 Full-range mechanical behavior study of FRP-to-concrete interface for pull-pull bonded joints, *Composites Part B: Engineering* **164** 333-344
[30] Garrido M, Madeira J FA, Proença M and Correia J R 2019 Multi-objective optimization of pultruded composite sandwich panels for building floor rehabilitation, *Construction and Building Materials* **198** 465-478
[31] ACI Committee 440 2008 *Guide for the Design and Construction of Externally Bonded FRP Systems for Strengthening Concrete Structures* ACI 440.2R-08 (Michigan: American Concrete Institute)
[32] Triantafillou T C and Antonopoulos C P 2000 Design of concrete flexural members strengthened in shear with FRP, *Journal of Composites for Construction* **4** 198-205
[33] Chen J F and Teng J G 2003 Shear capacity of fiber-reinforced polymer-strengthened reinforced concrete beams: fiber reinforced polymer rupture, *Journal of Structural Eng.* **129** 615-625
[34] Triantafillou T C 1998 Shear strengthening of reinforced concrete beams using epoxy bonded FRP composites, *ACI Structural Journal* **95(2)** 107-115
[35] CAN/CSA 2006 *Canadian Highway Bridge Design Code* S6-06 (Mississauga: Canadian Standards Association)
[36] Godat A, Aldaweela S, Aljaberi H, Al Tamimi N and Alghafri E 2021 Bond strength of FRP bars in recycled-aggregate concrete, *Construction and Building Materials* **267** 1-15

Acknowledgement
The authors would like to thank the Faculty of Civil Engineering and Built Environment, Universiti Tun Hussein Onn Malaysia (UTHM) for its support.

Strength of Modified Foam Concrete-Filled Hollow Section Using Fly Ash as Sand Replacement

Nurul 'Adilah Anati Murad[1], **Norashidah Abd Rahman**[1,2], **Ahmad Hazim Ahmad Firdaus**[1] **and Siti Amirah Azra Khairuddin**[1]

[1] Department of Civil Engineering, Faculty of Civil Engineering and Built Environment, Universiti Tun Hussein Onn Malaysia, Batu Pahat, 86400, Johor, Malaysia
[2] Composite Structural Engineering Technology Focus Group, Faculty of Civil Engineering and Built Environment, Universiti Tun Hussein Onn Malaysia, 8640, Johor, Malaysia
Corresponding author: nrashida@uthm.edu.my

Abstract: Hollow sections for columns, beams and trusses have been used in the steel construction industry for a decade. Concrete-filled hollow section (CFHS) has been widely used due to its aesthetic efficiency and to improve the load-carrying capacity. However, the use of normal concrete as infilled in steel hollow section has increased the dead load of structures. A modified foam concrete filled hollow section using fly ash as sand replacement (FCFHS-FA) is proposed to reduce the structure's dead weight. This study aims to determine the strength performance of FCFHS-FA structure by using two types of steel hollow section thickness and compare the strength between FCFHS and FCFHS-FA. Steel, preformed foam, and fly ash were used to increase the strength. Nine specimens were prepared and a compression test was conducted. The strength index was calculated to compare the strength of FCFHS with FCFHS-FA. Result shows that FCFHS-FA has a similar strength index compared with FCFHS.

1. Introduction

Steel Hollow sections (SHS) have been widely used in steel construction for a decade because the steel has strong durability and strength. The elongation of steel prevents structures to sudden failure. The rectangular hollow lattice structure shows the best mechanical compression resistance, and the highest bending properties are obtained by the truss-shaped prism structure [1]. Hollow steel sections have high strength, ductility and lightweight properties. However, they are prone to buckling. The ductility of a column decreases with the increase in depth-to-width ratio, concrete compressive strength and axial load level. Steel hollow sections are available in cold and hot rolled steel sections. Steel hollow sections have higher thermal conductivity than concrete structures [2]. Therefore,steel section filled with concrete is one of the solutions to reduce its buckling.

Concrete-filled hollow section (CFHS) is extensively used in the steel construction industry. CFHS is a steel hollow section with concrete infill. Concrete filler in steel hollow sections can provide flexural support, and steel hollow sections can influence concrete confinement to improve steel hollow performance. The performance of CFHS with various filling materials varies depending on the materials and cement used in the concrete infilled [2]. This approach decreases the deformation and buckling problem of the steel construction and increases the strength of CFHS. CFHS components are commonly used in various structural applications, including high-rise buildings, multitype bridges, towers and bracing systems [3]. The ability of CFHS to delay buckling of hollow sections [2, 4] increases the ductility and energy dissipation of the member. Steel hollow sections with concrete infill have a high local buckling and bending capacity and can be used to delay local buckling by placing the steel at the section's outer perimeter, where the moment inertia and radius of gyration are the greatest. However, studies show that the use of normal concrete raises a structure's dead weight [5], resulting in the expense of construction and lightweight concrete (LWC) [6].

According to Nurizaty et al. [2], CFHS outperforms reinforced concrete and steel section elements in terms of seismic performance. The flexural loading increases cyclically under constant axial load. Concentrially loaded circular CFHS columns may have better postyield behavior and stiffness than rectangular CFHS due to their larger diameter-to-thickness ratio. High-strength steel has a tendency to have a higher yield-to-ultimate stress ratio than mild steel but a lower strain ductility. The lighter and more slender the CFHS columns, the better they can replace traditional steel or reinforced columns in terms of resistance [4].

The application of CFHS has been improvised from time to time from NCFHS to lightweight LCFHSs. Thus, lightweight aggregate concrete may replace standard aggregate concrete in composite construction due to the former's low specific gravity and thermal conductivity levels, which can create lighter structures. LWC is approximately 20% to 30% lighter compared with normal weight concrete due to the composite action of LCFHS. The steel hollow section confines the LWC core and increases the strength, ductility and stiffness. The infill concrete protects the steel hollow sections and postpones or prevents local folding [3].

LCFHS is commonly used in buildings due to its density between 400 and 1,600 kg/m^3 that contributes to a decreased dead weight of a structure [5]. Khairuddin [6] used foamed concrete with a density of 40 kg/m^3 to 1,600 kg/m^3 and a strength of 10 MPa to 35 MPa. This lightweight foam concrete is made to strengthen the concrete by combining conventional cement, fine aggregate, foaming agent and compressed air [7-8]. The purpose of the foam is to provide a mechanism for producing a high ratio of air cells that produce a porous solid when mixed with cement [9]. However, the use of foamed concrete alone does not result in significant strength improvements [10].

1.1. Fly ash (FA)

Satpathy et al. [11] stated that the strength of structural LWC with FA cenospheres (FACs) and sintered FA aggregate (SFA) is less than the normal concrete but has a high density compared with the normal concrete. The concrete has an optimum strength with desired density, containing50% of FAC and 75% of SFA.

FA is an agricultural by-product used for renewable construction materials. It has an application density of 1,800 kg/m^3 without synthetic polyolefin fibres (macro or micro) for a power of 40 MPa [12]. According to Ahmaruzzaman [13], FA is typically grey in colour, abrasive in nature, often alkaline and refractory. Pozzolans, silica or silica and aluminium materials that form cement products at ambient temperatures along with water and calcium hydroxide are admixtures. In nature, FA usually is grey, abrasive, sometimes alkaline and refractory).

FA is the primary raw material needed for the production of SFA aggregates. One of the main issues associated with FA is the divergence in its fineness. With the increase in fineness of the FA, the average size of the manufactured aggregate increases, and the coarser FA particles require more moisture content to produce a fraction of the aggregate size. FA has various advantages, such as reduced the dead load, higher strength-to-weight ratio, low-cost construction, improved durability, higher tensile strain power, minim thermal expansion, greater insulation properties and improved fire resistance [14].

According to Golewski [15], Concrete with the addition of FA of up to 23% is distinguished above this limit in terms of normal fracture strength and low fracture toughness. The optimum FA content for foamed concrete is 20% to 25% to improve the higher short-term compressive strength. The compressive strength of structural lightweight foam concrete is 11.25 MPa, and the corresponding dry density value at 28 days is 1170 kg/m^3 [8]. According to Reddy and Dongkarar [16], the optimum FA used as a sand replacement is 70% and achieves the maximum value of compressive strength of 9.6 MPa. Therefore, a study on the foam concrete filled hollow section using FA as sand (FCFHS-FA) should be conducted.

1.2. Strength Index

Strength index (*SI*) is the ratio of ultimate load of the concrete-filled section column to the strengths of the composite column. Jamaluddin [17] suggest, SI values higher than 1.0 suggest have a positive interaction between the steel tube and concrete core and good concrete confinement . *SI* is described as Equation (1).

$$SI = N_e/N_u, \tag{1}$$

where N_e is the ultimate load of the experiment and N_u is the compression resistance of composite column as defined in EN 1994-1-1[18] as

$$N_u = {A_s f_{sk}}/{\gamma_{ms}} + {A_c f_{ck}}/{\gamma_{mc}}, \tag{2}$$

where Nu is the ultimate short-term axial load for short column; f_{sk} and f_{ck} are the characteristic strengths of steel and concrete, respectively; γ_{ms} and γ_{mc} are the material partial safety factors for steel and concrete, respectively. BS EN 1994-1-1:2004 [18] recommends that the values of characteristic strength of steel and concrete as $f_{sk} = f_y$ and $f_{ck} = 0.83 f_{cu}$, respectively. The values of f_y and f_{cu} refer to the yield strength of steel and cube compressive strength of concrete, respectively

2. Material and experimental work

2.1. Material preparation

The preparation of FCFHS-FA requires the following raw materials: FA, preformed foam, ordinary Portland cement (OPC), sand, water, foaming agent and additive of superplasticiser. The size of mould used for the modified foam concrete is 100 mm (length) × 100 mm (width) × 100 mm (height). The mixed design proportion can be referred in table 1. In this study, FA was used as a sand replacement for the modified foam concrete. The FA used is 60% [19] and 40% of sand.

The sand used for this concrete mix design is fine sand with a diameter less than or equal to 2.36 mm. The fine sand used for modified foam concrete (FC-FA) is dried under the sun to decrease its moisture content. The superplasticiser used is an additive material to increase the strength of concrete, and approximately 0.01 kg is used in the concrete mixture design. The volume method is used to find the weight of materials in kilogram (kg). The mix design for the foamed concrete is shown in table 1.

Table 1. Concrete mix design proportion.

Mixture	Foamed concrete	Modified foam concrete
Cement–sand ratio (C/S)	0.5	0.5
Foamed–cement ratio (F/C)	0.7	0.7
Water–content ratio (W/C)	0.55	0.55
FA (%)	-	60
Foamed agent	1:20	1:20

2.2. Specimen size of hollow section

The dimension of the hollow section used for this experiment is 100 mm (width) × 100 mm (height) × 350 mm (length) with a thickness of 2 and 4 mm, as shown in figure 1. The type of steel used is square hollow section steel.

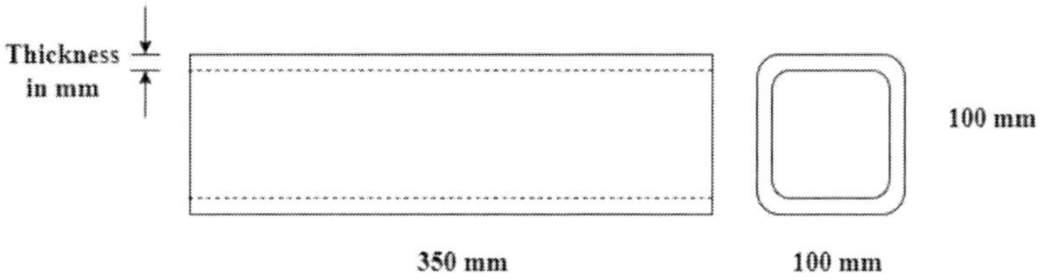

Figure 1. Dimension of hollow section.

2.3. Material properties

Cube tests were conducted to determine the strength of modified foam concrete at 28 days [20]. A tensile coupon test was performed to determine the strength of steel hollow sections in accordance with BS EN ISO 6892-1:2016 standard [21]. The strengths of modified foam concrete and FCFHS-FA are presented in table 2. The strength of the steel hollow section is presented in table 3.

Table 2: Material properties of lightweight cellular concrete and lightweight cellular concrete filled hollow section.

Material Properties	Average compressive strength (MPa)	Average load value (kN)
FC-FA	22.10	220.76

Table 3. Tensile strength of steel hollow section.

Type	Thickness (mm)	Yield strength, f_{yk} (MPa)
Steel hollow section I	2	323
Steel hollow section II	4	376

2.4. Experimental work

Six specimens of FCFHS-FA were prepared, with 3 specimens with 2 mm thicknesses and 3 specimens with 4 mm thicknesses. The steel hollow sections have the same dimension and outer diameter (figure 2).

(a) (b)

Figure 2. Steel hollow section preparation

The bottom of steel hollow sections is covered with plastics to avoid the concrete water flowing out from the bottom of steel hollow sections. The FCFHS-FA specimens were placed on the steel plate followed by concrete casting for the 6 specimens. The specimens were tested under compression load (figure 3).

Figure 3. Compression test on FCFHS-FA.

3. Result and discussion

3.1. Strength Index (SI)

Table 4 shows the result of the axial strength and *SI* of FCFHS-FA. *SI* is measured by using Equations (1) and (2). The experimental result and *SI* are compared with the FCFHS without RA from a previous study [23]. The strength of FCFHS-FA is lower compared with the result of previous study under the same thickness. However, the 4 mm specimen has higher strength compared with the 2 mm specimen. The strength of the FCFHS increases when the *b/t* is increased. According to Guler et al. [23], the specimens with thick steel hollow section walls exhibit increased strength improvement due to the strong confinement.

The strength of modified FCFHS using FA as sand replacement slightly decreases compared with the strength of modified FCFHS although the strength of the concrete is higher compared with the strength of FCFHS without FA. However, the *SI* is similar between the modified FCFHS-FA and FCFHS. The *SI* value of more than 1.00 indicates that the concrete prevents the local buckling of the steel tube [17].

Table 4. Strength of FCFHS.

Specimen	Thickness (mm)	*B/t* ratio	Compressive strength, Fc (MPa)	Theoretical value (kN)	Experimental value (kN)	*SI*
FCFHS-FA 21	2	50	22.10	313.08	366.07	1.00
FCFHS-FA 22	2	50	22.10	313.08	414.24	1.45
FCFHS-FA 23	2	50	22.10	313.08	418.85	1.50
FCFHS-FA 41	4	25	22.10	560.51	706.00	1.53
FCFHS-FA 42	4	25	22.10	560.51	697.73	1.49
FCFHS-FA 43	4	25	22.10	560.51	701.97	1.51
FCFHS 21[22]	2	50	14.50	323	495.00	1.53
FCFHS 22[22]	2	50	14.50	323	520.00	1.61
FCFHS 41[22]	4	25	14.50	608	870.00	1.44
FCFHS 42[22]	4	25	14.50	608	910.00	1.50

Note: '21', where 2 refers to 'thicknesses in mm', and 1 refers to 'Sample 1'.

3.2. Buckling

Figure 4 shows the local buckling failure appearance of the test specimens. In figure (a), FCFHS-FA 21 and (d), FCFHS-FA 41, the buckling mode appears at the top part of the steel hollow section structures. For specimens FCFHS-FA 22, FCFHS-FA 22, FCFHS-FA 42 and FCFHS-FA 43 in figures (b), (c), (e) and (f), the buckling mode appears at the bottom part of steel hollow section structures. Figures (b) and

(c) have the same specimen. For FCFHS-FA 22, buckling mode appears on two sides at the bottom part. Specimen FCFHS-FA 23 has no buckling mode. In this experiment, local buckling always occurs at the top part and bottom part. Figure 4 indicates that the buckling mode on steel hollow section structures with 2 mm thickness is more significant than that with 4 mm thickness. The buckling mode on these specimens is inconsistent. This appearance can be related to the bonding strength between the steel and the concrete. According to Abd Rahman [5], surface irregularities, such as roughness of steel section or surface of a concrete cross section along with specimen size, affect the bond strength.

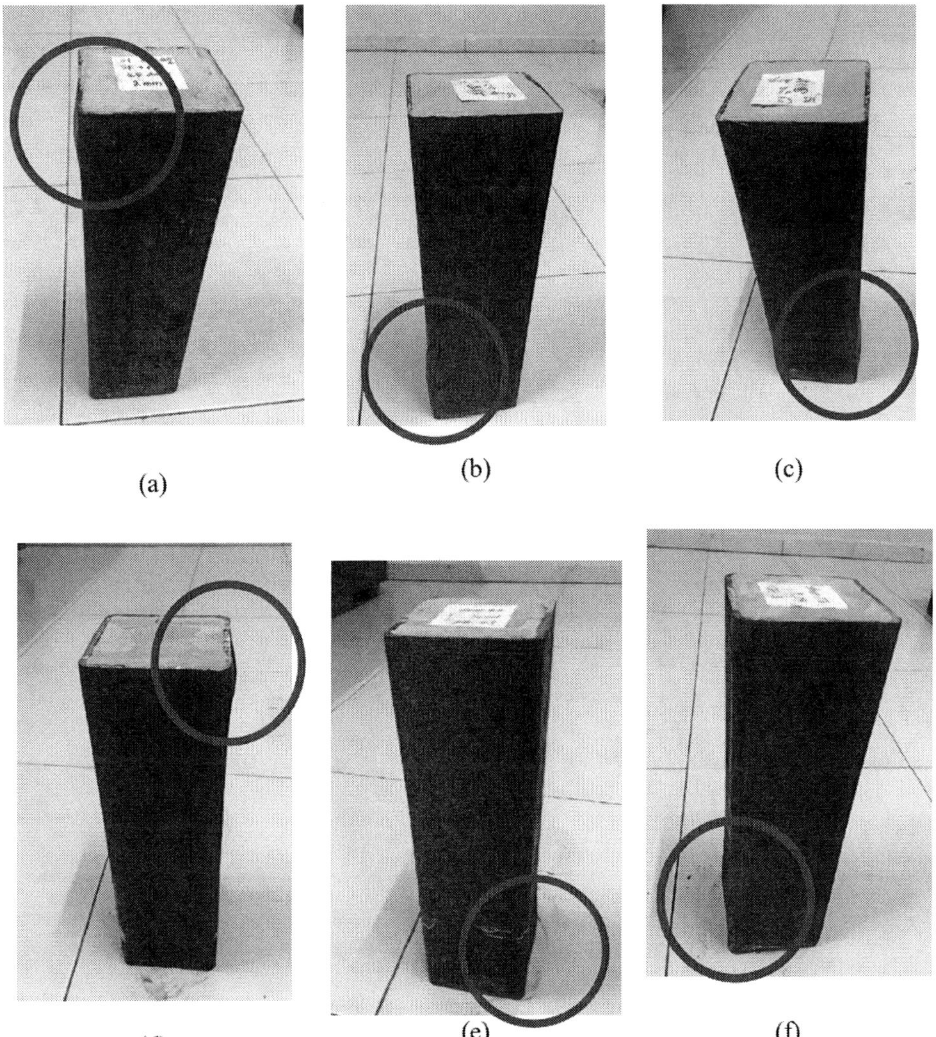

(a) (b) (c)

(d) (e) (f)

Figure 4. Typical local buckling failure of the test specimens. Figures (a), (b) and (c) are the test specimens with 2 mm thickness. Figures (d), (e) and (f) are the test specimens with 4 mm thickness.

4. Conclusion

The use of FA as sand replacement to foam concrete shows a similar strength although the strength of foam concrete using FA as sand replacement shows higher strength compared with normal foam concrete. Therefore, further study needs to be conducted by adding the fibre to improve the strength debonding of modified FCFHS using FA as sand replacement and by increasing the percentage of FA to 70% [16].The concrete and steel debonding will be explored to reduce the debonding issues and delay the local buckling mode.

5. References

[1] Wang J, Afshan S, Gkantou M, Theofanous M., Baniotopoulos C and Gardner L 2016 Flexural behaviour of hot-finished high strength steel square and rectangular hollow sections *Journal of Constructional Steel Research.* **27** 97 - 109

[2] Nurizaty Z, Mariyana A K, Shek P N, Rahman M S A, Najmi M M and Iqbal M K 2019 A review on concrete-filled steel hollow column subjected to fire and cyclic loadings *IOP Conf. Ser. Earth Environ. Sci.* **220** 012030

[3] Al-Shaar A A M and Göğüş M T 2018 Flexural behavior of lightweight concrete and self-compacting concrete-filled steel tube beams *J. Constr. Steel Res.* **149** 153–164

[4] Bagherinejad K, Hosseinpour E and Hosseini S H 2015 Evaluation of Rectangular Concrete-Filled Steel-Hollow Section Beam-Columns *J. Asian Sci. Res.* **5** 46–59

[5] Abd Rahman N, Khairuddin S A A, Faudzi M F, Mohd Asri M H I, Jamaluddin N and Mohd Jaini Z 2019 Axial strength of modified fibrous foamed concrete-filled hollow section *World J. Eng.***17** 175–181

[6] Khairuddin S A A, N Abd Rahman, Rossam M L and Mustafa A N 2020 Preliminary Study on Modified Foamed Concrete-Filled Hollow Section under Cyclic Load *IOP Conf. Ser. Mater. Sci. Eng.* **713** 012034

[7] Abd Rahman N, Khairuddin S A A, Jamaluddin N and Mohd Jaini Z 2018 Strength of reinforced fibrous foamed concrete-filled hollow section *Mater. Sci. Forum* **936** 219–223

[8] Gopalakrishnan R, Sounthararajan V M, Mohan A and Tholkapiyan M 2020 The strength and durability of fly ash and quarry dust light weight foam concrete *Mater. Today Proc.* **22** 1117–1124

[9] Chica L and Alzate A 2019 Cellular concrete review: New trends for application in construction *Constr. Build. Mater.* **200** 637–647

[10] Khairuddin S A A, Rahman N, Jamaluddin N, Mohd Jaini N, Elamin A and Rum R 2020 Bond strength of concrete-filled hollow section with modified fibrous foamed concrete *Arch. Civ. Eng.* **66** 97–108

[11] Satpathy H P, Patel K and Nayak A N 2019 Development of sustainable lightweight concrete using fly ash cenosphere and sintered fly ash aggregate *Constr. Build. Mater.* **202** 636–655

[12] Sahoo S, Selvaraju A K and Suriya Prakash S 2020 Mechanical characterization of structural lightweight aggregate concrete made with sintered fly ash aggregates and synthetic fibres *Cem. Concr. Compos.* **113** 103712

[13] Ahmaruzzaman M 2010 A review on the utilization of fly ash *Progress Energy Combust. Sci.* **36** 327–363

[14] Patel S K, Majhi R K, Satpathy H P and Nayak A N 2019 Durability and microstructural properties of lightweight concrete manufactured with fly ash cenosphere and sintered fly ash aggregate *Constr. Build. Mater.* **226** 579–590

[15] Golewski G L 2019 Estimation of the optimum content of fly ash in concrete composite based on the analysis of fracture toughness tests using various measuring systems *Constr. Build. Mater.*, **213** 142–155

[16] Reddy A B, Punith A and Dongarkar K 2017 Effect of Flyash on Strength Properties of Foam Concrete *Int. J. of Latest Eng. Res. Appl.* **2** 7-12

[17] Jamaluddin N, Lam D, Dai X H and Ye J 2013 An experimental study on elliptical concrete filled columns under axial compression, *Journal of Constructional Steel Research* **87** 6-16

[18] BS EN 1994-1-1 2004 : Design of composite steel and concrete structures - Part 1-1: General rules and rules for buildings (London:British Standard Institution)

[19] Bindiganavile V and Hoseini M 2008 *Developments in the Formulation and Reinforcement of Concrete : 11-Foamed concrete* (Sawston: Woodhead Publishing Limited) pp 231-255

[20] BS-EN-12390-3-2009 2009 Compressive strenght of test specimens (London:British Standard Institution)

[21] BS EN ISO 6892-1:2016 2016 Metallic materials. Tensile testing. Method of test at room temperature (London:British Standard Institution)
[22] Khairuddin S A A, Abd Rahman N, Jamaluddin N, Mohd Jaini Z and Ali N 2017 Ultimate strength capacity of a square hollow section filled with fibrous foamed concrete *IOP Conference Series: Materials Science and Engineering* **271** 012103
[23] Guler S, Lale E and Aydogan M 2013 Behaviour of steel fiber reinforce concrete filled steel tube *Advanced Steel Construction* **9** 14-25

Acknowledgments
Author would like to thank to Faculty of Civil Engineering and Built Environment, Universiti Tun Hussein Onn Malaysia and MALAKOFF Corporation SDN BHD, Tanjung Bin Energy Power Plant, Johor.

IOP Conf. Series: Materials Science and Engineering 1200 (2021) 012017 doi:10.1088/1757-899X/1200/1/012017

Flexural Capacity of RC Beams with Opening Strengthened using CFRP Sheet

M M Jusoh[1], N Ali[1], N A Abdul Hamid[1], N Salleh[1], S R Abdullah[1] and S Shahidan[1]

[1] Faculty of Civil Engineering & Built Environment, Universiti Tun Hussein Onn Malaysia, 86400 Parit Raja, Batu Pahat, Johor, MALAYSIA
Corresponding author: wirda@uthm.edu.my

Abstract. The use of Carbon Fibre Reinforced Polymer (CFRP) in strengthening has found to be an effective material which comprises of characteristic that comply to the requirement of structural component. CFRP was selected as strengthening material because of the capability to resist the corrosion and could regain the loss capacity due to presence of opening. The opening in structural member was essential in order to provide the route for the utility pipe, air conditioning, water supply and electrical conduit. However, the presence of opening has contributed to the reduction of stiffness, increase of deflection and extension of cracking of the beams. Therefore, this research was conducted to overcome the problem where the flexural capacity and the load deflection behavior of RC beam with opening strengthened by using CFRP sheet was analyzed. A total of five beam have been casted and tested. The specimens consist of beams with different type of opening which are rectangular and circular. The size of all specimen was 200 mm width, 250 mm height and 2000 mm for total length. The size of circular opening was 150 mm in diameter while rectangular opening was 150 x 200 mm. Bi-directional CFRP sheet were applied at the opening area as strengthening material and all beams were tested until failure. All of specimen were produced with the designed using 30 mm concrete cover, 6 mm link size and 10mm main bar size. The testing of specimens comprises of cube compressive test and four-point load for beam testing in order the determine the flexural strength of RC beam. The result from this research indicated that strengthened beam with circular opening which is SBOC-BI exhibit the highest ultimate load of 71.5 kN with flexural failure as the mode of failure.

1. Introduction

Consideration for mechanical and electrical system in a building was important for providing an efficiency of building services. Hence, providing the opening of beam was an essential factor to accommodate the services which comprise of the various system. Nowadays, the arrangement of installation for utility pipe, air conditioning ducts and water supply system seems not economical since it was hang at the ceiling or bolted at the wall surface [1]. This installation required the additional cover to achieve the aesthetic design which has contribute to increases in construction cost.

On the other hand, Chin et al. [2] mentioned that the presence of opening has contribute to the change in beam behaviour which may resulted in reduction of stiffness, increase of deflection and extension of cracking. These element was affected by the high concentration of load at the opening area. In additional, reduction of cross sectional area due to presence of opening have influence the actual beam capacity. Thus, strengthening at the opening area should be consider in order to re-gain the losses capacity of beam. The beam with opening could also lead to high in stress which can contribute to the presence of transverse crack [3]. In additional, the behavior of beam influenced the classification of opening whether large or small depends on the occurrence of flexural crack.

The diagonal reinforcement ratio was able to enhance the flexural strength at the opening area which can eliminate the premature shear failure [4]. Insufficient of reinforcement and large spacing of stirrup could lead to the decreasing of beam's serviceability. Other than that, the previous study stated that the amount of concrete thickness at the bottom and top chord of opening area should be considered in order to develop the flexural and shear strength of the beam [5]. Other than that, the depth of opening should not exceed 50% of overall depth for reducing the weakened area that effected from the presence of opening.

Ali et al. [6] stated that strengthening reinforced concrete structure by using conventional method which comprises of externally bonded steel plates or external post tensioning have been conducted over the past decades. However, this method has found to be not efficient because of existence in corrosion due to steel plate installation and contributed to several problem during maintenance process. Hence, application of FRP was found to be effective and able to contribute to reduce the maintenance process. FRP material was widely used because of its characteristics which are ease in installation, flexible and resistance to corrosion [7]. Furthermore, epoxy resin also has the potential for enhancing flexural and compression strength [8]. The epoxy consists of flexible element which can prevent the elongation of CFRP sheet before failure [9].

2. Materials and methods

Five specimens of beam were casted with the total length of 2000 mm which consist of 30 mm concrete cover, 6mm link size and 10 mm of main bar size. The opening size of rectangular and circular opening was 150 x 200 mm and 150 mm in diameter respectively. The area of the opening was around 60 % from the total area of the beam. Thus, figure 1 below illustrates the cross section of all specimens with difference type of opening.

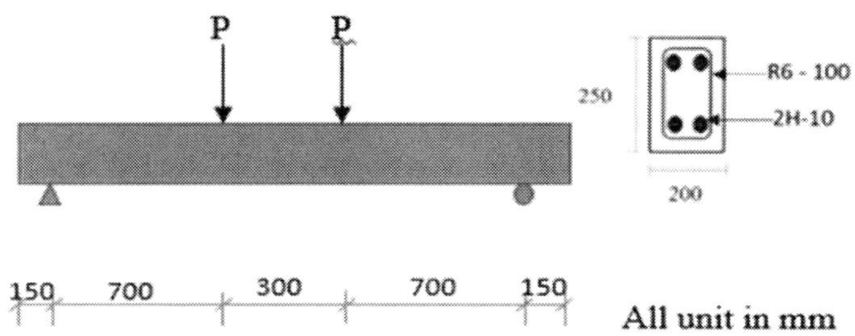

(a) Control beam (SB)

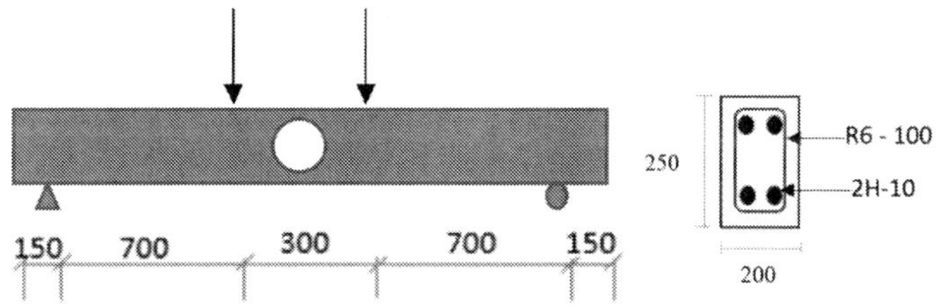

(b) Control beam with circular opening (CBOC)

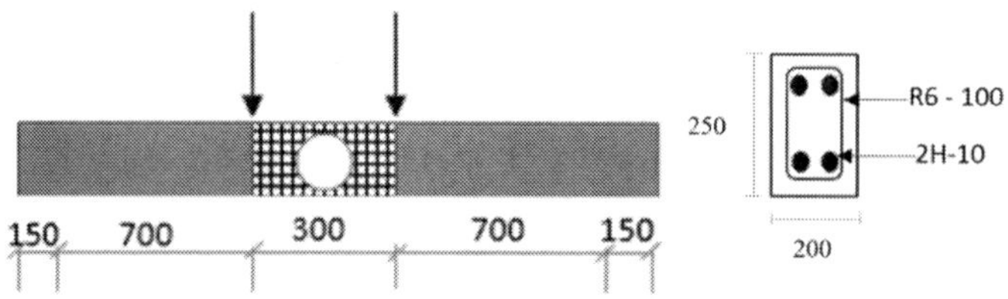

(c) Strengthen with biaxial CFRP sheet (SBOC-BI)

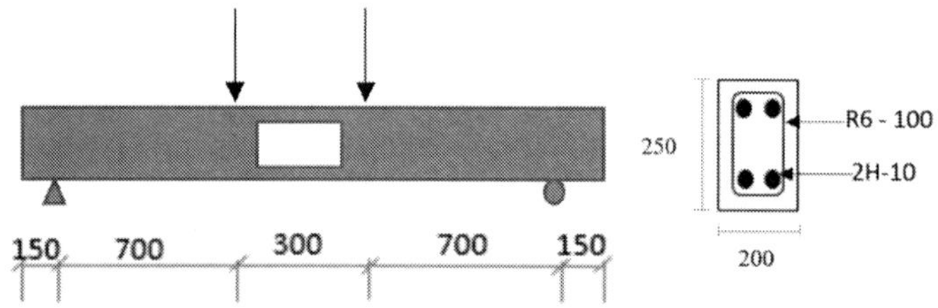

(d) Control beam with rectangular opening (CBOR)

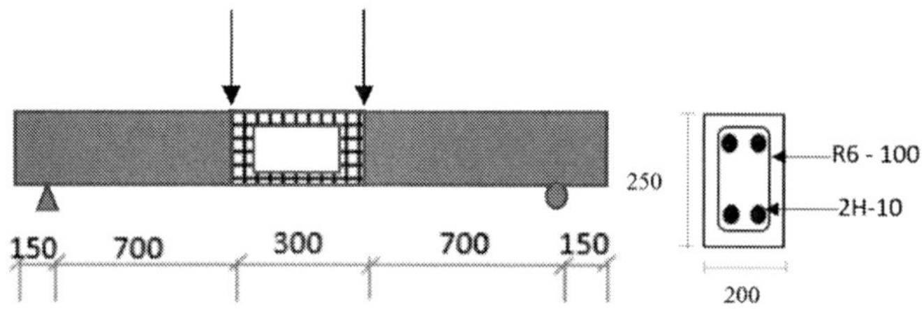

(e) Strengthen with biaxial CFRP sheet (SBOR-BI)

Figure 1. Specimen details.

2.1. Concrete

All of the specimens were fabricated by using ready mix concrete that obtained from local supplier. The concrete mixture with grade 30 has been used for conducting this research. The vibrator has been used during the compaction process of concrete into formwork which to avoid the presence of honeycomb.

2.2. Carbon Fibre Reinforced Polymer (CFRP)

Sika Wrap® - 160 BI C/15 bi-directional CFRP as shown in figure 2 below were applied for the two strengthened specimens with opening except for the control beam. The CFRP sheets were cut into 300 mm wide × 700 mm length. CFRP was applied at the side and bottom surface of opening area. Firstly, the surface of beam was prepared by using grinder and were cleaned by using brush which to remove the dust. Then, a thin layer of epoxy was applied at the concrete surface followed by installation of CFRP. Lastly, the second layer of epoxy mixture was applied by using roller in order to ensure the proper bonded between CFRP and surface of beam. Figure 3 shows the application of CFRP at the opening area while table 1 shows the properties of CFRP.

Figure 2. Bi-directional CFRP sheet.

Figure 3. Application CFRP sheet.

Table 1. Properties of CFRP sheet (Sika Wrap® - 160 BI C/15).

Density	1.75 g/cm³
Tensile strength	3800 N/mm² (nominal)
Tensile E-modulus	230000 N/mm² (nominal)
Elongation at break	1.5% (nominal)

2.3. Epoxy

Epoxy were applied at the surface of opening area which is the location of CFRP application. The type of epoxy used was Sikadur 330 and table 2 below shows the properties of Sikadur 330 while figure 4 shows the epoxy used in this research.

Table 2. Properties of epoxy (Sikadur-330).

Elongation at breaks	0.9 %
Flexural E-modulus	3800 N/mm²
Density	1.3 kg/L ± 0.1 kg/L
Tensile strength	30 N/mm²

Figure 4. Epoxy (Sikadur-330) which consist of Part A and Part B.

2.4. *Beam testing*

Figure 5 shows the four-point load test which shows the location of point load, supports and LVDT (Linear Variable Differential Transducer). The specimen was placed in the Magnus Frame before the testing process. Forklift and few manpower were needed during the testing process because the length of beam which approaching two meters. Two strain gauges have been installed on the concrete surface at the top of beam while the other two strain gauges were installed at the main reinforcement of the beam prior to concrete casting.

Figure 5. Four-point load test set-up.

3. Results and discussions

3.1. *Ultimate load and deflection*

Table 3 shows the data of ultimate load and deflection for all five specimens. From the results, it shows that SBOC-BI (strengthened beam with circular opening) exhibit the highest ultimate load with 71.5 kN compared to the other specimen. The application of CFRP as strengthening material for SBOC-BI specimen has contributed to 13.49 % of load incremental from SB. Other than that, SBOC-BI also shows the lowest deflection with maximum deflection of 15.005 mm and has proved the effectiveness of CFRP in reducing the deflection. On the other hand, the result of CBOR (beam with rectangular opening) indicated the lowest ultimate load with 58 kN and the highest deflection of 17.235 mm. This revealed that the sharp corner at the opening area was not able to accommodate the concentration of load applied.

Apart from that, SBOR-BI (strengthened beam with rectangular opening) shows the effectiveness of 19.83% CFRP contribution as strengthening material in reducing the deflection behavior of the beam and contribute to the inclination of load compared to CBOR. The load increment for all specimens were evaluated regarding to SB as the benchmark while the CFRP contribution was the comparison between strengthen and un-strengthened beam for each opening.

Table 3. Experimental results.

Specimen	Ultimate load (kN)	Load increment (%)	CFRP contribution (%)	Maximum Deflection (mm)
SB	63	-	-	16.235
CBOC	63.5	1	-	16.265
CBOR	58	8	-	17.235
SBOC-BI	71.5	14	12.60	15.005
SBOR-BI	69.5	10	19.83	13.250

3.2. Load-deflection behavior

Figure 6 illustrated the relationship between load and deflection for all specimens. From the graph, it can be seen that all specimen shows the same pattern behavior of load and deflection. It shows that SBOC-BI (strengthened beam with circular opening) indicated the ultimate load of 71.5 kN with 15.005 mm of maximum deflection value while CBOC (unstrengthened beam with circular opening) indicated the ultimate load of 63.5 kN with 16.265 mm of maximum deflection value. The comparison between these specimens revealed that SBOC-BI was able to reach higher ultimate load and lesser deflection value compared to the unstrengthened beam which is CBOC. This proved that the contribution of CFRP as strengthen material has improved the stiffness and load carrying capacity of the beam. The same pattern of the beam behavior also can be seen within CBOR (unstrengthened beam with rectangular opening) and SBOR-BI (strengthened beam with rectangular opening) curve in figure 6 where the application of CFRP has also influenced in improving the stiffness of the beam.

From figure 6, it shows that the curve of CBOC was close to the curve of SB (control beam). Although there was a small differences of ultimate load value between these two (CBOC and SB) specimens, CBOC has indicated the higher deflection value compared to the SB. This is due to the presence of the circular opening at the mid span which has reduced the actual capacity of the beam. CBOR shows the lowest load carrying capacity with the highest deflection value compared to the other specimens due to the stress concentration at the four edges of rectangular opening. It also shows that the presence of CFRP at the opening was able to improve the stiffness and reduced the deflection of the beam.

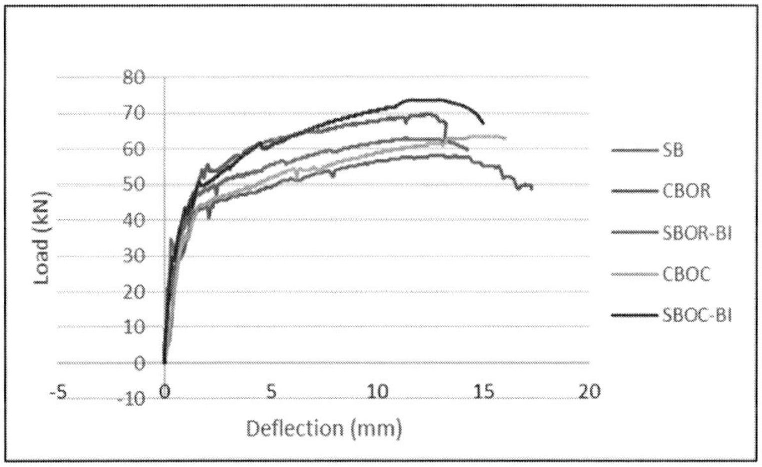

Figure 6. Load-deflection behaviour.

3.3. Crack pattern and modes of failure

Figure 7(a) shows the illustration of crack pattern while figure 7(b) shows the real cracking pattern of SB. It was observed that the first flexural crack was occurred at the mid span when the load reached 30 kN. As the load applied increased, the flexural crack occurred at the soffit of the beam and propagated to top of beam in vertical direction. From the observation, the width of flexural cracking at the mid span were widen before the failure which lead to the concrete crushing at the compression area of the beam. The beam failed at ultimate load of 63 kN and it can be seen that the modes of failure of SB was in flexural.

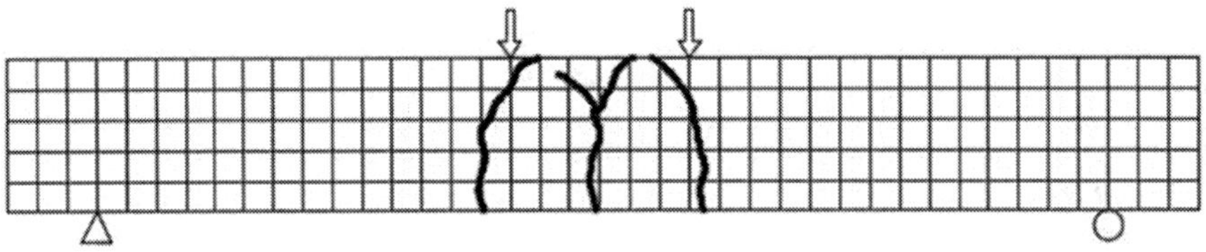

(a)

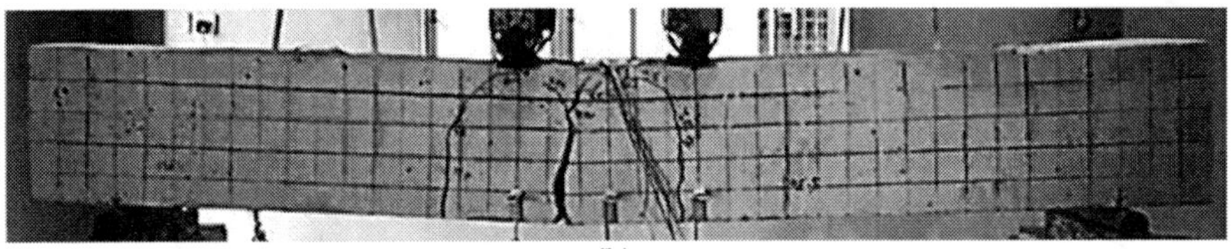

(b)

Figure 7. Illustration and real cracking pattern of beam SB.

On the other hand, figure 8 (a) shows the illustration of cracking pattern of CBOR while figure 8 (b) shows the real cracking pattern of CBOR. The first crack line that observed was appeared at the bottom chord of opening when the load exhibit 17.5 kN. When the load increased to 33.5 kN, a crack was occurred and extended to the left bottom corner of opening area. Meanwhile, the presence of the flexural crack was observed at the right bottom corner of opening area and a diagonal crack was existed afterwards when the load reach at 52 kN. The flexural crack along the bottom chord of opening became wider before the beam fail in flexural mode.

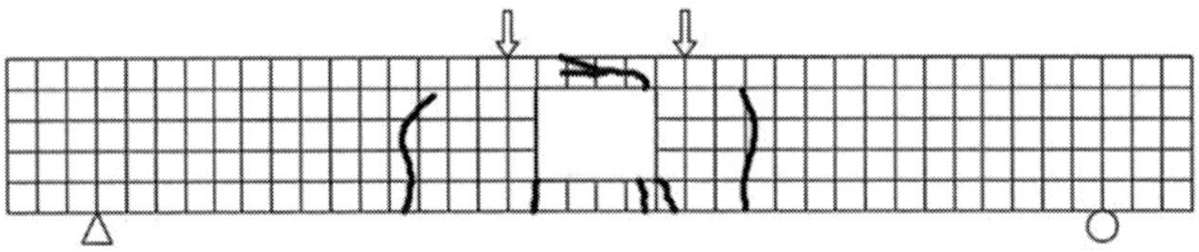

(a)

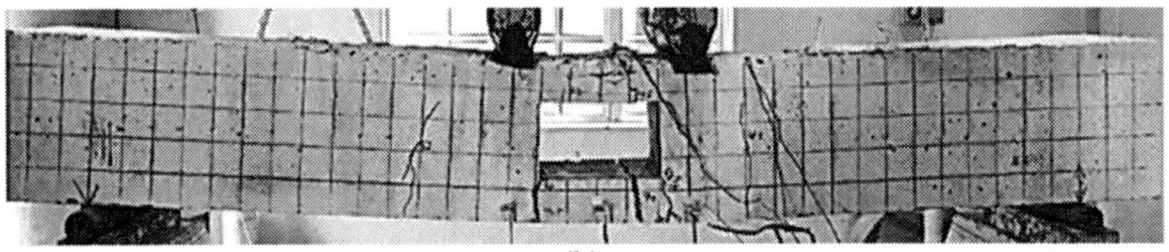

(b)

Figure 8. Illustration and real cracking pattern of beam CBOR.

Figure 9 (a) shows the crack pattern of beam SBOR-BI. From the observation, the initial crack was existed when the load approach to 41 kN . It was due to the application of CFRP at the opening which has influenced the initial value of cracking compared to the CBOR that indicated the value of 17.5 kN.

The increases of applied load has propagated the crack to the adjacent of point load location. In the same time, a flexural cracking was occurred at the left side near the opening area when the load reach to 51 kN. The beam failed when it exhibits the ultimate load of 69.5 kN. Figure 9 (b) shows that the flexural failure of the beam.

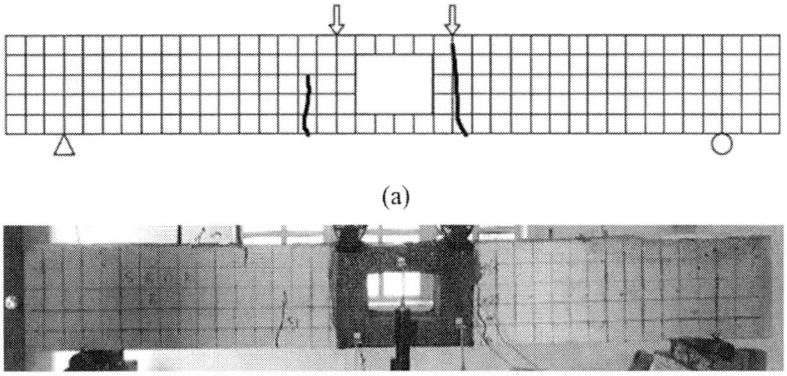

(a)

(b)

Figure 9. Illustration and real cracking pattern of beam SBOR-BI.

Figure 10 (a) shows the illustration of cracking pattern of CBOC while figure 10 (b) shows the real cracking pattern and failure mode of CBOC. Through the observation, the initial cracking occurred from the bottom of specimen which was incline to top when the load applied approached 44 kN. At the same time, another presence of flexural cracking has been observed to occur at the bottom chord of the opening which located at the mid span of specimen. When the load applied increased, the flexural cracking also existed at the top chord of the opening while another diagonal cracking occurred at the right side of opening with the same load approach. The cracking width at the mid of span were widen before the beam failure and the beam failed when the load reach 63.5 kN. From the observation, the beam fails in flexural mode.

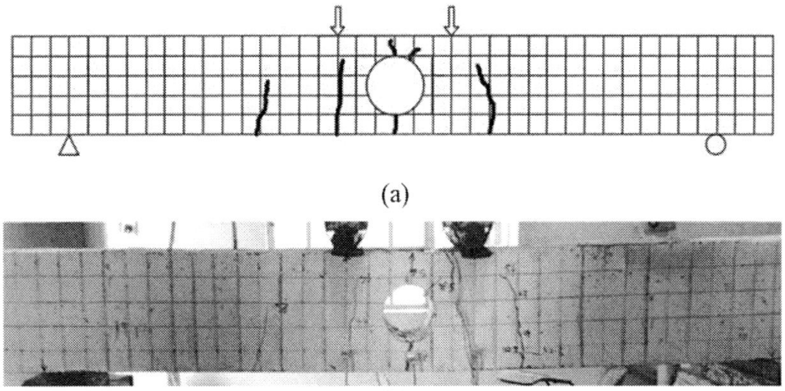

(a)

(b)

Figure 10. Illustration and real cracking pattern of beam CBOC.

Figure 11(a) shows the illustration of cracking pattern of SBOC-BI while figure 11 (b) shows the real cracking pattern and failure mode of SBOC-BI. The first flexural crack has been observed at the right side near the opening when the load reach at 35 kN. A diagonal cracking was presence when the load applied increased to 42.5. On the other hand, there were two diagonal cracking was observed which located at the right and left side of the opening when the load exhibit to 44 kN. The crack at the point load applied enlarged when the load increase to 55 kN. Lastly, the beam fails in flexural mode when the load approached 71.5 kN.

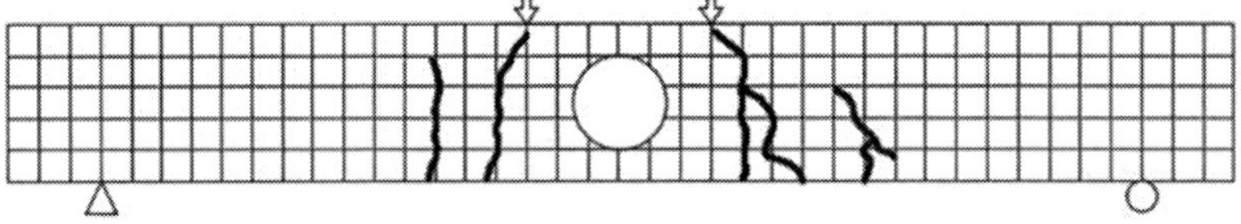

(a)

(b)

Figure 11. Illustration and real cracking pattern of beam SBOC-BI.

4. Conclusions

Through this research, all of five specimens were produced and successfully tested under four-point load testing. The result of all specimens have been analyzed in order to fulfill the objectives of this research. The application of CFRP as strengthening material were applied to each type of opening to evaluate the capacity and behavior of the strengthening specimen. The specimens without strengthening were considered as control specimen and solid beam was the benchmark for all specimens.

In conclusion, SBOC-BI has achieved the highest ultimate load among of all specimen. This is relevant to the expected result of this research which is the beam with circular opening was able to reach the higher load carrying capacity compared to the rectangular opening. SBOC-BI indicated 71.5 kN of ultimate load while SBOR-BI indicated 69.5 kN. This was due to the high stress concentration at the four edges of rectangular opening. From the comparison between control and strengthened specimen of each opening, the ultimate load for strengthened specimens increased due to the presence of the CFRP at the opening. This is followed by the deflection of the beam which has decreased due to the application of CFRP compared to the control beam. From the crack pattern analysis, the crack propagation of control specimen at the opening was extended to the top of specimen which lead to the crushing of concrete. In contrast, the crack pattern of strengthened beam was significantly decreased without presence of concrete crushing. It can be concluded that the application of CFRP as strengthening materials to enhance the flexural capacity for beams with opening as effective and gives promising results. However, more studies are needed to explore the possibilities of different opening location, opening size and opening shapes.

5. References

[1] Elsanadedy H M, Al-Salloum Y A, Almusallam T H and Alshenawy A O 2019 Experimental and numerical study on FRP-upgraded RC beams with large rectangular web openings in shear zones, *Construction and Building Materials* **194** 322–343

[2] Chin S C, Shafiq N and Nuruddin M F 2011 CFRP Composites for Strengthening RC Beams with Large Square Opening at CFRP composites for strengthening RC beams with large square opening at shear: FEM and Experimental *7th International Conference on Steel & Aluminium Structures* (Kuching)

[3] Aykac B, Kalkan I, Aykac S and Emre Y 2013 Flexural behavior of RC beams with regular square or circular web openings, *Engineering Structures* **56** 2165–2174

[4] Herrera L and Lemnitzer A 2017 Experimental performance of RC moment frame beams with rectangular openings, *Engineering Structures* **152** 149–167

[5] Ahmed A, Fayyadh M M, Naganathan S and Nasharuddin K 2012 Reinforced concrete beams with web openings : A state of the art review, *Journal of Material and Design* **40** 90–102

[6] Ali N, Samad A A A, Mohamad N and Jayaprakash J 2013 Shear Behaviour of Pre-cracked Continuous Beam Repaired using Externally Bonded CFRP Strips, *Procedia Engineering* **53** 129–144

[7] Salama A S D, Hawileh R A and Abdalla J A 2019 Performance of externally strengthened RC beams with side-bonded CFRP sheets, Composite Structures **212** 281–290

[8] Amran Y H M, Hung C, Alyousef R, Rashid R S M and Alabduljabbar H 2018 Properties and applications of FRP in strengthening RC structures: A review, *Structures* **16** 208–238

[9] Balamuralikrishnan R and Jeyasehar C A 2009 Flexural Behavior of RC Beams Strengthened with Carbon Fiber Reinforced Polymer (CFRP) Fabrics, *The Open Civil Engineering Journal* **3** 102–109

Investigation of Dynamic Factor, Rail Stress and Track Foundation Modulus of Ballasted Railway Track for Different Bed Conditions Using Numerical Methods

Shanmugasekar Thenappan[1]

[1] Department of Civil Engineering, University Technologi Malaysia (UTM), 81310 Skudai, Johor, Malaysia.
Email: shanmugasekar@gmail.com

Abstract. The track stiffness is the primary function of roadbeds thickness and subgrade characteristics. For this purpose, numerical scale track finite element technique representing the ballasted track with multi layered substructure founded on subgrade was simulated. The track deflection, stress was abstracted in static and dynamic conditions. The track significant design parameters: Foundation modulus, rail fatigue strength, rail bending stress and stress on subgrade levels were evaluated by using improved current track design numerical methods and compared against field test results which were carried out on part of MG Double track high speed main line (1600 km). Mathematical equations were developed to correlate the variables; ballast thickness, settlement, track stiffness, rail bending stress and rail fatigue strength on varying subgrade soil modulus. Incorporation of this parametric study will improve and optimise the conventional track design and maintenance standard. A simple improved track design was introduced by using single track stiffness parameter from conventional plate bearing test (PBT) on Force Displacement (FD) conventional curve method. The improved method with deriving equivalent track stiffness from rail pad and track substructure tested C value are accurate and simple. The current test method to determine the track stiffness in live track condition is expensive and unsafe with operational requirements. This PBT is simple, cost saving on labour, safe and without applying live train load.

1. Introduction

Vertical track stiffness is a function of the modulus of elasticity of different layers and components in the railway track systems [1]. Track compendium. PMC Media House. The properties included are rail pad modulus, rail flexural bending and roadbed modulus which is joint performance of ballast, sub ballast and subgrade below as shown in Figure 1.

A study by Pita et al. [2] showed that the optimum study recommends criteria to consider the global stiffness must have more than 52 kN/mm with limited settlement of 28 mm and for soft 32 kN/mm with 46 mm. The maximum bending stress evalauted in UK Rail is 110 MPa. Recent study by [3] concluded that the optimum value of track stiffness is depend on respective line specification and universally accepted value not applicable. The aim of this paper is to evaluate the response on the track foundation modulus C values, rail bending stress, fatigue strength to the varying ballast thickness on changing subgrade soil material value for the local train projects. Therefore,it is necessary to compare the experimental methods with numerical static and dynamic analysis taking into account the elasto plastic properties and non-linear behaviuor of track material and repetitive loading to the various speeds.

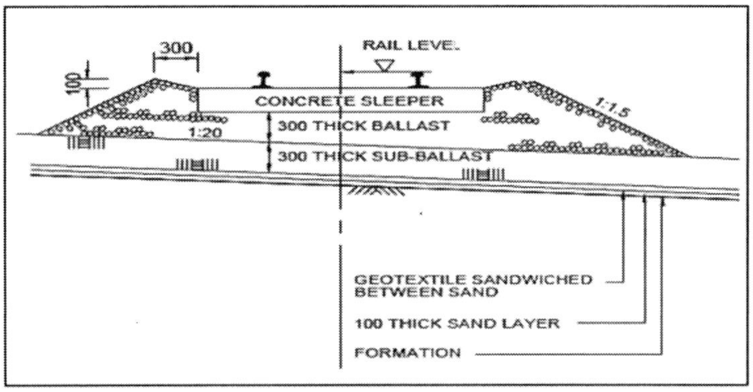

Figure 1. Typical Section of Ballsted Track in Main Line Malaysia [1].

In the present study, the author studied the response of rail fatigue and bending stress and its influences on C value (function of track stiffness and deformation) on different bed (ballast thickness) conditions on changing subgrade soil modulus. The ballast thickness (100 – 500 mm), sub ballast 300mm were chosen. The subgrade soil modulus was chosen to the range (25 – 48 MPa). The study focused on (PBT) plate bearing test method and deriving the track stiffness and response on rail stress evaluation as per project requirement.

1.1. The (C) definitions and influence

The (C) is track stiffness and called in different name in different textboo$_{ks}$ that are; Ballast modulus, Coefficient of Ballast, Track Foundation Modulus (C) [4]. It is denoted in letter (C). Here the author refers as (C). The (C) indicates at which value of (N/mm^2) surface pressure the sleeper subsides by 1 mm. The unit is N/mm^3 or Kg/cm^3.

2. Review methodology

This chapter briefs on how the review was carried out:

i. Rail fatigue and bending stress evaluation from the Field Test result (Plate Bearing Test -(C) value determination test.
ii. Evaluation of (C) Test and its response for varying ballast thickness on varying subgrade soil condition with numerical modelling.

3. Current design concept of ballasted track structure

The track is designed utilizing load bearing approach to sustain the static and dynamic loads which are transferred to the subgrade through sleeper ballast by ensuring that the strength of the components is not exceeding their capacity and to limit allowable rail bending stress, sleeper bending stress ballast pressure and subgrade pressure [5].

Current approach on Winkler model (WM) and discrete support model (DSM) are Semi empirical. The dynamic response on the characteristic of the track components is not well understood to form the basis of rational design. It relies on observed dynamic response to an equivalent static response by making use of various load factors. It is understood that ballast and subgrade properties need to be examined and included.

3.1. Track structure design approach

The steps involved are: -

i. The (C) value parameter derivation from PBT results.
ii. Mean Rail bending stress (σ_{mean})
iii. Determination of dynamic factor (DF)
iv. Calculation of Maximum Bending Stress in rail and Verification of bending stresses against fatigue strength using SMITH diagram.

3.1.1. (C) Derivation methods. The (C) is the track parameter and derivative of track stiffness. The common methods adopted in railway industry are classified into 3 Major groups, that are theoretical, theoretical-experimental, and experimental. Theoretical method is questionable on validity, experimental methods about their field test which leads to time, cost, speed restrictions and safety concern on live track. Hence not possible for all rail roads in all places. Considering the practicality, constructability, minimisation of train traffic, safety, cost, and labour, one of the simple experimental tests is (PBT) on two cycles.

3.1.2. The (C) value using Plate Bearing Test (PBT). The (C) value is derived from the settlement using the (PBT). The conventional (PBT) is carried out on the prepared roadbed surface (which is completed to lay the sleeper) and measured the total settlement by applying rail seat load in two cycles. The basic principle is the intensity of pressure (kPa) exerted by the bearing plate to the resultant deformation provides the total elasticity modulus of the roadbed and subgrade. The plate bearing test with two cycles confirming to the British (BS 1377) [6] and German (DIN 18134:2012-04) [7] standards. The PBT consists of applying static load in uniform increment on a circular plate that is placed on close contact (roadbed) and measuring the deflection for each increment in two cycles. During the first cycle at initial stage, the voids in ballast are closed similar to what happens in actual track. The behaviour on load deformation is linear as shown in Figure 2a. The second cycle represents the track simulation stage on accumulative load wherein the track ballast particles further densified and create a low –voids medium. The behaviour is non-linear due to accumulative load and properties of subgrade and roadbed material [8]. The idealised theoretical and experimental test load settlement behaviour of track simulation immediately after construction i.e., at cycle 1 and after tamping i.e., cycle 2 are shown in figure 2(a) and (b). From the FD developed graph, the slope (s) is obtained from applied load and residual settlement in second cycle. The C value is derived by dividing the area of the loading plate (A) from (s) value as shown in Equation (1).

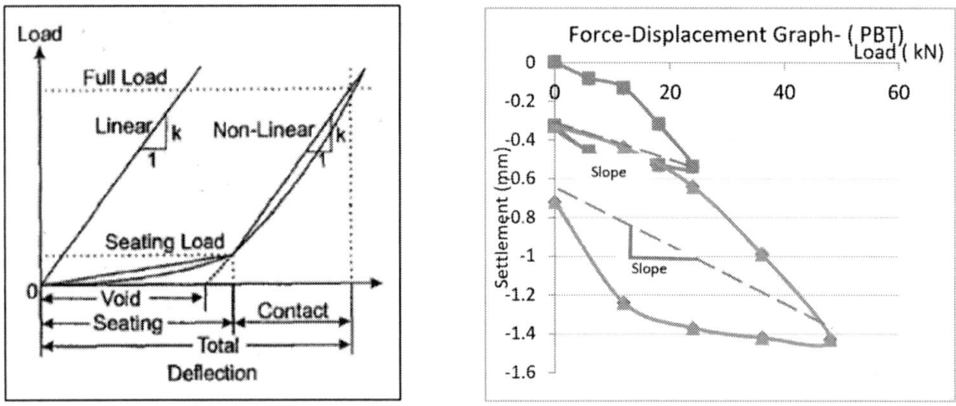

(a) (b)

Figure 2. (a) Idealised theoretical load-settlement graph [9], (b) Idealised field experimental test load-settlement graph.

The C value was derived from load deformation behaviour graph on second cycle.

$$C = \frac{s}{A} \tag{1}$$

For load evaluation main line passenger cum freight network operated by KTM trains (Locomotive and EMU) on metre gauge Class 26 Blue Tiger with maximum 20-ton axle load arranged in 1.850m front axle spacing (figure 3) was considered. The rail section UIC 54 kg/m, concrete sleeper on 600 mm spacing were considered. The maximum rail seat load transfers to the immediate sleeper ranging 0.43 to 0.65 of (p) [10] i.e., wheel load and transfers to 3 adjacent sleepers. The maximum rail seat load of 50 kN (0.5p) was considered.

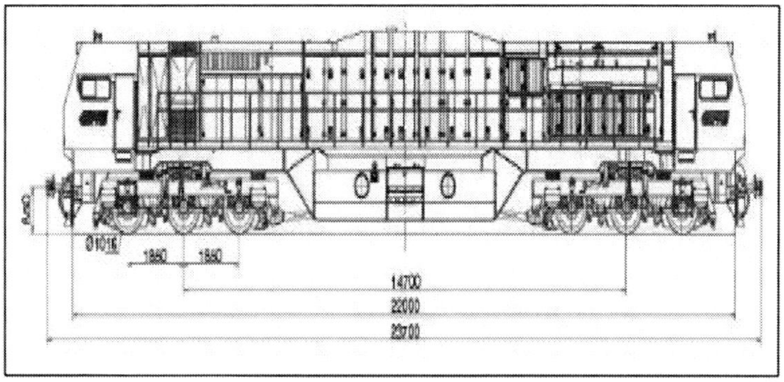

Figure 3. DE AC 33 C- HP 3300- KTMB 26 class-Blue Tiger Locomotive: axle load 20 tonnes.

3.2. Development of C value - rail bending stress working equations

In the current practise, the rail design is based on the theory developed by Winkler and later by Zimmerman [4]. Based on recent study [11], negligible differences were evaluated between two track models, that are one on continuous support Winkler Model (WM) assuming elastic foundation and another is discrete supports models (DM) which comprises rail pad sleeper ballast and subgrade stiffness. The WM slightly increase in DM and negligibly decreases in rail deflection [12].

Conventional track consists basically of two parallel continuous beams, the rails, which are fixed at regular intervals onto sleepers' support. The ballast bed rests on a formation. Winkler hypothesis applies to track support. At each point of support the compressive stress is proportional to the local compression. This relation is,

$$\alpha = Cw \tag{2}$$

Where,

α =Local compressive stress on the support (N/m^2)
w= local subsidence of the support (m)
C= Foundation Modulus (N/m^3)
In discreet support
Kd= C A_{rs}
Kd= Sum F/Sum w
K= kd/a
a= Spacing between centres of discrete support

Zimmermann's method is appropriate which considered the influence of adjacent wheel hogging effect on deflection, bending moment and characteristic length. The total Equivalent stiffness (k_{eq}) was derived with tested standard rail pad stiffness (k_p) with ballast subgrade stiffness (ks).

The rail mean (α_mean) is derived using general equation (3) and (4) with developed moving three-wheel load influence envelope from the Zimmermann Analysis [10] (figure 4).

$$Mmax = \frac{QL}{4} \tag{3}$$

$$L = \sqrt[4]{4EI/k} \tag{4}$$

Where,
w=Rail deflection (m)
Q= wheel load (N),
EI= Rail Bending stiffness, Nm^2
L= Characteristic length, (m)
k= Foundation coefficient / Track modulus

(N/m/m)
k = Support stiffness
k = kd/a,
kd = Spring constant of the support= C*A

The Dynamic Factor (DF) is applied to the static load to consider dynamic impact on stresses and deflection in the railroad bed and sub grades. as in equation (5).

$$\alpha_{max} = DF\ \alpha_{mean} \tag{5}$$

The Eisenmann's DF (ϕ) influences on three factors as in Equation (6)

$$\phi = 1 + t\ ¢\ (1 + (V-60)/140) \tag{6}$$

Where,
t= multiplication factor of standard deviation 1for 68.3 Probability occurrences, 2 for 95.4 and 3 for 99.7
¢= factor depending on track quality 0.1 for very good, 0.2 good and, 0.3 bad condition
s = Speed Factor, V= Train speed (kmph)

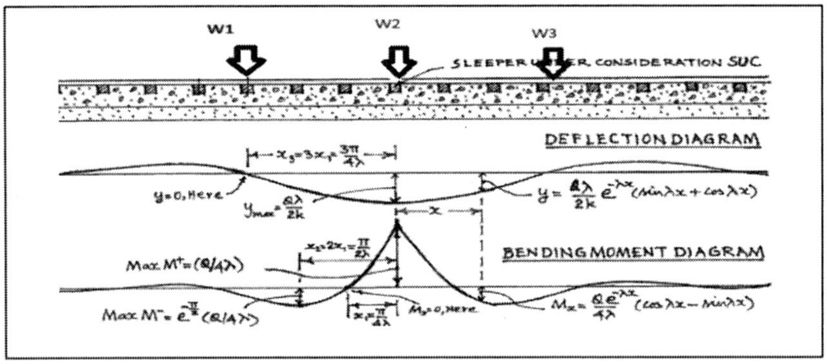

Figure 4. Moving load bending moment and deflection profile-Zimmermann analysis [10].

3.3. Verification of bending stresses against fatigue bending strength using SMITH diagram envelop.
From the Smith Diagram [4] the permitted limit of stress to the fatigue condition for UIC 54 kg UTS 900 was derived. The ultimate tensile strength is 900 N/mm^2. The upper limit should not exceed yield limit FY (580 N/mm^2). The permissible range of stress fluctuation is 280 N/mm^2 (DK) σ_q (Live Load stress) while the constant background stress of 180 N/mm^2. For any mean stress along the line OA the safe stress range is AP and AQ. Apart from this additional constant background stress in the rail are due to temperature variation and residual stress while manufacturing process. These are derived with maximum thermal stress of 50 N/mm^2 as per project requirement and residual stress on manufacturing process of 130 N/mm^2. The developed fatigue bending strength smith diagram is shown in figure 5.

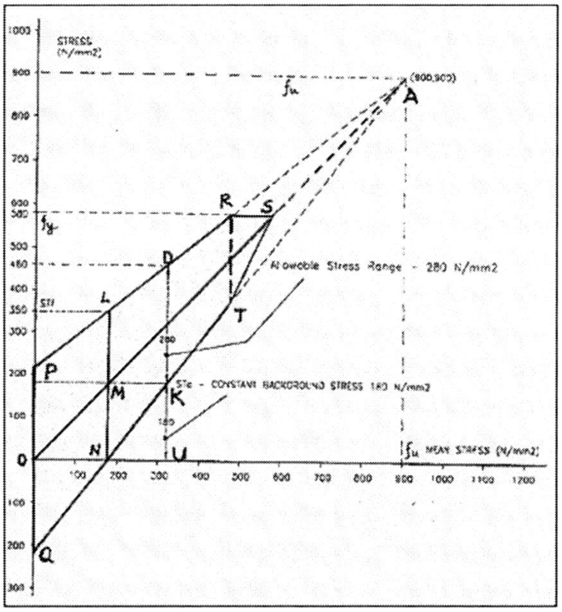

Figure 5. Developed Smith Diagram- the limit of fatigue strength range UIC 54 Rail UTS 900 [4].

4. Field test results evaluation and rail bending stress derivation

The C value test on completed double track were carried out on Two sites. Three locations along Site 1 i.e Rawang- Ipoh –Padang Besar Track project (509 kM) and four locations along Site 2 i,e Gemas-Johore Bahru Track Project (200 kM). The test carried out on 14 m width 30m length along the track area over ballast. The plate of size range 450-600 mm dia. with 25 mm thick were used. The increment of loading in 2 cycles are 24 and 48 kN at Site One and increment loading of 150 kPa and 300 kPa in Site 2. The settlement was measured.

Table 1. The summary of C value, rail bending stress derived from plate bearing test.

Subgrade Modulus N/mm^2	Vertical Displacement (mm)	Ballast Thickness (mm)	Derived C Value kg/cm^3	Derived Mean rail Stress N/mm^2
48	-0.861	100	36.526	52.734
	-0.843	200	37.308	52.726
	-0.838	250	37.507	53.137
	-1.057	300	29.765	52.722
	-0.836	350	37.600	52.726
	-0.838	400	37.509	52.732
	-0.842	450	37.368	52.739
	-0.846	500	37.181	52.726
35	-0.996	100	31.571	55.060
	-0.960	200	32.763	51.620
	-0.949	250	32.763	52.000
	-1.147	300	33.149	53.170
	-0.937	350	27.408	51.000
	-0.936	400	33.149	52.000
	-0.936	450	33.585	53.000
	-0.939	500	33.606	54.000
25	-1.185	100	26.539	56.139
	-1.117	200	28.162	53.631

	-1.100	250	28.162	53.224
	-1.294	300	28.581	56.139
	-1.077	350	24.309	56.140
	-1.072	400	28.581	56.139
	-1.070	450	29.378	52.782
	-1.070	500	29.342	53.138
Range (25-48) MPa	0.83-1.29	100-500	24.3-37.6	56-59

Figure 6. Test set up on prepared track ballast - plate bearing test (field photograph).

4.1. Evaluation of plate bearing test

The field plate bearing tests (PBT) load -settlement graphs were developed. The K_{eq} was derived from tested Rail Pad stiffness of 100 kN/mm in project requirement. The rail means bending stress $\sigma_{mean} = M/Z$ was evaluated. The summary of derived C value and Mean Rail stress σ_{mean} are shown in table 2.

1-1: Site 1 in location 1

2-1: Site 2 in location 2

5. Numerical analysis

The numerical model of plate bearing test similar to experimental test on track within tank of size (1.5m x 1.5m x 1.2 m) was simulated. The track substructure (ballast sub-ballast and subgrade) was simulated like field test using ABAQUS programme. The static and dynamic repeated load to the frequency range (1.5-31 Hz) were applied and evaluated. The load frequency of a train is derived as f=V/L, where V is the speed (10 to 200 kmph) and L is the distance between axles. The front axle distance of 1.80m for 20t axle load locomotive was assumed. The response on overall deformation and stress distribution were evaluated to determine the C value on different bed (ballast thickness) conditions on changing subgrade soil modulus [13].

Table 2. Material properties used in numerical analysis.

Item	Density (t/mm^3)	Elastic Modulus (N/mm^2)	Poison	Friction	ø	Cohesion (Cu) N/mm^2	Thickness (mm)
Steel Plate	7E-9	20000	0.35	-	-	-	450 dia 25
Ballast	2E-9	127	0.35	45	15	0.001	Varies (100-500)
Sub ballast	1.9E-9	127	0.35	45	15	0.001	300
Sub-grade	1.8E-9	25-48 Varies	0.35	-	-	-	600

5.1. Analysis - determination of C value

The numerical model consists of three layers (substructure and subgrade) the top layer is 100-500 mm thick ballast (varying), middle layer is 300 mm thick sub ballast and bottom layer is 600 mm thick subgrade. The subgrade base is fixed support as founded on rigid base. The model is shown Figure 7a. The parameters of materials assumed are shown in table 3. The aim of static analysis is to evaluate the total vertical settlement, static stress on track bed layers for different roadbeds thickness on varying subgrade conditions. From the deformation value the (c) value and rail stress were derived.

Two cycles loadings applied incrementally similar to the field plate bearing test. In first cycle 25 kN and 50 kN in second cycle. In second stage the dynamic analysis were carried out. The aim of dynamic analysis is to evaluate the dynamic factor (DF) i.e., the ratio of dynamic stress and static stress on subgrade.

The vertical settlement and von mises stress distribution are shown in figure 7 (a), (b), (c) and (d).

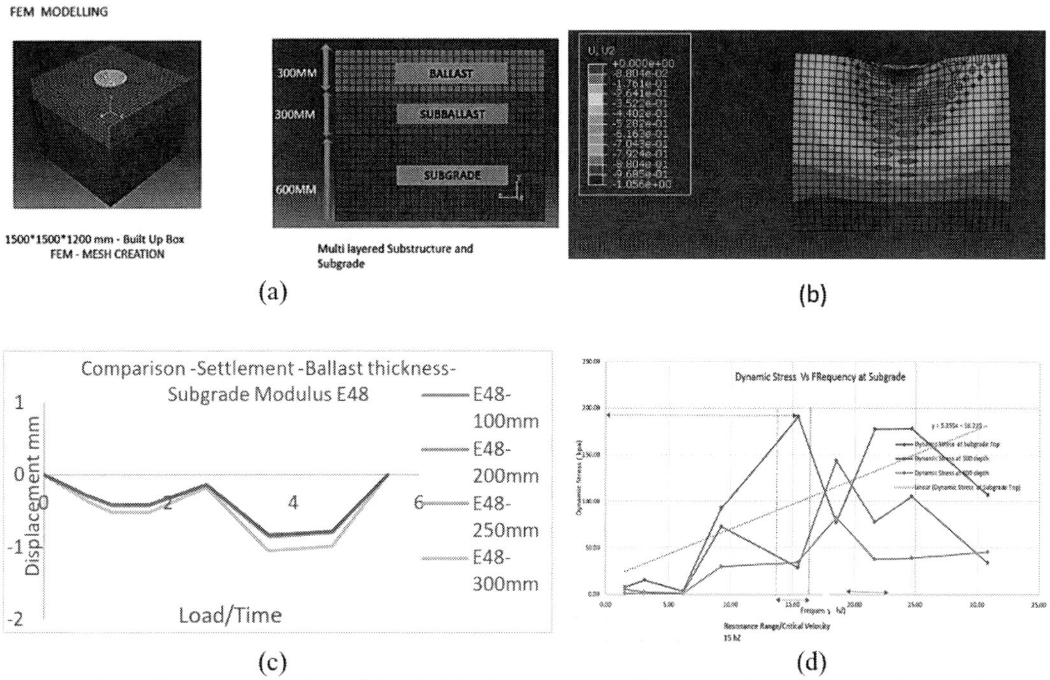

Figure 7. (a) FEM simulated model, (b) Displacement at ballast level in E45 subgrade maximum 1.05mm, (c) Load displacement on two cycles for E 47 sub grade modulus, (d) Dynamic stress on subgrade.

Table 3. Mathematical equation and dynamic factor.

Soil Subgrade	Subgrade Modulus (MPa)	Settlement From Ballast Thickness	C Value From Ballast Thickness	Rail Bending Stress From Ballast Thickness
High	47	Y= 0.0002x-1.0268	Y=0.005x+29.444	Y=-0.0015*58.714
Medium	35	Y = 6e-5x-0.8902	Y=0.005x+29.444	Y=-0.0009*57.716
Low	25	Y = 0.0003x-1.2163	Y=0.005x+29.444	Y=0.0004x+56.928

Variable and derivative	X= Ballast thickness 100 < X < 500 Y= Settlement (mm)	X= Ballast thickness (mm), 100 < X < 500 Y= C value (kg/cm3)	X= Ballast thickness (mm), Y= 100 < X < 500 Rail stress (MPa)

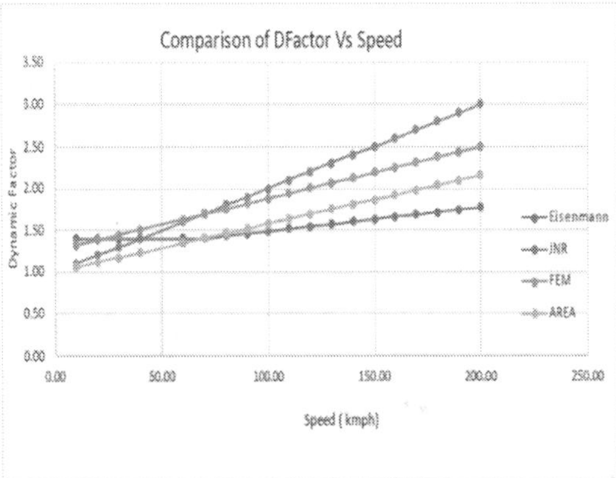

(a)

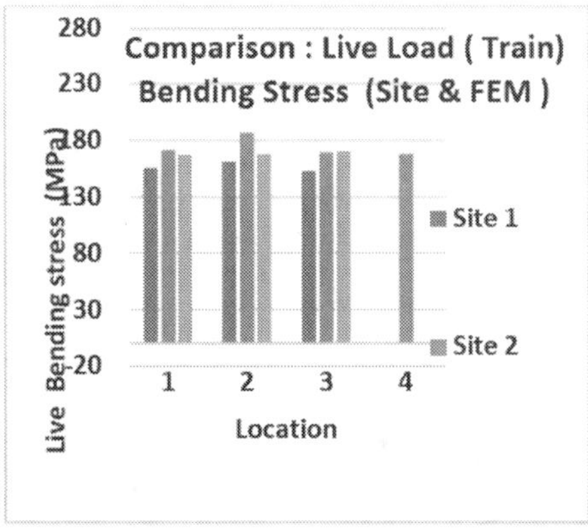

(b)

Figure 8. (a) Comparison of dynamic factor by various; (b) Comparison- fatigue rail stress (field methods. Vs FEM on 300 thick ballast).

5.2. Results - C value rail bending stress dynamic factor derivation

From the results the linear relationship was observed. The mathematical expressions were obtained for various subgrade with different sub ballast thickness on C value and rail bending stress. The speed factor(s) was derived on mathematical equation, 0.0155X+0.646, X is speed in kmph. The DF Factor of 2.9 for maximum speed of 160 kmph was derived (figure 8(a)).

5.3. Comparison of field and numerical values (settlement, C value, rail fatigue and bending stress)

The (C) value was derived from the PBT load -deformation data. The ballast thickness of 300 mm thick was constructed in both sites. Hence the comparison between FEM and Field is focused on 300 mm ballast thickness only.

Table 4. The summary of settlement, C value, rail stress, fatigue stress derived from site and FEM.

Location/ Model	Settlement (mm)	C Value (kg/cm^3)	Mean Rail Stress (MPa) (σ_{mean})	DF	Dynamic Stress N/mm^2 ($\sigma_{Q\ Live\ Load}$)	Total Stress (MPa) (Live+ Temp+Residual)	Free Stress (MPa) ($\sigma_{Free\ Stress}$)	Fatigue Limit stress Utilization %
1-1	0.46	65.43	53.99	2.9	156	336	124	73
1-2	0.71	42.39	55.82	2.9	161	341	119	74
1-3	0.35	86.00	53.06	2.9	153	333	127	72
2-1	1.37	21.90	59.49	2.9	172	352	108	77
2-2	2.83	10.60	64.72	2.9	187	367	93	80
2-3	1.20	25.00	58.67	2.9	170	350	110	76
2-4	1.14	26.32	58.36	2.9	169	349	111	76
FEM (47 MPa) Subgrade	1.06	28.30	57.93	2.9	167	347	113	76

FEM (35 MPa) Subgrade	1.15	26.09	58.36	2.9	169	349	111	76
FEM (25 MPa) Subgrade	1.29	23.26	59.09	2.9	171	351	109	76

6. Evaluation and discussions

The displacement decreases with increase of subgrade modulus. It implies that soil subgrade modulus is the dominant factor influencing on rail deflection vertical stress on subgrade, ballast, sub ballast levels. The (track stiffness) C increases with decrease in rail bending stress.

Increase in ballast has two significant effects firstly increase the thickness secondly the distance increases from sleeper base. Hence the stress spreading resulting the reduction on deviatory stress [14] on subgrade. The mathematical linear equations were evaluated relating the variables ballast thickness, C value, rail stress on varying subgrade based on deformations results.

In FEM average settlement of 1.30mm and static stress at subgrade 44 kPa were evaluated. These values are closure to 4mm settlement and 40 kPa stress estimated by Harry Tan. It implies FEM is reliable and realistic.

From the Dynamic Analysis, the maximum dynamic stress was evaluated 190 kPa at the subgrade at the resonance frequency of 15 Hz. The first resonance frequency occurred at 15 hZ. (Refer Figure 8b. It is close to the few literatures evaluated railway frequency value of 16 hZ [15, 16], 15-25 hZ [16].

On comparison of the maximum settlement, Site 1 (0.7mm) is less than Site 2 (2.8mm) Relatively the C value higher, resulting the rail experiences less bending stress. It implies that Site 1 is much stiff than Site 2. Consequently, the Fatigue limit Derived from smith Diagram (Refer Figure 8b). Site 2 reaches close to the limit 280 MPa when comparing to Site 1. It implies for 50 MPa temperature and 130 MPa residual stress, the allowable live load Stress is 280 MPa. The maximum utilization on Site 1 (161MPa-74% of 280MPa), Site 2, (187 MPa - 80 % of 280 MPa). The low stiffness in Site 2 resulted in 93 MPa (20 % of 280 MPa), remaining free strength to reach the limit of 280 MPa. It shows that Site 2 has the tendency to experience more fatigue stress resulting deterioration rate shall be higher than Site 1. Consequently, rail life significantly shall reduce and, maintenance cost shall increase.

7. Conclusions

In this paper, the response of static and dynamic analysis of ballasted track structure for different railroad bed condition on changing subgrade was evaluated by using numerical methods on taking in to account the elasto-plastic properties of subgrade soil and ballast, to consider the non-linear behaviour and to evaluate the plastic deformation due to train repetitive loading. The track stiffness comparison between two sites were evaluated and implies that. Site 1 much stiff than Site 2when comparing settlement, C value and rail stress. The response of rail bending and fatigue strength in different ballast thickness and different subgrade conditions were evaluated. The impact of rail bending stress was influenced by deformation and shows linear variation on Track foundation modulus. Based on reserve capacity of rail fatigue strength, the rail life, deterioration rate could be assessed. A simple improved track design was introduced by using single track stiffness parameter from conventional plate bearing test (PBT) on Force Displacement (FD) conventional curve method. The improved method with deriving equivalent track stiffness from rail pad and track substructure tested C value are accurate and simple. The current test method to determine the track stiffness in live track condition is expensive and unsafe with operational requirements. This PBT is simple, cost saving on labour, safe and without applying live train load.

Mathematical equations were developed to correlate the variables, ballast thickness settlement, track stiffness, rail fatigue strength on varying subgrade soil modulus. Incorporation of this equations and DF will improve the accuracy and reliability of the conventional track design methods and maintenance standards. Besides, it could assist to explore and assess on the quality control and quality assurance on functionality of the track system in the field.

8. References

[1] Lichtberger B 2005 *Track Compendium: Formation, Permanent Way, Maintenance. Economics, 1.* (Hamburg: Eurailpress Tetzlaff-Hestra GmbH & Co)

[2] Pita A L, Teixeira P F & Robusté F 2004 High speed and track deterioration: the role of vertical stiffness of the track, P*roceedings of the Institution of Mechanical Engineers, Part F: Journal of Rail and Rapid Transit* **218** 31-40

[3] Puzavac L, Popović Z and Lazarević L 2012 Influence of track stiffness on track behaviour under vertical load, *Promet-Traffic&Transportation* **24** 405-412

[4] Esveld C 2001 *Modern railway track (Vol. 385)* (Zaltbommel: MRT-Production.

[5] Burrow M P N, Bowness D and Ghataora G S 2007 A comparison of railway track foundation design methods, *Proceedings of the Institution of Mechanical Engineers, Part F: Journal of Rail and Rapid Transit* **221** 1-12.

[6] Standard B 1999 *BS 1377-7, Methods of test for Soils for civil engineering purposes-Part 7: Shear strength tests (total stress)* (London: UK: British Standard Institute)

[7] DIN D 2012 *18134: Soil–testing procedures and testing equipment–plate load test* (Berlin, Germany).

[8] Sadeghi J M 2008 Experimental evaluation of accuracy of current practices in analysis and design of railway track sleepers, *Canadian Journal of Civil Engineering* **35** 881-893

[9] Berggren E 2009 *Railway track stiffness: dynamic measurements and evaluation for efficient maintenance* PhD Thesis (Stockholm: KTH)

[10] Sadeghi J M 2005 Investigation on the accuracy of the current practices in analysis of railway track concrete sleepers, *International Journal of Civil Engineering* **3** 31-45

[11] Sadeghi J & Barati P 2010 Evaluation of conventional methods in Analysis and Design of Railway Track System, *International Journal of Civil Engineering* **8** 44-56

[12] Kouroussis G, Gazetas G, Anastasopoulos I, Conti C and Verlinden O J S D 2011 Discrete modelling of vertical track–soil coupling for vehicle–track dynamics, *Soil Dynamics and Earthquake Engineering* **31** 1711-1723

[13] Mulugeta M 2015 Dynamic Simulation of T-Track: Under Moving Loads *Advances in Intelligent Systems and Computing* **334** 147-161

[14] Sayeed M 2016 *Design of ballasted railway track foundations using numerical modelling with special reference to high-speed trains* PhD Thesis (Western Australia: Curtin University)

[15] Li Z and Wu T X 2008 Modelling and analysis of force transmission in floating-slab track for railways, *Proceedings of the Institution of Mechanical Engineers, Part F: Journal of Rail and Rapid Transit* **222** 45-57

[16] Chen R Zhao X Wang Z Jiang H and Bian X 2013 Experimental study on dynamic load magnification factor for ballastless track-subgrade of high-speed railway, *Journal of Rock Mechanics and Geotechnical Engineering* **5** 306-311

Initial Implementation of Structural Health Monitoring System of a Railway Bridge

D A Purnomo[1], W A N Aspar[1], W Barasa[1], S M Harjono[1], P Sukamdo[2] and T Fiantika[1]

[1]Center of Technology for Transportation System and Infrastructure, BPPT, Jakarta, Indonesia 10340
[2] Department of Civil Engineering, Universitas Mercu Buana, Indonesia, 11650
Corresponding author: dwi.agus@bppt.go.id

Abstract. In order to determine the actual condition of the railway bridge structure in the field, predictive monitoring is needed by installing a structural health monitoring system (SHMS). In the process of applying the SHMS, a bridge design review was applied to have railway bridge characteristics. The purpose of conducting this design review is to determine the allowable threshold for deflection and vibration of the bridge. This paper will present the analysis of the steel frame structure; with a span of 51.60 meters, 4.45 meters wide, of 5.00 meters high, respectively. According to the applicable standards, the loads used following the function of the bridge on the railroad tracks are calculated. The purpose of this paper is to (1) analyze the strength of the attached profile against the working forces, especially the live load of the rail line, (2) to know the deflection that occurs, (3) to know the natural frequency that occurs, and (4) to develop expert systems. The simulation results are used as the basis for placing sensors on the bridge and as the basis for determining the threshold for the railway bridge SHMS.

1. Introduction

The maintenance of a railway infrastructure plays an important role in the operation of transportation systems and infrastructures. Such activity aims to ensure the safety of operations and the availability of railroads and related equipment for traffic control. Maintenance is one of the main costs of operating rail transport. Furthermore, the increasingly fierce competition in the traffic market demands maintenance improvements, which aim to reduce maintenance costs while maintaining safe operations. This problem is expected to be overcome by the methodology presented in this paper. The first step of the methodology consists of a cluster-based approach to equipment reliability analysis. Its purpose is to identify groups of railway items that can be given the same reliability target. The second step is to pay attention to the level of service required in the transportation system. This needs to be done by identifying the most critical items in an effort to develop a railway system reliability model. This second step system includes implementing SHMS (structural health monitoring systems) for railway infrastructures such as SHMS on railroads and railway bridges.

As widely known, SHMS is the process of assessing the health/damage condition of a civil building structure using instruments. The term SHMS is familiar to the civil construction industry and has been widely installed in tall buildings, bridges, tunnels, dams, and other important civilian buildings. The SHMS, in general, serves as an early warning to increase the security and reliability of the infrastructure system by detecting damage before it reaches a critical state and allowing an assessment of the condition of the infrastructure being observed with the SHMS facility.

The benefits of SHMS include early detection of problems or damage to the structure of a building object which can prevent fatal damage, such as the collapse of the main structure or the settlement of the foundation. In that sense, SHMS can monitor in real-time the state of the building object. The implementation of the SHMS for the BH77 railway bridge in Lampung starts with determining until obtaining adequate structural modeling through a review of the bridge design. The step of implementing the SHMS on the BH77 railway bridge with collecting the dimensions of structural bridge members because the as-built drawing of the bridge was not available. Furthermore, deflection calculations as the effect of loads and natural frequency are carried out. Once overall displacement data were obtained, IoT Ecosystem will be followed by an intelligent decision-making process. The expected output from the development of this expert system is to obtain the health condition of the bridge structure in real-time through the reading dashboard to the authorities in taking necessary actions.

2. Literature study

In 2012 Macchi et al. [1] stated that the maintenance of railroad infrastructures plays an important role in railway transportation. This aims to ensure the safe operation and availability of railroads and related equipment for traffic control. There has been an increasing demand for improved conditions of transport infrastructures in Indonesia. Bridges are important facilities of transportation infrastructures and require such monitoring. Davis and Goldberg [2] reported that there were 66,405 structurally deficient bridges in the United States in 2013. The structurally deficient bridges were more than 11 percent of all bridges. Moreover, most of them were more than 65 years old. This condition is similar to that in Europe [3], in which the majority of bridges were built in the post-war era period from 1945 to 1965. In Indonesia, many railway bridges were built more than 100 years ago. The railway bridges were built in connection with the construction of railway lines. The first rail line was built by the Staats Spoorwegen (SS) company. The railway was constructed along 115 kilometers, located between Surabaya and Pasuruan. The railway was completed and inaugurated on May 16, 1878 [4]. Subsequently, the construction of several bridges and railroads began [5][6].

In Indonesia, the loading conditions of railway bridges have changed in recent decades due to increased freight volumes [7]. Besides, many of those railway bridges have been structurally deficient due to gradual deterioration with time. Rehabilitation and extension of the lives of these structures raise important maintenance and safety issues. Practically, bridge maintenance has been performed utilizing visual inspection methods, which are highly variable and lack resolution. It can only detect damage when it is visible. In addition, structurally deficient bridges may be left undiscovered. Several railway bridges have collapsed due to a lack of structural capacity information and routine maintenance. Chupanit and Phromsorn [8] suggested that a bridge health monitoring technique is necessary rather than visual inspection alone, and it may not be adequate.

Performance evaluation of existing railway bridges is usually by load testing method. The method covers retrofitting or strengthening old bridges, bridges that are in service already, and new railway bridges for post-construction maintenance. Duvnjak et al. [9] applied proof and diagnostic load testing when monitored and diagnostic a damaged railway bridge. Proof load testing is very useful for the evaluation of bridges when information related to the capacity of the bridge is insufficient. However, most existing railway bridges past their design life and require have much heavier loads during the design period. For this condition, it is suggested that assessment and prediction of the current conditions and safety of the remaining life on older railway bridges are necessary as a result of increased loads on locomotives [10]. Therefore, Mirza et al. [11] evaluated the effect of wheel loading in the rail component system between railheads and sleepers interaction.

In special conditions, there are often crossings with slopes greater than the determining ramps. These special conditions are referred to as steep ramps with a length of ramps that must meet the applicable provisions [12]. Various methods are engineered to make the rail slope within a safe threshold by constructing a bridge to overcome gradients. Railway bridges are bridges specially designed to be crossed by trains. Planning these bridges from the railway tracks, free space of the bridge until the load received by the bridge is adjusted to the trains that cross the bridge.

Axle loads used as a basis for planning must be following the classification of the path, and the largest load operated [13]. Bridge inspection systems and monitoring procedures have to be in line with the bridge management system. Improved bridge inspection addressing a structurally deficient bridge has substantial cost and practical implications. Various countries [14] have developed bridge inspection measurements integrated with the bridge management systems. The development of the bridge monitoring systems in Japan has been highly influenced by geographic and socioeconomic conditions [15]. Japan, geographically, is pronely subjected to natural disasters. Therefore, civil infrastructure monitoring has to be performed regularly. Damage detection techniques were developed and implemented so that the service life of these structures can be monitored beyond the design basis service life [16]. The process of damage detection techniques for civil engineering infrastructures is referred to as Structural Health Monitoring [17]. Malekjafarian et al. (2015) performed indirect bridge monitoring utilizing an instrumented vehicle [18]. The system is called the "drive-by bridge health monitoring." The car was instrumented with sensors, consisting of the most common accelerometers, and fitted on its axles. Li et al. [19] conducted an extensive study regarding vertical vehicle-track interaction at pivotal points on railway bridge approaches and lateral misalignment issues associated with bridges located in sharp curves.

The dynamic interaction between wheels and rails is increasing due to the accelerated speed of the train and the additional heavy axle loads, especially in the vertical direction [20]. Therefore, there is a need to improve and inspect existing railway bridges to meet the service requirement. The purpose of this study is to (1) analyze the strength of the attached profile against the working forces, especially the live load of the rail line, (2) to know the deflection that occurs, (3) to know the natural frequency that occurs, and (4) to develop expert systems.

3. Research methodology

3.1. Study activities

The study of this program was to evaluate the existing steel railway bridge structure. The railway truss bridge is a Warren truss configuration with a vertical height of 5.00 m, a single span of 51.6 m, and a width of 4.45 m. In order to have a comprehensive load-carrying capacity, loading combination which acts to the railway bridge structure were numerically simulated. Three-dimensional (3D) detailed models of the railway bridge were established. Finite element analysis was applied for the analytical calculation of the railway bridge in this study. The finite element analysis was performed by means of SAP2000® software Ver. 6.1 [21]. As mentioned earlier, due to lack of documentation, existing dimensions should be directly measured at the bridge location. Numerical simulation was then performed to understand the bridge characteristics, especially related to its deflection, stress distribution, and natural frequency. The flow of the study diagram and the effect of increased loading and loading standards on this railway bridge can be seen in figure 1.

3.2. Loading combination of the railway bridge

Before carrying out a numerical analysis of the loading combination, it is necessary to test the numerical model with a simpler problem. The initial step is to equalize all the truss geometries of the railway bridge to obtain a completely symmetrical and uniform bridge. Furthermore, the symmetrically distributed load is applied, for example, the combined magnitude of the dead load and the live load evenly. This is to check whether the resulting deflection and stresses are symmetrical, and the magnitude should be smaller than the applicable standard [22].

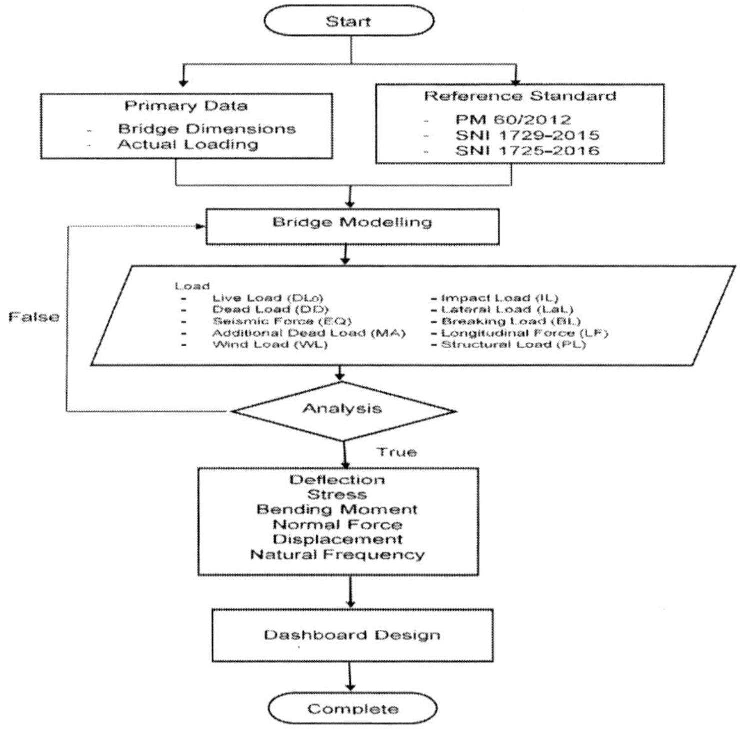

Figure 1. Flowchart of the study.

Existing bridge elements shall exhibit reliable load-carrying capacities. It should, therefore, be verified and determined by using applicable standards with appropriate safety factors. All main provisions for loading actions on the railway bridge should be considered.

Characteristic values of the bridge loading conditions should be according to Indonesian National Standard [22] and following the transportation minister regulation number PM 60 of 2012 [23] regarding railway line technical requirements. In many cases, EN 1991:2-2003 [24] regarding Technical Requirements of Railway Lines and Actions on Structures and Traffic Loads on Bridges should be considered. The reliability of the existing bridge structural members and their cross-sections should also be verified, and the load-carrying capacity should be determined by means applicable standard [25][26].

In order to have a comprehensive load-carrying capacity, a loading combination based on the code was applied to the railway bridge structure. Load factors implemented for this numerical simulation follow SNI 1725-2016 [22] and are presented in table 1. Table 1 represents the applied loading combination and becomes the input for numerical simulation modeling of the railway bridge. Strong mean that loading combination takes into account the forces caused on the bridge under normal state without taking into account the wind load. In this boundary state, all the normal forces that occur are multiplied by the appropriate load factor. Roman numerals, I through V, indicate the loading type acting on the bridge. Extreme conditions are loading combination that considers the combination of live loads reduced by loads due to ship collision, vehicle collisions, flooding, or other hydraulic loads, except for cases of loading due to vehicle collisions. Loading cases due to flooding should not be combined with loads due to vehicle collisions and ship collisions.

Serviceability load is loading related to bridge operation with all loads having a nominal value and taking into account the wind load speed of 90 km/hour to 126 km/hour. This combination is also used to control deflection in steel culverts, tunnel lining plates, thermoplastic pipes, control the crack width of reinforced concrete structures, and tensile stress analysis in segmental concrete cross-sections. This loading combination should also be used for the investigation of slope stability [22].

Meanwhile, fatigue load is a combination of fatigue and fracture loads in relation to fatigue life due to the induction of loads that have unlimited time. The results of the numbers listed in this table 1 represent the load factors used as input for numerical simulation modeling of the railway bridge [27]. Three-dimensional (3D) detailed models of the railway bridge were established, and the models were shown in figure 2.

Table 1. Loading Combination Acting on the Railway Bridge [22][27].

No	Load limits	DL	MA	LL	IL	LaF	LF	PL	BL	WR	WL	QX	QY
1	Strong I	1.1	2.0	1.8	1.0	1.0	1.0	1.0	1.8	0.0	0.0	0.0	0.0
2	Strong II	1.1	2.0	1.4	1.0	1.0	1.0	1.0	1.4	0.0	0.0	0.0	0.0
3	Strong III R	1.1	2.0	0.0	1.0	1.0	1.0	1.0	0.0	1.4	0.0	0.0	0.0
	Strong III L	1.1	2.0	0.0	1.0	1.0	1.0	1.0	0.0	0.0	1.4	0.0	0.0
4	Strong IV	1.1	2.0	0.0	1.0	1.0	1.0	1.0	0.0	0.0	0.0	0.0	0.0
5	Strong V R	1.1	2.0	0.0	1.0	1.0	1.0	1.0	0.0	0.4	0.0	0.0	0.0
	Strong V L	1.1	2.0	0.0	1.0	1.0	1.0	1.0	0.0	0.0	0.4	0.0	0.0
6	Extreme I X	1.1	2.0	0.3	1.0	1.0	1.0	1.0	0.3	0.0	0.0	1.0	0.3
	Extreme I Y	1.1	2.0	0.3	1.0	1.0	1.0	1.0	0.3	0.0	0.0	0.3	1.0
7	Extreme II	1.1	2.0	0.5	1.0	1.0	1.0	1.0	0.5	0.0	0.0	0.0	0.0
8	Serviceability I R	1.0	1.0	1.0	1.0	1.0	1.0	1.0	1.0	0.3	0.0	0.0	0.0
	Serviceability I L	1.0	1.0	1.3	1.0	1.0	1.0	1.0	1.0	0.0	0.3	0.0	0.0
9	Serviceability II	1.0	1.0	0.8	1.0	1.0	1.0	1.0	1.3	0.0	0.0	0.0	0.0
10	Serviceability III	1.0	1.0	0.0	1.0	1.0	1.0	1.0	0.8	0.0	0.0	0.0	0.0
11	Serviceability IV R	1.0	1.0	0.0	1.0	1.0	1.0	1.0	0.0	0.7	0.0	0.0	0.0
	Serviceability IV L	1.0	1.0	0.75	1.0	1.0	1.0	1.0	0.0	0.0	0.7	0.0	0.0
12	Fatigue	0.0	0.0	0.75	0.75	0.0	0.0	0.0	0.75	0.0	0.0	0.0	0.0

DL = dead load; IL = impact load; PL = structural load; WL = left wind load
MA = additional dead load; LaF = lateral force; BL = braking load; QX = seismic force X
LL = live load; LF = longitudinal force; WR = right wind load; QY = seismic force Y

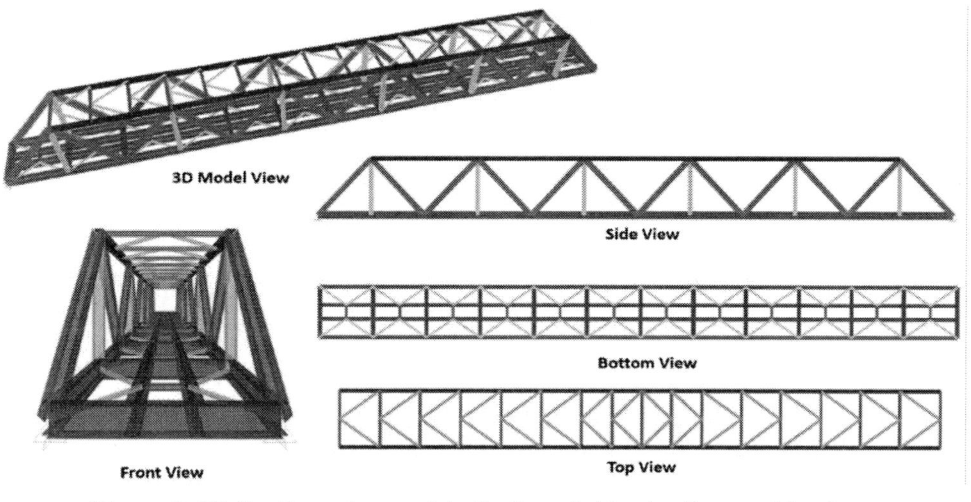

Figure 2. 3D Configuration model of railway bridge loading combinations.

The material property used for numerical simulations was taken as steel with quality material grade BJ55 with a modulus of elasticity of 200,000 MPa and yield stress of 410 MPa. However, justification was applied regarding diagnostic investigation and inspection in the field since the as-built drawing was not available. In table 1, the dead load (DL) is the mass of the rails and sleepers installed, which was 0.938 kN/m'.

The forces (PL) were loads acting on the bridge structure during the construction process, including all the forces that occur due to changes in the segmental statics of the construction, and were taken as 3.00 kN/m'. The impact load (IL) applied to the structure was a running load of 50 kN/m'. The live load was taken to be 90 kN for one axle. The braking load (BL) designed on the structure was 22.5 kN in the direction of the X-axis or longitudinal direction. A seismic load was also considered during numerical simulation. Seismic loads applied to the railway bridge structure were related to the bridge location. Acceleration of soil parameters of $Ss = 0.571$, and $S_1 = 0.236$ was considered in the design. The numbers are based on the spectral response of the peak ground acceleration. The amplification is determined based on the soil classification in the seismic design criteria for a structure at ground level.

3.3. Proposed transfer data sensor

An expert system is a system that tries to adapt human knowledge into computers. Computers are attempted to solve problems as usually executed by experts. A good expert system is designed to solve a particular problem by imitating the work of experts. Likewise, for the implementation of SHMS on the BH77 railway bridge, an expert system will be developed through the use of big data. This concept is associated with Artificial Intelligent (AI) and the Internet of Things (IoT). Figure 3 shows the schematic sensor data transfer flow chart for the SHMS implementation on the BH77 railway bridge. In general, the developed expert system includes four main things. (a) Managers use the user interface to enter instructions and information from the system. The input methods used by managers include menus, commands, natural language, expert system output using two forms of explanation: explanation of questions and description of problem solutions. Via the user interface, the conditions of the BH77 railway bridge can easily be monitored using sensors. (b) Knowledge Base consists of facts that describe the problem domain and presentation techniques that use facts according to logic. Rules are details in an unchanging situation. True and false conditions, where the action will be taken if the conditions are true. (c) The engine interface is part of an expert system that forms reasoning using the knowledge base contents in a certain order. (d) The development engine is used to build rule sets with an approach to programming languages and the expert systems section for the system development process.

Bridge - SHMS covers the components of the data retrieval process via a sensor system with parameters of deflection, stress on bridges, and vibration. Via the Bus System, the results of sensor monitoring are processed and sent directly to the Data Acquisition Module (DAQ) in the bridge location area. The results of DAQ data collection are sent and stored through the Gateway into Cloud Storage. The infrastructure control center can utilize the actual data on Cloud Storage to monitor bridge conditions online in a dashboard. Based on the data set on the cloud storage and the set of views of bridge experts in the near future, a bridge health prediction system can be developed.

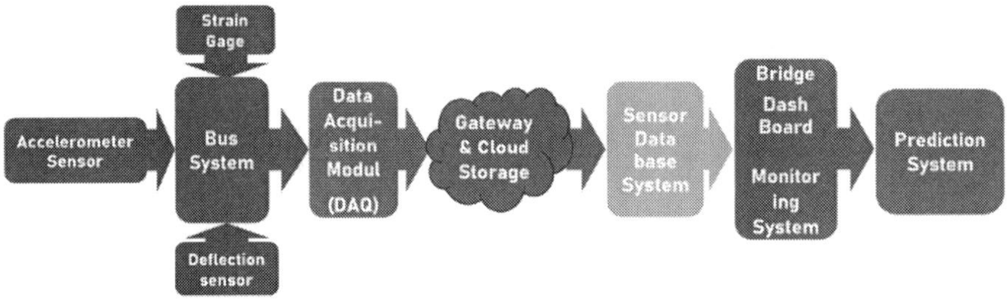

Figure 3. Sensor data transfer flow chart.

4. Results and discussion

4.1. Structural characteristics maximum capacity

Indonesian National Standard [22], SNI 1725-2016 on loads of bridges requires that loads act on the bridge must not be greater than the bridge's capacity. Therefore, the bridge's capacity ratio and the applied load shall not be greater than 1. Figure 4 shows the capacity ratio on the BH77 railway bridge.

Based on the numerical simulation results, the bridge capacity of each structural member was visualized using color degradation from grey, blue, green, yellow, to red. The grey color represents the capacity of the bridge is much greater than the capacity of the bridge as the loads pass over the bridge, and it is categorized as very safe. The blue color indicates that the bridge is still in the very safe category, and the green color is the safe category. The yellow color is in the warning category, while the orange color is in the very critical category, whereas the red color indicates that the bridge will collapse.

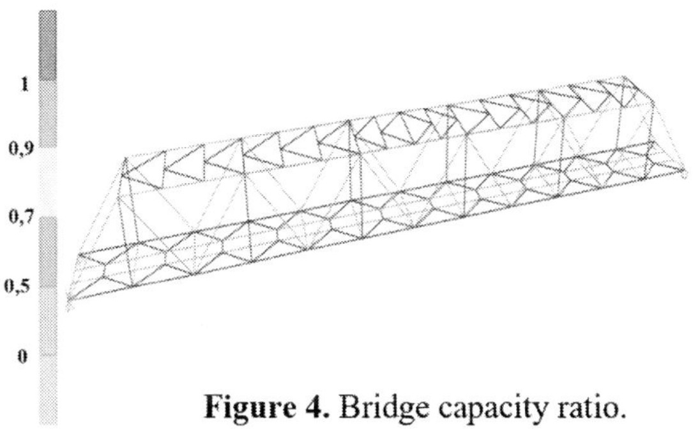

Figure 4. Bridge capacity ratio.

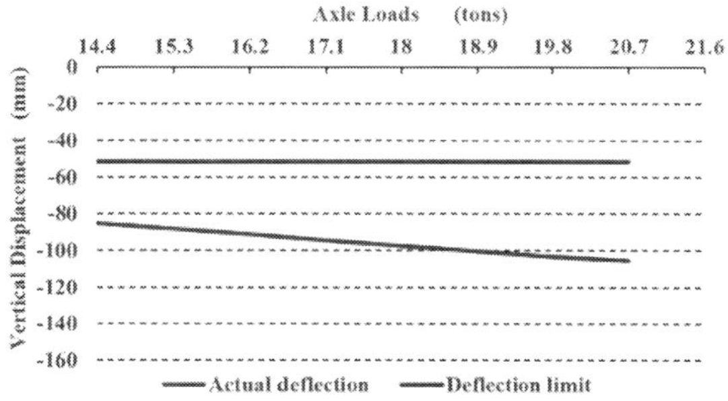

Figure 5. The maximum deflection of the BH77 railway bridge.

4.2. Maximum bridge deflection

Indonesian transportation minister regulation number PM 60 of 2012 regarding Technical Requirements for Railway Tracks [23] requires that the maximum deflection limit for the frame is L/1000 or equal to 51.6 mm, in which L is span length. Figure 5 shows load – maximum deflection characteristics of the BH 77 railway bridge. It can be seen from figure 5 that the maximum deflection is 93.551 mm, and it is greater than the required maximum deflection. This implies the structure fails to meet the criteria for maximum deflection in either the frame or the girder components. In summary, the existing deflection control criteria are not sufficient to address serviceability concerns. Exception of the girder 116, the deflection values of all bridges were significantly less than those mentioned earlier. However, this maximum deflection occurs at the strong load combination I, in which the calculated loading combination during operation never happens. Meanwhile, the calculated operation that occurs is a combination of 1 DL plus 1 LL simultaneously. This combination of loading results in a maximum deflection of 41,925 mm, which is smaller than the combination of strong loading I, and this is still acceptable.

4.3. Maximum stresses and natural frequency

SNI 1729-2015: Indonesia National Standard on Specifications for structural steel buildings requires that maximum allowable stress is $0.66 f_y$. In the design process, the yield stress of the steel is taken to be 410 MPa. Therefore, the maximum stress that occurred in any part of the bridge should not exceed 270.6 MPa. The maximum stress that occurs in the numerical modeling is 100.4 MPa. Therefore, it can be stated that the bridge structure is safe in terms of stress because it is significantly smaller than the allowable stress. The distribution of maximum stress works on the BH77 railway bridge can be seen in figure 6. Furthermore, it can be observed that the minimum and maximum frequency equal $23.28L^{-0.592}$ Hz, and $94.76L^{-0.748}$ Hz [24], respectively, where L represents the span length of the railway bridge. This boundary results in a natural frequency of about 2.283 Hz – 4.961 Hz. The analyzed frequency is the frequency at the first time the bending occurs in the structure with mass participation must be greater than 90%.

After reviewing the simulation output results, it can be seen from figure 6 that the first bending occurs in mode 4 with the natural frequency and mass participation of 4.162 Hz and 93.639%, respectively. In summary, the identified dynamic characteristics showed typical values for the type of railway bridge structure.

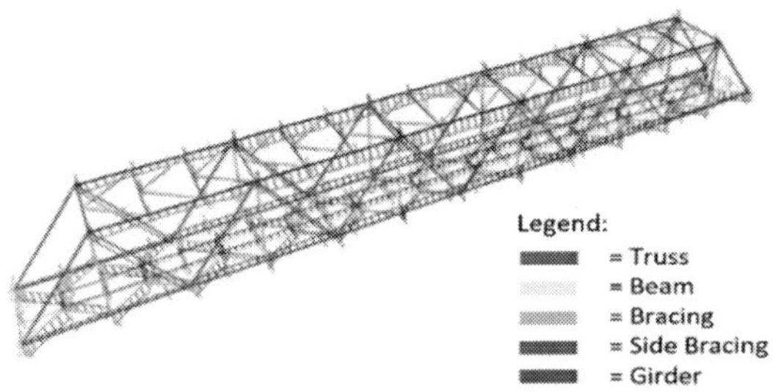

Figure 6. The maximum stress of the bridge structure.

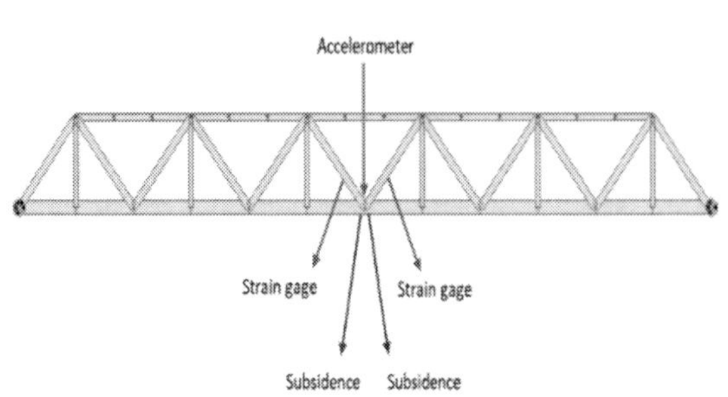

Figure 7. Illustration of sensor installation location.

4.4. Components of structural railway bridge monitoring system

The expert system adopted for application on this BH77 railway bridge SHMS is the Brickwork Expert [28] cluster system, an Expert System for civil building design structure and is commonly called BERT. BERT is used to examine a building design, furthermore provides some recommendations for improvement. The input can be in the form of images or other formats as outlined on the reading dashboard. If a bridge construction, for example, wants to be monitored in terms of bridge vibration, then a vibration sensor will be used, which will be connected to the transfer instrument. Furthermore, the data transferred through the transfer instrument will enter the data interface and document control. Therefore, through the expert system installed in the Control Center, analysis can be carried out automatically, the results of which will appear on the reading dashboard for use by stakeholders or policymakers in taking actions and policies on the results of these readings. This expert system is adjusted for engineering transportation purposes [29]. Figure 7 shows an illustration of the data sensors installed at the railway bridge. The plan for the location of the sensor installed on the bridge is as recommended by Fiantika [30]. The accelerometer sensor is used to measure vibrations on the bridge. Measurement of vibration on the bridge will be useful to reduce the risk of further damage to the bridge due to vibration and frequency is too high. In this bridge, the accelerometer sensor is also used as an alternative to measuring deflection on the bridge. Based on the analysis using SAP2000, the largest deflection is in the middle of the bridge span with a position just below the girder above which there is a rail. Therefore, the accelerometer sensor is placed in the middle of the span to measure the deflection and vibration of the bridge, especially the BH 77 bridge model is a double symmetrical bridge, meaning that it is symmetrical between the x and y axes.

5. Concluding Remarks

From the results of the review of the BH77 railway bridge design and numerical analysis, the facts are obtained and are presented as follows:

This paper has described various aspects of the initial effort to build an SHMS for a railway bridge. As an initial phase, the BH77 railway bridge was used as a case study of SHMS development. The numerical model shows that the critical boundary conditions are determined by deflection, while the natural stresses and frequencies are within safe limits. Maximum deflection occurs at frame 116 (girder). Other than frame 116, the deflection is still very far from the maximum deflection. In terms of service load, a deflection that occurs is still in a safe condition, provided bracing frames are dimensionally upgraded. These results can be used to provide an initial picture of the readings for deflection and strain measuring equipment that will soon be installed in the field.

The maximum stress that occurs is 100.4 MPa. Therefore, this structure can be stated that the bridge structure is still safe under stress control because it is smaller than the allowable stress. The maximum natural frequency allowed is a minimum of $23.28L^{-0.592}$ Hz and a maximum of $94.76L^{-0.748}$ Hz or about 2.283 Hz – 4.961 Hz. The frequency to be reviewed is the frequency of the first bending of the structure, and the mass participation must be greater than 90%.

The numerical analysis results show that the first bending occurs in mode 4 with the frequency and mass participation of 4.162 Hz and 93.639%, respectively. The results of this frequency value are also used as the basis for selecting the type of sensor to be installed on the bridge. From the analysis of the bridge capacity, maximum deflection, maximum stresses, and maximum frequencies, the type of sensor installed in the field can be determined and the location of its installation. This can be used for sampling against the installed sensor. Bridge-mounted sensors include accelerometers, strain gauges, and displacement leveling. Sensor sampling is carried out every second when a train is passing and every 15-20 minutes when no train is passing. Data from sensor sampling are collected by the acquisition instrument data and sent to the data processing center via instrument transfer, which is then used by the adopted expert system, which can be read in the reading dashboard of the bridge structure.

6. References

[1] Macchi M, Garetti M, Centrone D, Fumagalli L and Piero Pavirani G 2012 Maintenance management of railway infrastructures based on reliability analysis, *Reliability Eng. and Sys. Safety.* **104(C)** 71-83

[2] Davis S L and Goldberg D 2013 *The Fix We're In For: The State of Our Nation's Bridges 2013* (Washington: Transportation for America) the USA

[3] Znidaric A, Pakrashi V, O'Brien E and O'Connor A 2011 A review of road structure data in six european countries, *Proc. of the Inst. of Civ. Eng.: Urban Design and Planning* (Yantai, China)

[3] Gitgita 2018 *What is the History of Railway in Indonesia*, Retrieved on January 13, 2020, from http://sejarahunik.net/sejarah/ bagaimana-sejarah-kereta-api-indonesia.html

[4] Subarkah I 1981 *Jalan Kereta Api* (Bandung: Idea Dharma)

[5] Subarkah I 1992 *Sekilas 125 Tahun Kereta Api Kita 1867 - 1992* PERUMKA (Bandung: Yayasan Pusaka)

[6] Barasa W, Fiantika T, Purnomo D A, and Aspar W A N 2020 The comparative study of railway bridge design load between PM 60/2012 and EN 1991, *Majalah Ilmiah Pengkajian Industri* **14** 107-116

[7] Chupanit P, and Phromsorn C 2012 The importance of bridge health monitoring, *World Acad. of Sci., Eng. and Tech., Int. J. of Civ. and Envi. Eng.* **6** 389-392

[8]` Duvnjak I, Damjanovic D, Bartolac M, Francic Smrkic M and Skender A 2019 Monitoring and diagnostic load testing of a damaged railway bridge: Monitoring and diagnostic load testing of a damaged railway bridge, *Frontiers in Built Envi.* **5** 108

[9] Chikermane S A 2011 Structural health monitoring of a steel railway bridge for increased axle loads, *Struct. Eng. Int.* **21** 210–216

[10] Mirza O, Hassan M K, Pham T H, Assaf E and Stanshall G 2017 The effect of wheel loading on a railway bridge, *Proc. of the 8th Australian Small Bridges Conf.* (Queensland)

[11] Rosyidi S A P 2015 *Rekayasa Jalan Kereta Api (Tinjauan Struktur Jalan Rel)*, Penerbit Lembaga Penelitian, Publikasi dan Pengabdian Masyarakat Universitas Muhammdiyah Yogyakarta (Yogyakarta: LP3M UMY)

[12] Muspitasari T, Sulistyowati I and Kumara W 2017 Evaluasi peraturan pembebanan gandar kereta api di pulau jawa terhadap kondisi aktual, *Jurnal Teknik Sipil* **14** 13

[13] Akgul F 2012 A BMS development project with an integrated inspection program *Proc. of the 6th Int. Conf. on Bridge Maintenance, Safety and Management* (Lake Come) 2659–2665

[14] Fujino Y and Siringoringo D M 2011 Bridge monitoring in japan: the needs and strategies, *Structure and Infra. Eng.*, **7** 597–611

[15] Farrar C R and Worden K 2012 *Structural Health Monitoring: A Machine Learning Perspective* (New Jersey: John Wiley & Sons Inc.)

[16] Farrar C R and Worden K 2007 An introduction to structural health monitoring, *Phil. Trans. of the Royal Soc. of London A: Math. Phys. and Eng. Sci.* **365** , 303-315

[17] Malekjafarian A, Patrick J, McGetrick P and O'Brien E J 2015 A review of indirect bridge monitoring using passing vehicles, *Shock and Vib. J.* **2015** 286139,

[18] Li D, Otter D and Carr G 2010 Railway bridge approaches under heavy axle load traffic: Problems, causes, and remedies, *Proc. of the Inst. of Mech. Eng. Part F: J. of Rail and Rapid Transit* **224** 383–390
[20] Zhai W and Sun X 1994 A detailed model for investigating vertical interaction between railway vehicle and track, *Int. J. of Veh. Mech. and Mob.* **23** 603–615
[21] SAP2000$^{(R)}$ 1997 *Integrated Finite Elements Analysis and Design of Structures* TUTORIAL MANUAL Ver. 6.1 (Berkeley: Computers and Structures, Inc)
[22] SNI 1725-2016 2016 *Indonesia National Standard: Loads for Bridges* (Jakarta: National Standardization of Indonesia)
[23] Minister of Transportation 2012 *Regulation no. 60 of 2012: Railway Technical Requirements* (Jakarta: Minister of Transportation)
[24] European Standard 1991:2 2003 *Traffic Loads on Bridge. European Committee for Standardization* (Brussels: European Standard)
[25] SNI 1729-2015 2015 *Indonesia National Standard: Specifications for Structural Steel Buildings* (Jakarta: National Standardization of Indonesia)
[26] SNI 1726-2012 2012 *Indonesia National Standard: Seismic Resistance of Planning Procedure for Building and Non-building Structures* (Jakarta: National Standardization of Indonesia)
[27] Barasa W, Aspar W A N, Purnomo D A, Nadi M A B and Fiantika T 2021 Assessment of an existing steel railway bridge structure *The 2nd Int. Sem. on Civ. and Envi. Eng. (2nd ISCEE)* (Bogor)
[28] Bowen J N, Cornick S P and Bull S P 1987 BERT-an expert system for brickwork design *Proc. of Expert Sys. '86* (Brighton)
[29] Mohammed A A, Ambak K, Mosa A M and Syamsunur D 2019 Expert Sys. in Eng. Trans.: A review, *J. of Eng. Sci. and Tech.* **14** 229–252
[30] Fiantika T, Aspar W A N, Purnomo D A, Barasa W and Harjono S M 2021 A review of structural health monitoring system and application for railway bridge *The 6th Ann. Appl. Sci. and Eng. Conf.* (Kuala Lumpur)

Acknowledgments
This paper is part of the research performed in the Center of Technology for Transportation Systems and Infrastructure - Agency for the Assessment and Application of Technology, Indonesia. The study can be conducted with support from BPPT and Kemenristek/BRIN via LPDP grants. Without financial support from those agencies, this study would have been impossible. The authors would like to express their gratitude for the financial support received from the institutions. The authors would also like to thank the reviewers who actively provide direction and suggestions. The authors declare that there is no conflict of interest regarding the publication of this paper.

Numerical Analysis of Crack Propagation in Tubular Glass Column

A W Ahmed-Abdullamohamed[1], M K Kamarudin[1] and M M. Yussof[2]

[1] Universiti Teknologi MARA (UiTM), School of Civil Engineering, College of Engineering, Shah Alam, 40450, Selangor, Malaysia
[2] School of Civil Engineering, Universiti Sains Malaysia, Nibong Tebal, 14300 Pulau Pinang, Malaysia
Corresponding author: mkhairul3965@uitm.edu.my

Abstract. The demand for transparency has increased in the construction industry and contemporary architecture over the last decade. The prior researchers focused on glass columns because their uniqueness and transparent characteristics generate an impressive visual feature. Past studies on structural glass entailed numerous experimental investigations, but FEA was applied in a few investigation exercises. This study aims to validate the experimental data and analyse the crack in the tubular glass column and determine the effectiveness of different slenderness ratios of the glass column. This study investigated the column structural behaviour under compression with different geometrical dimensions of hollow section laminated glass columns to determine their load-carrying capacity, buckling performance, and failure mechanism. Finite element analysis using the explicit method was performed by using ABAQUS. The study found that the failure mechanisms depend on the slenderness ratio classified into two failure modes, either buckling or crushing. The glass column failed due to buckling when the slenderness ratio is more than 40, while it failed due to crushing when the slenderness ratio is less than 40. The finite element analysis did not correlate perfectly with the experimental data since the FEA underestimating the glass performance.

1. Introduction

Structural glass is widely used in current structures and in a variety of significant scientific and scholarly works. There has been a growth of glass application in the construction sector, and this has a partial relationship with the glass columns application. Presently, glass columns do not have significant use in the building industry, but their progress is fascinating. For the utilization and optimally exploiting glass columns as construction materials, the columns should offer excellent structural functionality and a simple fabrication process. Still, above all, they should guarantee safety. Modern buildings often use architectural glass, and there has been an increase in the use of structural glass in recent years because it has excellent artistic characteristics. In addition, structural glass possesses high action as a compressive component in the architectural building. As a result of this advantage, glass plays a significant role and performs a substantial function of an element that bears the load. Therefore, structural engineers need to understand the response of glass to loads significantly.

Glass research in the early years mainly focused on recognizing the material characteristics of the different kinds of glass. These include annealed glass, heat-strengthened glass, and fully toughened glass. Laminated glass (LG) was introduced as a discovery to promote the efficiency and strength of structural glass. Besides, LG is two or more glass layers, joined using an interlayer that originates from the glasses above. Generally, two glass kinds of glasses are made; Borosilicate glass, often applied in laboratory apparatus or Float glass, is appropriate for the building sector. While safety concerns are important aspects of structural glass, the optimal advantage of glass use should be received to mitigate risks to the public. Therefore, laminated glass has a prospective preference since it has exhibited excellent building performance [1].

An intensive study of glass structures has been undertaken by Veer and Pastunink [2]. They initially focused on beams and plates of glass. Nevertheless, research on how glass column behaves began much later, particularly the laminated tubular glass column. The laminated tubular glass column concentrated on evaluating the capacity to carry loads coupled with the interlayer's impact to ensure the glass component's cohesion. There was an extension of the study to include the safe failure attribute of the glass column that is laminated after the initial fracture in the loading process [3]. Besides, according to Overend et al. [4], varying dimensions of the cruciform glass columns that are laminated have been examined. This type of cruciform cross-shaped glass column easily connects the beam-column, but it shows a low torsional rigidity as load-bearing in structures. Therefore, the tubular glass column is concerned with attaining higher torsional strength and has structural effectiveness for glass applications. The buckling performance of a single laminated tubular glass column has been investigated by Kamarudin et al. [5]. The research revealed that the laminated glass buckling performance depended on the ratio of slenderness. The failure mechanism of the glass column happens either through the buckling or a deforming crush. Nevertheless, Kamarudin et al. proposes an FEA to be utilized to investigate fracture development in the laminated tubular glass column under axial load compression.

Studies on laminated glass are vital for pursuing advanced glass technology. Specifically, more studies on building applications of adhesive bonding are needed. Even though structural silicone glazing has been used for years, there are still problems with improper processing and workmanship [6]. In addition, during the composite action of the laminated glass, when one outer glass layer is broken, the interlocking efficiency of fragments in touch enables the cracked glass to act as if it were still intact [7]. On the other hand, the residual static load carrying capacity and stiffness of laminated glass are known to be limited following glass fracture [8]. Furthermore, a significant finding shows that the bearing capacity and secant stiffness of tempered glass panels increase as the thickness of each layer of the panel increases [9,10].

This study investigated the propagation of cracks in laminated tubular glass columns based on experimental and numerical findings, including additional simulations conducted in ABAQUS. Since previous studies from Kamarudin *et al* did not incorporate the fracture behaviour in the analysis modelling, hence this study adopted the explicit method in ABAQUS/Explicit to analyse the fracture behaviour. ABAQUS is a robust application to reproduce the glass fracture behaviour by using the brittle cracking material model. Furthermore, this investigation portrays the mechanism of glass failure since it determines the initiation of the first crack on the glass surface. Hence, it is paramount to investigate the structural characteristics of tubular glass columns of different dimensions by FEA. The main objectives of this study are to determine the crack propagation in tubular glass columns by using ABAQUS and to carry out a parametric study in looking at the effect of different slenderness ratios of glass columns.

2. Numerical model

In this study, ABAQUS/Explicit was used to determine the crack propagation in the glass column. Data were obtained from the experiment to achieve correlation and verify the obtained results in this study. The finite element method is one of the handiest methods to study the behaviour of a structure. It gives solutions for complicated geometry very readily. ABAQUS/Explicit is a finite element analysis-based software; It gives many solutions to structural analysis such as fracture, failure, and damage of a structure. The analysis modelled in this study to simulate the behaviour of the tubular glass column under compressive loads was used ABAQUS version 6.15.

IOP Conf. Series: Materials Science and Engineering 1200 (2021) 012020 doi:10.1088/1757-899X/1200/1/012020

2.1. Element type
The element type used in this study was S4R to construct a tubular glass column modelled by 3D extrusion deformable solid shell element. S4R is a double-curved linear quadrilateral with four notes, as displayed in figure 1. Thus, this element can provide a curved geometry.

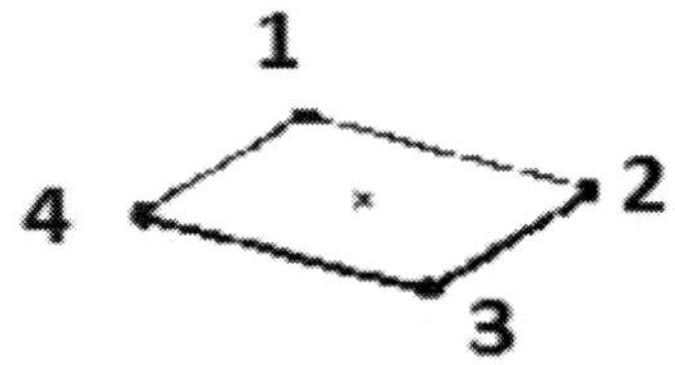

Figure 1. S4R – Shell, 4 nodes, reduced integration (ABAQUS Manual).

2.2. Geometry
The structural behaviour of six types of tubular glass columns was investigated. All the specimens have the same height of 1500 mm with different outer diameters, named; 60 mm diameter with 7.0 mm thickness (TGC-60-7.0), 24 mm diameter with 2.5 mm thickness (TGC-24-2.5), 100 mm diameter with 7.0 mm thickness (TGC-100-7.0), 200 mm diameter with 7.0 mm thickness (TGC-200-7.0), 300 mm diameter with 7.0 mm thickness (TGC-300-7.0) and 400 mm diameter with 7.0 mm thickness (TGC-400-7.0). The identification of the specimens, e.g. TGC-60-7.0 is TGC means the "tubular glass column", then the following number means the outer diameter of the TGC, and the last number means the thickness of the TGC. The details of all specimens are shown in table 1.

Table 1. Details of specimens for TGC.

No.	Test	Specimen	Height, *L* (mm)	Outer Diameter, *do* (mm)	Inner Diameter, *di*	Thickness, *t* (mm)
1	T1	(TGC-60-7.0)	1500	60	46	7
2	T2	(TGC-24-2.5)	1500	24	19	2.5
3	T3	(TGC-100-7.0)	1500	100	186	7
4	T4	(TGC-200-7.0)	1500	200	186	7
5	T5	(TGC-300-7.0)	1500	300	286	7
6	T6	(TGC-400-7.0)	1500	400	386	7

2.3. Material properties
In this study, the tubular glass column was used as linear elastic. Therefore, the model was built to respond elastically. The material Young's modulus of the glass used as the experiment which is 6400 N/mm^2 and 0.2 Poisson's ratio. Tables 2 to 4, present the material properties of the glass column.

Table 2. The material properties of the glass.

Properties	Symbol	Unit	Value
Glass density	ρ	kg/m^3	2490
Young's modulus of Glass	E	MPa	64000
Poisson's ratio	v	-	0.2

Table 3. Brittle cracking properties.

Direct stress after cracking	Direct cracking strain
138	0
0	1E-006

Table 4. Brittle shear and failure properties.

Brittle Shear		Brittle Failure
Shear retention factor	Crack opening strain	Direct cracking failure strain
1	0	
0	1E-006	1E-006

2.4. Load and Boundary Conditions

A concentrated load has been defined at the top of the tubular glass column with 200 kN. Two different types of support for fixed-defined boundary conditions were used in this study. The boundary conditions at the top and bottom are presented in figure 2 and table 5. All three directions (U) and the rotational degrees of freedom (UR) were restrained to be zero at the bottom fixed-end. At the top movable end, all three directions (U) and the rotational degrees of freedom (UR) were restricted to be zero except that the U3 was free to move 7 mm with a tabular amplitude as shown in table 6.

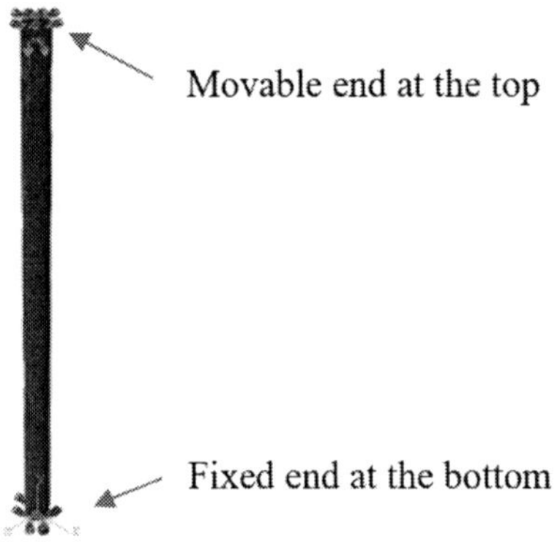

Figure 2. Load and boundary conditions in the analysis model.

Table 5. Boundary conditions at top and bottom of the TGC.

Boundary Conditions	At Top	At Bottom
U1	0	0
U2	0	0
U3	7	0
UR1	0	0
UR2	0	0
UR3	0	0

Table 6. Amplitude.

Time/Frequency	Amplitude
0	0
0.001	0.001
1	1

2.5. Mesh Density

In the meshing step, it is necessary to get good accuracy results by using optimum mesh density. A refinement mesh element was carried out to reach the optimum mesh density. The mesh element built by using a various number of edge seed, as shown in table 7. To achieve the best mesh density, an Eigenvalue analysis for buckling performed by comparing the theory buckling load with the buckling load in the FEA test. By using the equation 1 obtained from ABAQUS guidebook, the critical buckling load calculated for TGC-60-7 and TGC-24-2.5 from the FEM analysis.

$$P_{cr} = E \times a \times \frac{\Delta L}{L} \times A \quad (1)$$

P_{cr} is the critical buckling load, E is Young's modulus, a is the eigenvalue, ΔL is the displacement input, L is the column height and A is the cross-section area. The buckling load in theory calculated by using equation 2.

$$P_{cr} = \frac{\pi^2 EI}{L_e{}^2} \quad (2)$$

The comparison between the buckling load in theory and the buckling load in FEM analysis for different mesh densities are shown in Table 7. The optimal mesh size for tubular glass column analysis was determined using 40 elements, as seen in figure 3.

Table 7. Mesh quality.

TGC-60-7				TGC-24-2.5			
No. of elements	FE Pcr (kN)	Theoretical Pcr (kN)	% error	No. of elements	FE Pcr (kN)	Theoretical Pcr (kN)	% error
10	403.5	467.58	14	10	9.82	11.10	12
20	438.64	"	6	20	10.67	"	4
30	445.31	"	5	30	10.83	"	2
40	447.65	"	4	40	10.89	"	2
50	448.74	"	4	50	10.92	"	2
60	449.33	"	4	60	10.93	"	2
70	449.71	"	4	70	10.94	"	1
80	449.95	"	4	80	10.95	"	1
90	450.1	"	4	90	10.95	"	1
100	450.22	"	4	100	10.95	"	1

Figure 3. An example of 40 numbers of mesh density elements on the edge of the column for TGC-60-7.

2.6. Methods

To study the crack propagation in laminated tubular glass columns, a geometrically Linear Elastic Fracture Mechanic (LEFM) analysis was carried out by using the Explicit method in Abaqus/Explicit. The brittle cracking model was used to analyse the crack in the tubular glass columns. The brittle cracking model is the perfect method to analyse the applications where the brittle behaviour subjects. This model assumes that the material is linear elastic in compression. Therefore, the brittle cracking model can be used for glass and other materials such as concrete and brittle rocks or ceramics.

3. Failure mechanism in different slenderness ratio

Six specimens of tubular glass column constituted of various geometrical dimensions (diameters and thickness) were analysed under a displacement rate of 7 mm, to investigate the effect of different slenderness ratios in the tubular glass column.

Based on table 8, the study found that the failure mechanisms of the specimens analysed were failed catastrophically due to crushing, except TGC-24-2.5 specimen failed due to buckling deformation. The type of failure can be divided depend on the slenderness ratio into two rates, either low or high. The column with a low slenderness ratio fails due to crushing, and the column with a high slenderness ratio fail due to buckling. Figure 4 shows the deformed shape of tubular glass columns that failed by crushing at TGC-100-7, TGC-200-7, TGC-300-7, and TGC-400-7.

Figure 5 shows the buckling mode shape of the tubular glass column TGC-24-2.5. Figure 4(a) and 4(b) shows the crack propagation and the final crack during the experimental tests, the crack was highlighted in red colour, and it failed due to buckling. The fracture that occurred through the whole height interestingly ended up by breaking the glass column into many large parts in the middle, as shown in figures 4(c) and 4(d). Furthermore, there was no fracture at the top and bottom end of the column.

Table 8. Summary of glass columns experimental results of TGC specimens.

No.	Specimen	Column slenderness ratio, λ	Maximum failure load (kN)	Axial displacement failure (mm)	Type of failure mechanism
1	(TGC-60-7.0)	40	40	1.5	Crushing
2	(TGC-100-7.0)	23	94.7	3.9	Crushing
3	(TGC-200-7.0)	11	92.7	3.8	Crushing
4	(TGC-300-7.0)	7	92.3	3.8	Crushing
5	(TGC-400-7.0)	5	92.2	3.8	Crushing
6	(TGC-24-2.5)	98	25	0.9	Buckling

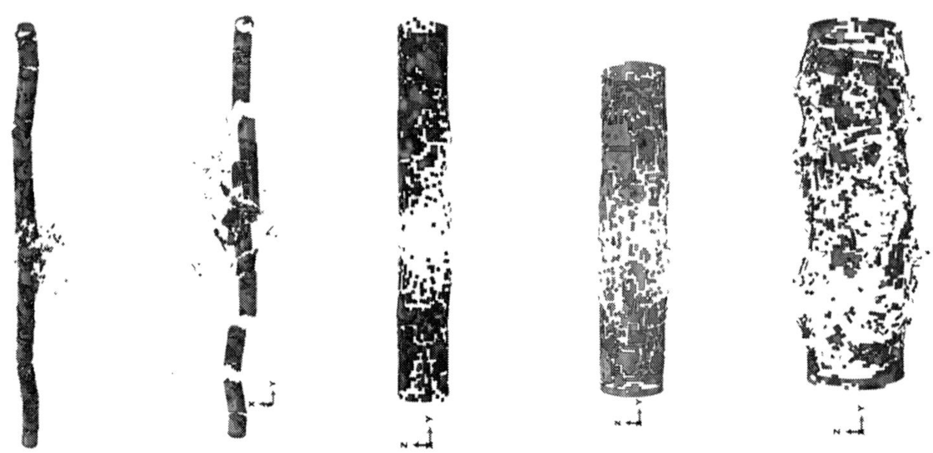

a) TGC-60-7 b) TGC-100-7 c) TGC-200-7 d) TGC-300-7 e) TGC-400-7

Figure 4. Deformed shape of tubular glass columns failed by crushing.

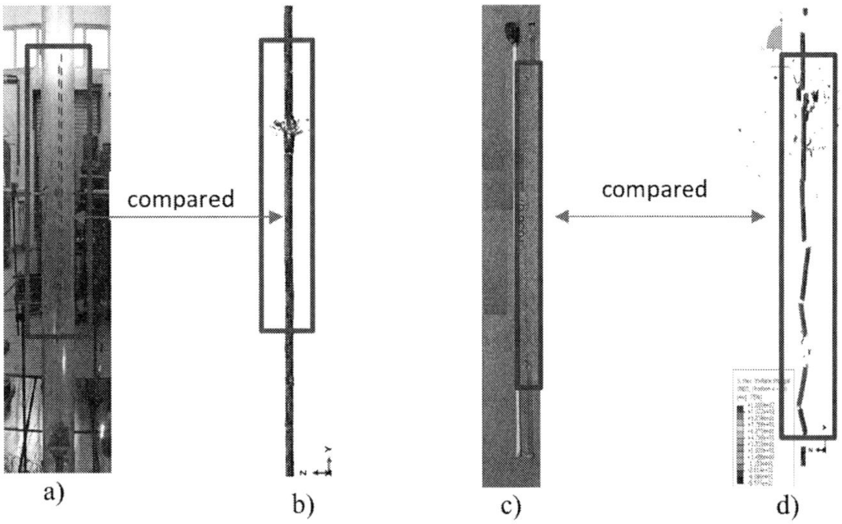

Figure 5. The deformed shape of tubular glass columns failed by buckling with compared between this study and an experimental done by Kamarudin et al. [5] in TGC-24-2.5; a) First crack initiate from the experimental. b) First crack from this study, c) Final crack from the experimental. d) Final FEA results, large glass fracture parts in the middle in this study.

3.1. Comparison and validation of the results:

A comparison of the FEA results in this study and experimental results from previous researchers is presented in table 9. It was noticed that all the specimens with a slenderness ratio less or equal to 40 mm failed due to crushing. In contrast, all specimens with a slenderness ratio of more than 40 mm failed due to buckling.

Table 9. Comparison between this study and the previous study tests with different slenderness ratio on glass columns.

Year	Researcher name	Column height (mm)	Outer diameter (mm)	Thickness (mm)	Support condition	Column slenderness ratio, λ	Maximum failure load (kN)	Type of failure mechanism
2021	This study	1500	400	7	Fixed	5	75	Crushing
2021	This study	1500	300	7	Fixed	7	75	Crushing
2021	This study	1500	200	7	Fixed	11	79	Crushing
1999	Veer	550	40	4	Fixed	22	35	Crushing
2021	This study	1500	100	7	Fixed	23	68	Crushing
2018	Kamarudin	1500	60	2	Fixed	37	*64	Crushing
2018	Kamarudin	1500	60	7	Fixed	40	*105	Crushing
2021	This study	1500	60	7	Fixed	40	40	Crushing
1999	Veer	550	40	4	Pinned	43	110	Buckling
2018	Kamarudin	1500	50	1.8	Fixed	44	*50	Buckling
2005	Overend	1080	50	7	Pinned	70	50	Buckling
2018	Kamarudin	1500	24	2.5	Fixed	98	*3	Buckling
2021	This study	1500	24	2.5	Fixed	98	25	Buckling
2018	Kamarudin	1500	20	1.8	Fixed	116	*1	Buckling
2007	Veer & Bos	555	12	3.5	Pinned	170	2	Buckling

*average value from the experimental work [5]

3.2. Load versus Axial Displacement Behavior

Figure 6 shows the axial load versus the displacement of TGC-60-7 in comparison between the experimental work and FE work (by Risk Method) carried out by Kamarudin et al. [5] and result from Brittle Cracking analysis. FEA by Riks Method tends to overestimate the glass tube performance but still showing an acceptable load-axial displacement behaviour and the load-lateral displacement behavior. FEA by Brittle Cracking Method has shown a consistent gradient line for the specimen's stiffness, however the strength prediction is showing an underestimate value compared to the experimental results.

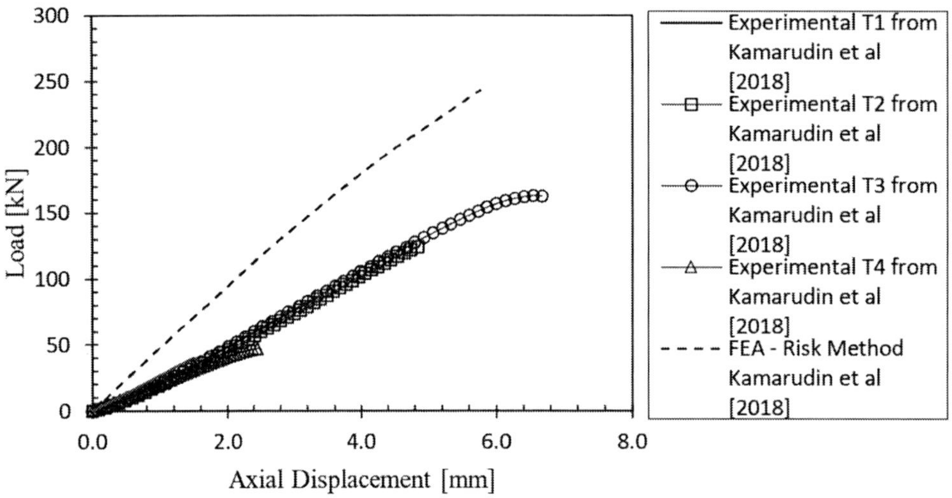

Figure 6. Load vs Displacement behavior of all glass column types.

4. Conclusions

The purpose of this study is to analyse the crack behavior of the tubular glass column under the compressive load. The structural performance of the tubular glass column was evaluated and investigated the effect of different geometrical dimensions. The explicit method in ABAQUS/Explicit was used to analyse the structural behaviour of tubular glass columns. The main conclusions of this study are as follows:

(i) The outcome of this study validates the previous researchers' theory that the strength of the tubular glass column depends on the slenderness ratio. The failure mechanisms depend on the slenderness ratio that has been classified into two failure modes, either buckling or crushing. If the glass column has a slenderness ratio of more than 40mm can be classified as a slender column that is subjected to buckling failure mode. While if it is less than 40mm can be classified as a short column that is subjected to crushing failure mode.

(ii) The relationships between the load and the axial displacements indicate a linear behaviour of the glass with an instantaneous failure without plastic deformation.

(iii) The Brittle Cracking method managed to analyse the linear behaviour of the glass. However, it tends to underestimate the glass strength. Nevertheless, when the analysis is run based on the Riks method, the linear behaviour is scarcely simulated compared to the experimental work. Therefore, the best way to analyse the glass strength is by using the Brittle Cracking method compared to the Riks method.

(iv) Glass strength cannot be justified by mean strength as compared to concrete and steel since glass inherits unique flaws though it came from a similar batch of geometrical dimensions. The same dimension of the tubular glass column will not give a similar ultimate load and displacement.

(v) Most FE results did not correlate perfectly between experimental data and FEA since the FEA leads to underestimating the glass performance. Nevertheless, the FEA gave acceptable behaviour at the lateral load-displacement and axial load-displacement.
(vi) The crack propagation related to the failure mode can be classified into two considerations either by buckling or crushing. The glass column will probably fail due to buckling in the slender column, and it will fail due to crushing in the short column.

In conclusion, based on the results of the finding, all the objectives of this study were achieved. The crack propagated in the tubular glass column has been found by using ABAQUS and shows the effect of different slenderness ratios on the tubular glass column. Consequently, the results have been validated with the previous researchers.

5. References

[1] Jacob L, Davies P S, Ray A and Yang J 2005 Safety glass performance in buildings-a review of a current test methods and selection criteria *9th International Conference on Automotive and Architectural Glass* (Tampere, Finland)
[2] Veer F A and Pastunink J R 1999 Developing a transparent tubular laminated column *6th International Conference on Automotive and Architectural Glass* (Tampere, Finland)
[3] Nieuwenhuijzen EJ, Bos F P and Veer F A 2005 The laminated glass column *9th International Conference on Automotive and Architectural Glass* (Tampere, Finland)
[4] Overend M, Vasallo C and Camillieri K 2005 The design, assembly and performance of glass columns, *9th International Conference on Automotive and Architectural Glass* (Tampere, Finland)
[5] Kamarudin M K, Yussof M M, Disney P and Parke G.A.R 2018 Experimental and numerical investigation of the buckling performance of tubular glass columns under compression, *Structures* **15** 355-369
[6] Machalickaa K, Eliasovab M 2017 Adhesive joints in glass structures: effects of various materials in the connection, thickness of the adhesive layer, and ageing *International Journal of Adhesion and Adhesives*. **72** 10-22
[7] Wang X, Yang J, Chong W T A, Qiao P, Peng S, Huang X 2020 Post-fracture performance of laminated glass panels under consecutive hard body impacts, *Composite Structures*. **254** 112777
[8] Wang X, Yang J, Liu Q, Zhao C 2018 Experimental investigations into SGP laminated glass under low velocity impact, *International Journal of Impact Engineering* **122** 91–108
[9] Campione G, Orlando F, Fileccia G, Pauletta M 2019 Bond characterization of monolithic and layered glass panels and ultrasonic tests to control glued surfaces, *Engineering Structures*. **198** 109545
[10] Chen Y, He K, Han S, Yuan Y 2019 Experimental study on three-layer tempered glass panel on quadrilateral simple bearing under local uniformly distributed load, *Thin-Walled Structures*. **139** 294–309

Prediction of Traffic Vibration Effect on Heritage Building at Muar, Johor, Malaysia

J S Ng[1], T N T Chik[1] and N A Yusoff[2]

[1] Jamilus Research Centre, Faculty of Civil Engineering and Built Environment, Universiti Tun Hussein Onn Malaysia (UTHM), Parit Raja, 86400, Batu Pahat, Johor, Malaysia
[2] RECESS Centre, Faculty of Civil Engineering and Built Environment, Universiti Tun Hussein Onn Malaysia, Parit Raja, 86400, Batu Pahat, Johor
Corresponding author: thayati@uthm.edu.my

Abstract. The vibration produced by road traffic can contribute to the long-term adverse effects on heritage structures. The objectives of this study were to analyze the dynamic response of the heritage building under time-dependent loads caused by traffic vibration by carried out the transient analysis in ANSYS and to determine the level of Vibration Criterion of the heritage building due to traffic vibration by using VSATs in MATLAB software. This research was to predict the traffic vibration effect on the old building of Telecom Muar at Johor, Malaysia. A field testing was carried out using a mobile application, iDynamics, to obtain the vibration signal induced by road traffic. The data obtained were analyzed using ANSYS and MATLAB software. Two types of analysis were carried out using ANSYS, which are modal analysis and transient analysis. MATLAB was used to obtain the vibration criteria plot (VC) for the building. The natural frequency of the fundamental mode shape was 2.57Hz. The natural frequency of the building is acceptable as it is below the human sensitivity frequency range. On the other hand, the level of vibration criteria of the building falls on VC-E, which is the lowest level in the Generic Vibration Criteria Plot implemented by Gordon (1991). In conclusion, both natural frequency and the vibration criteria of the selected building is acceptable according to the results obtained from the analysis conducted.

1. Introduction

Vibration is one of the frequent issues which lead to the failure of buildings. Normally, building structures are exposed to the two most important vibration sources which are the internal and external sources [1]. The internal sources of vibration come from the usage of machinery, elevators, HVAC systems, and the activities of a human inside a building; while the external sources of vibration originate from an earthquake, rail and road traffic, construction activities, and wind [2].

The traffic-induced vibration in Malaysia has increased significantly since the population in Malaysia increased year by year and the amount of newly registered vehicles has also increased drastically in Malaysia [3]. Thus, the vibration produced by road traffic needs to be focused on as it has the possibility to contribute to the long-term adverse effects on historic structures [2]. Traffic vibration is often caused by heavy vehicles that drive at relatively high speed and contact with the uneven road surface. The contact between the wheels of vehicles with the road surface causes a dynamic excitation and creates waves that disperse in the soil. The dispersion of waves will impact the foundations of the surrounding structure [4] and cause unwanted structural damage or also excessive vibration on floor at building [5].

The aim of the study was to predict the effect of traffic vibration on the heritage building, the old building of Telecom Muar at Johor, Malaysia. Software analysis using ANSYS and MATLAB were carried out in this study. The Finite Element Modelling (FEM) software, ANSYS, was used to model the heritage building and carried out the transient analysis. VSATs and ModalV in the MATLAB software were used to obtain the vibration criteria level of the heritage building. The vibration criteria obtained will then be used to compare with the Generic Vibration Criteria guideline implemented by Gordon [6].

2. Generic vibration criteria

In this study, generic vibration criteria were used as the guideline to determine the traffic vibration effect on the heritage building. There is a total of five types of criterion curves, which are curves VC-A through VC-E. The curves are taken from the form of a set of one-third octave band velocity spectra, together with the International Standard Organization (ISO) guidelines of the impact of vibration on people in buildings [6]. Figure 1 shows the graph for the generic vibration criterion curve and table 1 shows the description for the generic vibration curves.

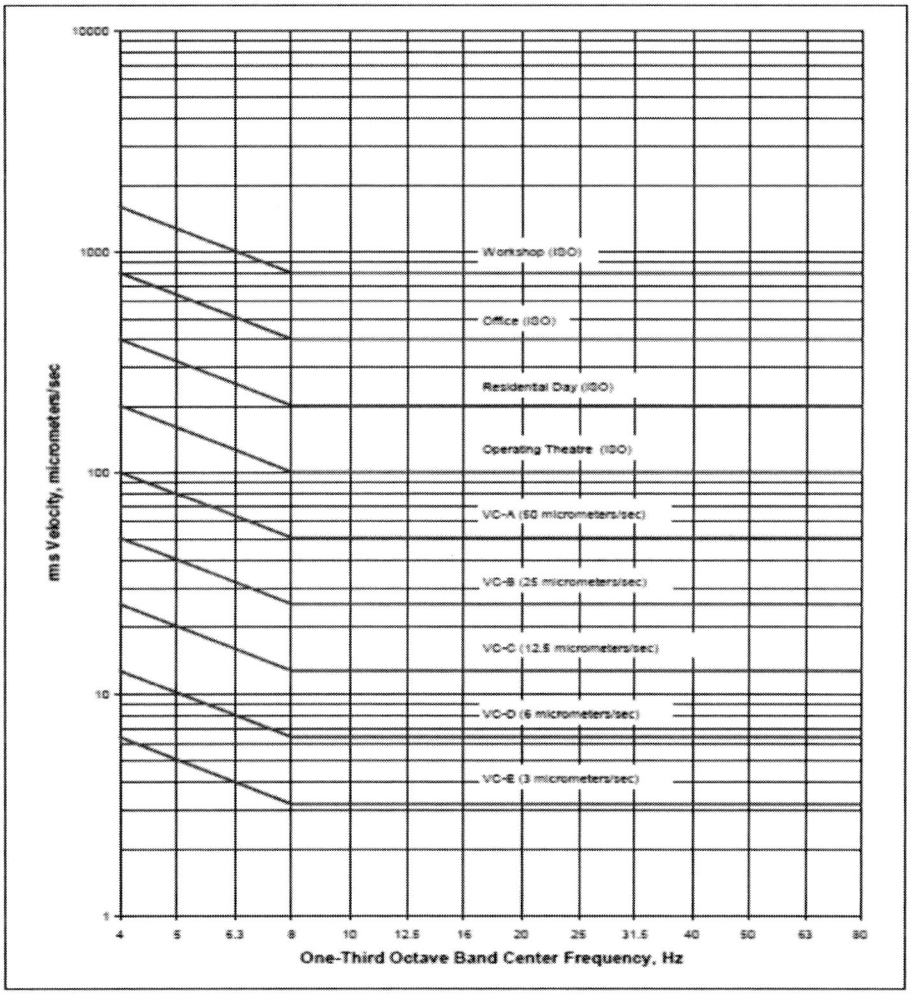

Figure 1. Generic Vibration Criterion (VC) Curves by Gordon [6].

Table 1. Description of Generic Vibration Criterion (VC) Curves [6].

Criterion curve	Max Level (μm/ s)	Detail Size (μm)	Description of use
Workshop (ISO)	800	N/A	Distinctly feelable vibration. Appropriate to workshops and non-sensitive areas
Office (ISO)	400	N/A	Feelable vibration. Appropriate to offices and non-sensitive areas.
Residential day (ISO)	200	75	Barely feelable vibration. Appropriate to sleep areas in most instances. Probably adequate for computer equipment, probe test equipment and low-power (to 20X) microscopes.
Operating theatre (ISO)	100	25	Vibration not feelable. Suitable for sensitive sleep areas. Suitable in most instances for microscopes to 100X and for other equipment of low sensitivity.
VC-A	50	8	Adequate in most instances for optical microscopes to 400X, microbalances, optical balances, proximity and projection aligners, etc.
VC-B	25	3	An appropriate standard for optical microscopes to 1000X, inspection and lithography equipment (including steppers) to 3 micron line widths.
VC-C	12.5	1	A good standard for most lithography and inspection equipment to 1 micron detail size.
VC-D	6	0.3	Suitable in most instances for the most demanding equipment including electron microscopes (TEMs and SEMs) and E-Beam systems, operating to the limits of their capability.
VC-E	3	0.1	A difficult criterion to achieve in most instances. Assumed to be adequate for the most demanding of sensitive systems including long path, laser-based, small target systems and other systems requiring extraordinary dynamic stability.

3. Methodology

The old building of Telecom Muar at Johor, Malaysia as shown in figure 2 was selected as the case study location for this study. The in-situ field testing was carried out at the surrounding of the building by using the mobile application, iDynamics. The two-storey old Telecom Muar building was modelled using ANSYS to carry out Finite Element Analysis. Besides, ANSYS was also used to carry out a transient analysis. Vibration criterion analysis was carried out using VSATs and Modal V in MATLAB.

3.1. Modal analysis

Vibration characteristics in terms of natural frequencies and mode shape of the structure can be determined through modal analysis. The parameter and the geometry of the building were defined in the analysis. The building consists of two-storey buildings with a dimension of 32.9m x 22.8m, and a height of 7.5m. The frame structure of the building includes a reinforced concrete frame system and also a timber concrete composite slab structure for the first-floor slab.

Figure 2. Front view of the old building of Telecom Muar.

Table 2 shows the properties of the building structure and table 3 shows the sizes of beams and thickness of a slab for each floor of the building. The concrete graded at 40 N/mm^2 was assumed to be used for the concrete structure.

Table 2. Properties of the building structure.

Materials	Density (kg/m^3)	Elastic Modulus (GPa)	Poisson Ratio
Reinforced concrete	2500	38	0.2
Concrete	2400	38	0.2
Timber	1000	10	0.2

Table 3. Sizes of beam and thickness of the slab.

Floor	Beam size (mm)		Slab Thickness (mm)
Ground Floor	Primary beam	300 x 600	300
	Secondary beam	300 x 500	
First Floor	Primary beam	300 x 600	130
	Secondary beam	300 x 500	
	Timber joist	43 x 400 @ 750mm c/c	
Roof	Primary beam	200 x 600	125

3.2. Transient analysis

Transient dynamic analysis is a technique used to determine the time history dynamic response of a structure to arbitrary forces varying in time. This analysis provides the displacement, stress, strain, and force-time history response of a structure to any other combination of harmonic or transient loads [7].

3.3. Vibration criteria analysis

Vibration criteria analysis was carried out using VSATs and ModalV in MATLAB. The results generated using VSATs analyses were performed and characterized by colors. The results obtained from

ModalV were formed as a plotted graph, then were compared with the Generic Vibration Criterion (VC) curve by Gordon [6].

4. Results and Discussion

The results obtained from the modal analysis, transient analysis and vibration criteria analysis were presented and discussed in this section. For all the analysis, the ground floor was not be taken in the analysis. The ground floor was assumed to be fixed to the ground; while the upper floor was set to be free. Besides, only vertical displacement (z-axis) of the floor structure was investigated. The horizontal displacement was not considered in this study.

4.1. Result for modal analysis

The result of the modal analysis was generated through ANSYS, where the result shows that the natural frequency of the fundamental mode shape of the building was 2.57 Hz. According to Siddika et al. [8], the first mode for free vibration is considered critical and has a high risk to cause resonance to occur. The other modes of free vibration can be critical for high-rise structures and assumed as less critical than the natural period for low-rise structures. Hence, the fundamental mode obtained from the modal analysis is the most important mode to reflect the behavior of the building.

According to the ISO base curve, as shown in figure 3, the human sensitivity is at its maximum when the frequencies are between 4Hz to 8Hz; after that, the sensitivity will decrease as the frequency increases [9]. Hence, the natural frequency for the first mode shape can be accepted as it is below the range between 4Hz to 8Hz. The mode shape and natural frequencies for the first ten-mode are shown in table 4. Green color indicates that the model does not undergo any displacement; blue color means that only minimum displacement is faced by the model; yellow color shows the model has moderate displacement, and red color means the model is under its maximum displacement.

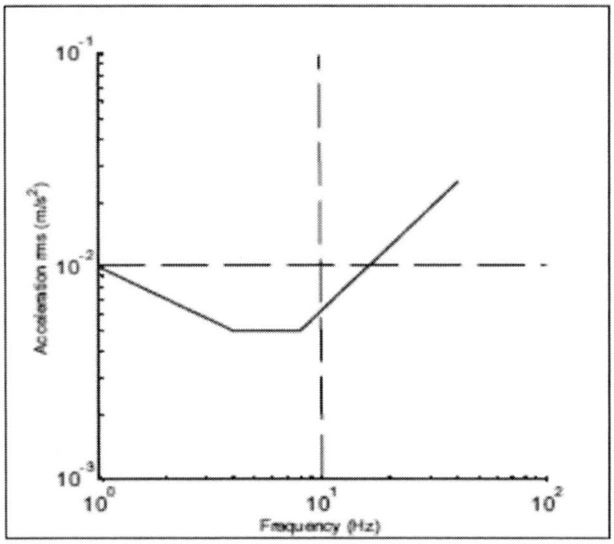

Figure 3: ISO base curve [9].

Table 4. Model of the first ten modes and their natural frequencies.

Mode Shape (ANSYS)	
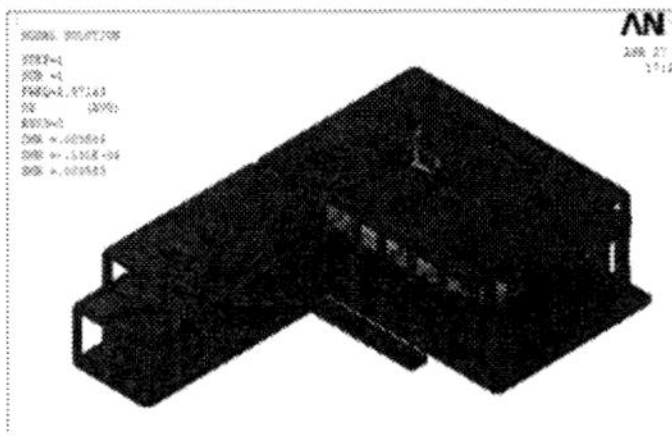	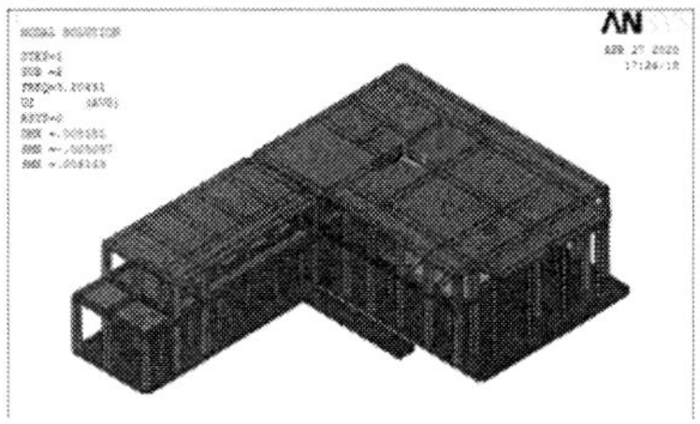
Mode 1, 2.57 Hz	Mode 2, 3.20 Hz

Mode 3, 3.36 Hz	Mode 4, 3.89 Hz
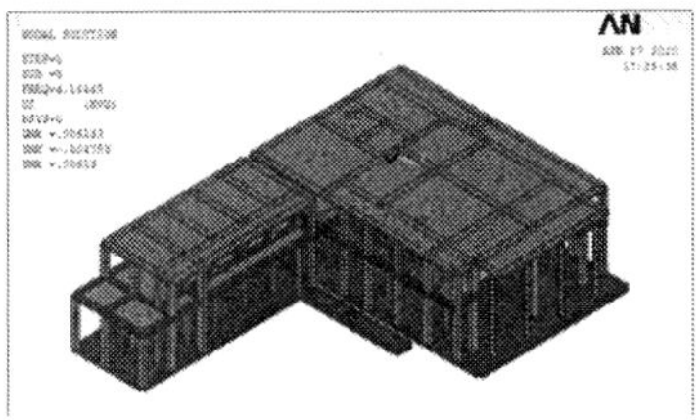	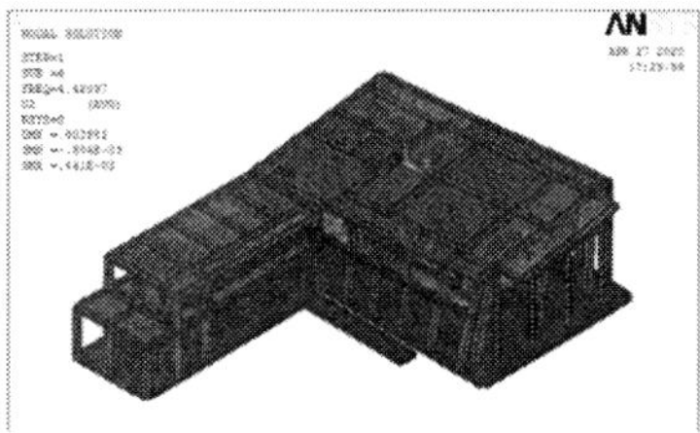
Mode 5, 4.14 Hz	Mode 6, 4.42 Hz
	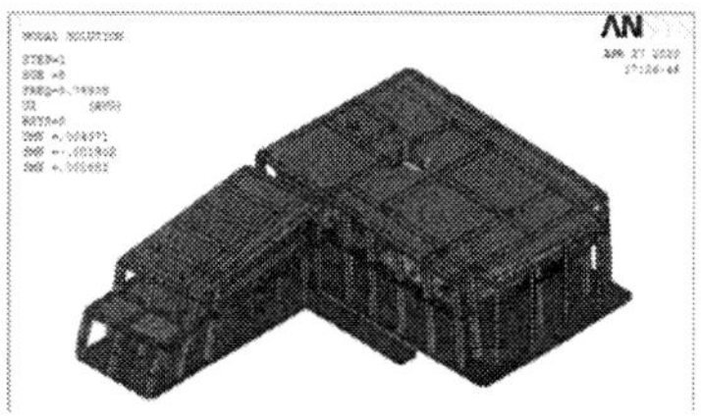
Mode 7, 5.75 Hz	Mode 8, 5.80 Hz
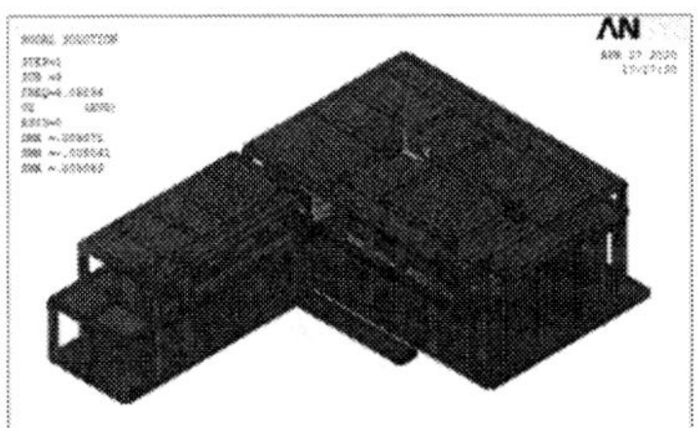	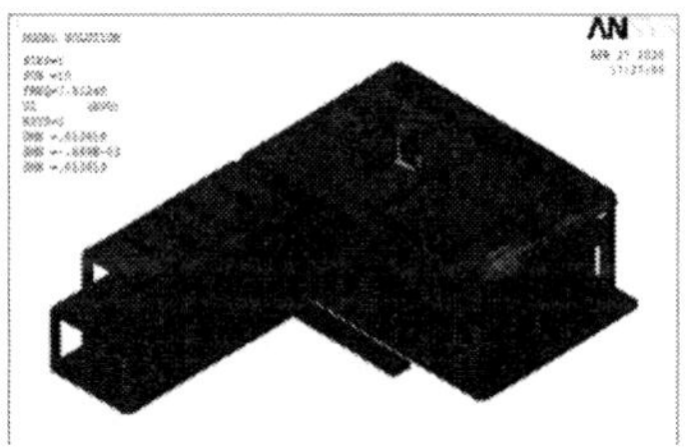
Mode 9, 6.89 Hz	Mode 10, 7.51 Hz

4.2. Result for transient analysis

Transient analysis, also known as time history analysis, was conducted using ANSYS software. This analysis was used to determine the dynamic response of the building under the time-dependent loads. The vibration input location in ANSYS is shown in figure 4. The results obtained from the transient analysis were greatly affected by the distance of the vibration input location in ANSYS and signal input obtained from iDynamics. The vibration input location in ANSYS was set in the middle of the building, while the vibration signal was obtained outside of the building (point 1 and point 2). The dynamic soil-structure interaction caused the vibration wave to be dissipated when transferring through the soil structure to the building structure [10]. This was shown in the result obtained in this analysis. The vibration waves received by the building structure decrease to 0 m/s^2 when it reaches the middle of the building. Figure 4 indicated the location of the Point 1 and Point 2 from the main road. From the results obtained, it can be concluded that the traffic-induced vibration which acts onto the building was negligible as the acceleration for the first floor was 0 m/s^2. Table 5 shows the result obtained from the transient analysis.

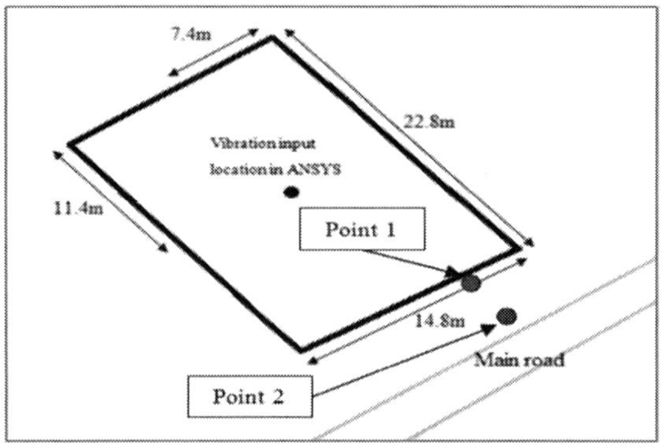

Figure 4. Vibration input location in ANSYS.

Table 5. Results of transient analysis for traffic-induced vibration.

Comparison between the vibration signal input and output results for traffic-induced vibration for point 1 and point 2 in the z-direction.		
	Input (from iDynamics)	Output (from ANSYS)
Point 1	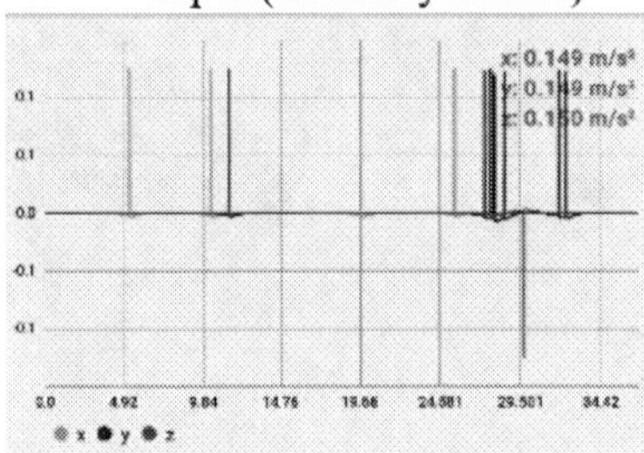	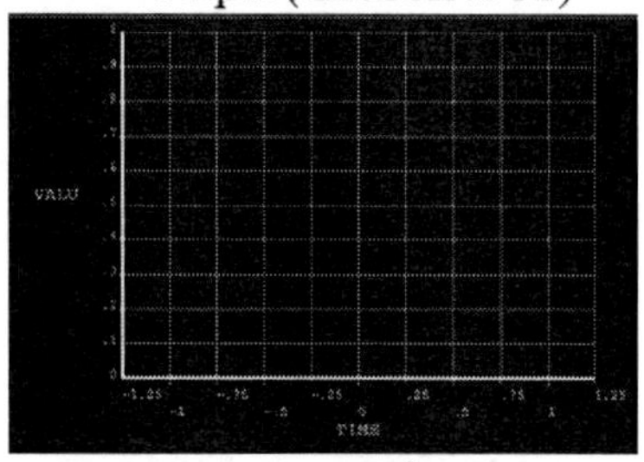
Acceleration	0.150 m/s^2	0 m/s^2
Point 2	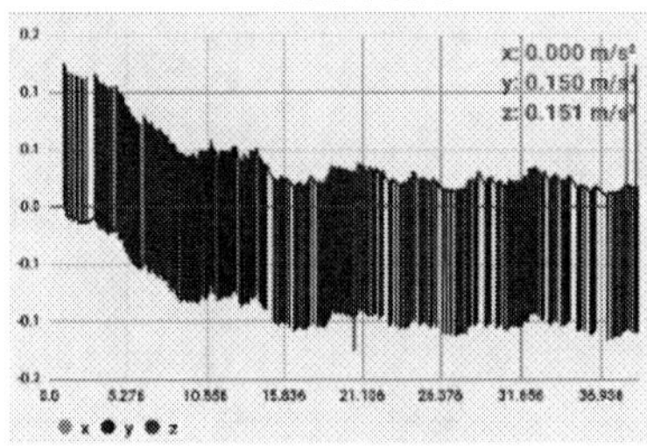	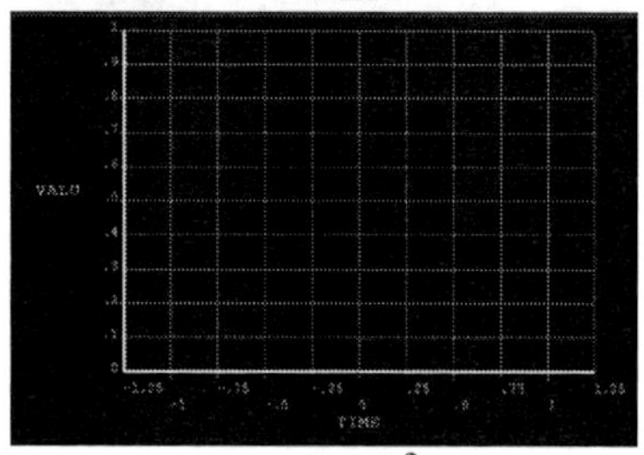
Acceleration	0.151 m/s^2	0 m/s^2

4.3. Result for vibration criteria analysis
Vibration criteria analysis was carried out using VSATs and ModalV in MATLAB. In this analysis, the vibration input location was the same as the transient analysis. The vibration criterion level of the building fell on curve VC-E for both of the analyses. Further discussion will be carried out as follows.

4.3.1. Vibration serviceability assessment tools (VSATs). By using this tool, the vibration criterion for the overall structure can be obtained. The level of vibration response is being characterized by the color pattern as shown in figure 5. The vibration criteria analysis carried out using VSATs in MATLAB shows that the first-floor structure was categorized under VC-E. According to Gordon (1991), a building structure under the VC-E category is assumed to be suitable for placing sensitive equipment and also other systems which require extra dynamic stability [6]. Hence, the traffic vibration around the old building of Telecom Muar did not give any significant effect on the building structure. Table 6 shows the results obtained from VSATs.

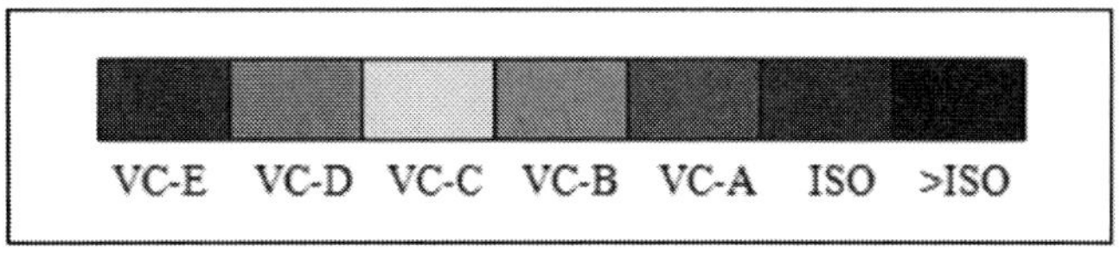

Figure 5. Characteristic of color pattern for vibration response for VC plot.

Table 6. Vibration criteria analysis for first-floor structure of the old building of Telecom Muar with vibration input from Point 1 and Point 2

Vibration input	Vibration response
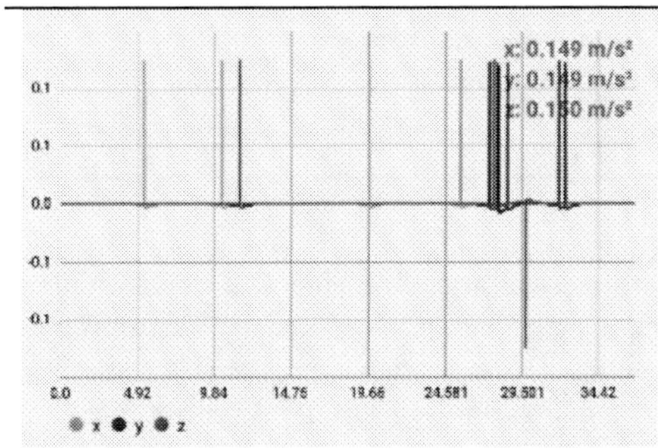	
Vibration input for Point 1	
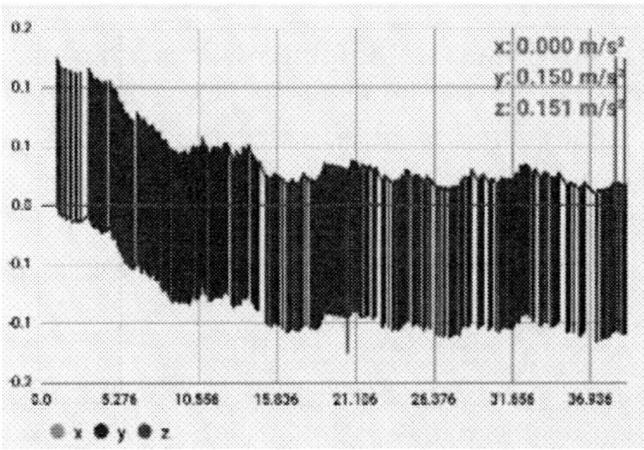	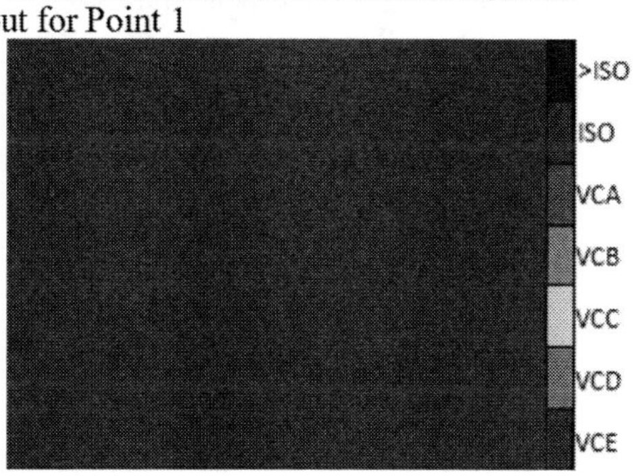
Vibration input for Point 2	

4.3.2. ModalV. The vibration criteria plot obtained from the analysis shows that the vibration level of the old building of Telecom Muar was 0 μm/s. It can be assumed that the vibration level of the building drops at the lowest level (VC-E) according to the generic vibration criteria plot implemented by Gordon [6]. Hence, it can be said that this building did not undergo any significant effect on traffic vibration that occurs around the building. Table 7 shows the result obtained from ModalV.

Table 7. Vibration criteria plot for the input data from Point 1 and Point 2

Input data	Vibration Criteria Plot
Point 1	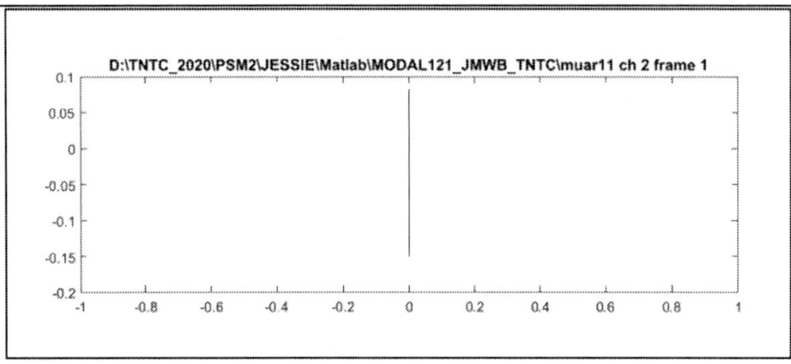
Point 2	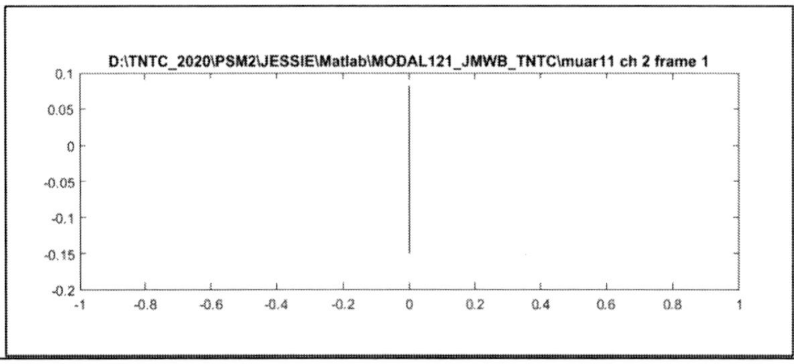

5. Conclusion

In conclusion, both objectives have been achieved by the researcher at the end of this study. Overall, the displacement of the floor structure for the first ten-mode shape can be accepted. The natural frequency of the building is acceptable as the frequency of the fundamental mode shape, 2.57Hz, falls below the range as stated in the ISO base curve. On the other hand, the output acceleration obtained from the transient analysis in ANSYS was 0 m/s^2. This result indicates that the traffic-induced vibration which acts onto the old building of Telecom Muar was negligible.

For the vibration criteria analysis, the level of vibration of the old building of Telecom Muar fell on VC-E, the lowest level in the Generic Vibration Criteria Plot implemented by Gordon [6]. This has shown that the traffic-induced vibration does not show any significant effect on the building structure of the old Telecom Muar building. Hence, it can be concluded that the old building of Telecom Muar was not affected by the traffic vibration that occurs around the building.

There are some suggestions provided as follows to enhance the quality of the research:

i. The accelerometer is recommended to be used to determine the traffic-induced vibration. In order to obtain the vibration signal more accurately compared to the mobile applications.
ii. Besides, further checking needs to be carried out to determine the accuracy of the vibration measurement using mobile applications.
iii. The road condition, type of vehicles passing and the distance from the main road to the building are also important factors to be included in the research.

iv. Lastly, the type of soil around the building should also be taken into consideration as the propagation of vibration waves will be affected by the type of soil.

6. References

[1] Chik T N T, Ahyer M Z, Yusoff N A, Ilias M E, Wan Ibrahim M W and Ghazali M I 2014 Vibration response on MiNT-SRC building due to ground-borne vibrations from humans using finite element modelling, *Applied Mechanics and Materials* **660** 536–540
[2] Hunaidi O 2000 Traffic vibrations in buildings, *Construction Technology Update* **39** 1–6
[3] Almselati A S I, Rahmat R A O K and Jaafar O 2011 An overview of urban transport in Malaysia, The *Social Sciences* **6** 24–33
[4] Lombaert G and Degrande G 2001 *Study on Determining Factors for Traffic Induced Vibrations in Buildings* Final Report BWM-2001-06, DWTC Research Programme Sustainable Mobility Research Project MC/01/040 (Leuven: Department of Civil Engineering, Katholieke Universiteit Leuven)
[5] Chik T N T, Yew T A, Ibrahim M H W and Yusoff N A 2019 Vibration criteria on floor at laboratory room *IOP Conference Series: Earth and Environmental Science* **220** 012021
[6] Gordon C G 1999 Generic vibration criteria for vibration sensitive equipment, *SPIE Conference on Current Development in Vibration Control* **3786** 22–33
[7] Wang E and Nelson T 2002 Structural dynamic capabilities of ANSYS *Proc ANSYS 2002 Conference* (Pennsylvania)
[8] Siddika A, Awall M R, Al Mamun M A and Humyra T 2019 Study on natural frequency of frame structures, *Journal of Computational Engineering and Physical Modeling* **2** 38–49
[9] Ljunggren F 2006 *Floor Vibration - Dynamic Properties and Subjective Perception* Doctoral Thesis (Lulea: Lulea University of Technology)
[10] François S, Pyl L, Masoumi H R and Degrande G 2007 The influence of dynamic soil-structure interaction on traffic-induced vibrations in buildings, *Soil Dynamics and Earthquake Engineering* **27** 655–674

Acknowledgement
Authors wishing to acknowledge assistance or encouragement from their colleague for valuable guidance, support and throughout this study.

Numerical Simulation of Modified Rubberized Concrete Block Under Impact Loads

T Y Pei[1], S N Mokhatar[2], N A N A Mutalib[3] and S J S Hakim[4]

[2,4] Structural Dynamics and Computational Engineering Focus Group (StrucDyCE), Faculty of Civil Engineering and Built Environment, Universiti Tun Hussein Onn Malaysia (UTHM), 86400 Parit Raja, Johor, Malaysia
[1,3] Faculty of Civil Engineering and Built Environment, Universiti Tun Hussein Onn Malaysia (UTHM), 86400 Parit Raja, Johor, Malaysia
Corresponding author: shahruln@uthm.edu.my

Abstract. Rubberized concrete was innovated by many researchers to enhance energy absorption under impact load and by reusing scrap tires. Thus, this research was aims to develop the numerical procedure using the Finite Element Method (FEM) to simulate modified rubberized concrete under impact loads and predict its energy absorption under different impact loads. Three existing constitutive models: Concrete Damage Plasticity (CDP), Drucker-Prager (DP), and Modified Drucker-Prager Cap (MDPC) available in ABAQUS software were used to replicate the rubberized concrete with 10% of Rice Hush Ash (RHA) as cement substitution and different percentages (0%, 5%, 10%, 15%, and 20%) of crumb rubber as sand replacement. All three models produced successful FEM results with reasonable modelling assumption, and the CDP model was more effective in simulating rubberized concrete under impact to predict energy absorption than DP and MDPC models. Further, it was concluded that crumb rubber could enhance the energy absorption of concrete. Generally, the energy absorption of the concrete increased as the crumb rubber increase. However, the strength decreased as the crumb rubber increased, but 10% of RHA in concrete mix can maintain the concrete strength. Overall, this study reveals that FEM incorporated with the CDP model is able to predict the impact response of modified crumb rubber as an application of concrete road barrier.

1. Introduction

Concrete, a human-made material that is the most frequently used on the earth. Concrete is designed to withstand impact loads, but its energy absorption is not perfect for a structure requiring high energy absorption capacity. For example, a concrete road barrier with high capacity energy absorption can reduce or minimize impact force from uncontrolled vehicles to protect the passengers in the car [1]. On the other hand, waste tire disposal becomes a significant issue to be considered in the world due to the vehicle industry's development. There are around 1.6 billion new tires generates per year and 1 billion waste tires each year worldwide. The waste tire is a great volume and a lot of void space that require valuable landfill space and need more than a hundred years to decompose. Hence, rubberized concrete that replaces sand aggregate with crumb rubber was produced, which can reutilizing scrap tires and enhance energy absorption under impact load at the same time. A more recent study investigated rubberized concrete consisting of fine aggregate with crumb rubber has similar workability to normal concrete. The rubber enhances the flexural tensile strength, ductility, and damping ratio of concrete due to brittle concrete [2].

The impact resistance and energy dissipation of rubberized concrete also increased compared with normal concrete. However, it will reduce the concrete's density and compressive strength because the rubber is low bonding adhesion. Besides, the crumb rubber was more effective in increasing the strength of concrete than other forms of rubber, and the percentage of rubber replaced also affects the strength of concrete [3]. This research aims to develop the numerical procedure using the finite element method to effectively simulate modified rubberized concrete under impact loads. Rubberized concrete blocks with 10% of Rice Hush Ash (RHA) for cement substitution and crumb rubber replacement volume of sand with different percentages (0, 5, 10, 15, and 20%) were successfully modelled. The technology keeps moving, and numerical techniques became a popular method rather than experimental approaches. Finite element analysis (FEA) is the modelling of products and systems by setting up a virtual environment or condition to predict how the product reacts to the real-world forces, vibration, heat, fluid flow, and other physical effects. FEA is the application of the FEM to solve real engineering problems. This numerical simulation tool has good performance in time, energy, and cost-effectiveness.

The finite element model will connect these system points that are nodes to form the design's shape. However, the accuracy of the result is based on the model mesh, material descriptions, and structural properties [4]. Therefore, this study focuses on the material descriptions of Drucker-Prager (DP), Concrete Damage Plasticity (CDP), and Modified Drucker-Prager Cap (MDPC). The rubberized concrete mixes the characteristics of concrete and rubber particles. So, the material described by DP and MDPC for ductility of the crumb rubber, and CDP for the brittle cracking of concrete. The crumb rubber has a good performance to dissipate energy due to its elasticity characteristic compared with sand, gravel, and cement. Therefore, the rubberized concrete increases the energy absorption of concrete under impact load by reducing stress concentration and the brittleness of concrete, restraining or postponing the occurrence as well as the development of cracks [5]. Concrete with high energy absorption can apply in structure require high energy absorption such as road barrier to reduce the injuries and deaths during the accident.

2. Numerical modelling

In this study, the numerical simulation of modified rubberized concrete under the impact loading was modelled using ABAQUS software. The practical work concerning the energy absorption of the modified rubberized concrete block under low-velocity impact loads was carried out by other researchers at the Universiti Tun Hussein Onn Malaysia. The size of the test specimen was 300 mm × 300 mm × 200 mm. A cylindrical projectile circular cross-section with a diameter of 40mm, a length of 800mm, and a weight of 80kg are used as an impactor of the test. The steel projectile was imposed on concrete specimens from a height of 1.6m.

2.1. Geometrical design of model

A three-dimension deformable model for rubberized concrete and steel projectile were constructed for analysis in ABAQUS. The 8-noded brick continuum elements (C3D8R) were created for the modified rubberized concrete. The element mesh with C3D8R and by symmetry revolved 360° to produce the cylindrical shape created for steel projectile to prescribe impact load.

2.2. Material property of model

The rubberized concrete element adopting three different material descriptions where are DP, MDPC and CDP. Table 1 shown the material properties of rubberized concrete with varying percentages of crumb rubber. In contrast, other parameter data and materials properties of steel projectile were taken from the previous study [6] shown in tables 2 and 3.

Table 1. Material properties of rubberized concrete.

RHA (%)	Crumb Rubber (%)	Mass Density, ρ_{rc} (kg/m^3)	Compression Strength, f_{cu} (N/mm^2)	Tensile Strength, f_{st} (N/mm^2)	Young's Modulus, E (N/m^2)	Poisson's Ration, ν_c
0	0	2156	40.67	3.03	3.17E+6	0.20
10	5	2137	30.33	2.75	2.75E+6	0.17
10	10	2114	26.00	1.93	2.49E+6	0.16
10	15	2100	22.67	1.84	2.37E+6	0.15
10	20	2057	17.67	1.46	2.08 E+6	0.14

Table 2. Parameter of DP, MDP, and CDP [6].

CDP model parameters					
Dilation angle	Eccentricity, ε	σ_{bo}/σ_{co}	K_c	Viscosity parameter,μ	
38°	1	1.12	1	0.666	
DP model parameters					
Angle of friction, β		Flow stress ratio, K		Dilation angle, ψ	
30°		1		20°	
MDPC model parameters					
Material cohesion, *d* [N/m2]	Material angle of friction, *β*	Cap eccentricity parameter, *R*	Initial cap yield surface position	Flow stress ratio, *K*	Strain rate
4.71E+00	51°	0.65	0.0011	1	1.5

Table 3. Material properties of steel projectile [6].

Young's modulus, E_s [N/m2]	Poisson's ratio, ν_s	Density, ρ_s [kg/m3]	Yield stress, f_y [N/m2]	Ultimate stress, f_u [N/m2]
2.10E+11	0.29	7800	5.60E+08	6.30E+08

2.3. Initial and boundary conditions and load

The initial velocity assigns to the penetration direction to simulate rubberized concrete's motion under impact loading. The boundary condition is assigned to the bottom node of the rubberized concrete block to avoid displacement or rotation, as figure 1.

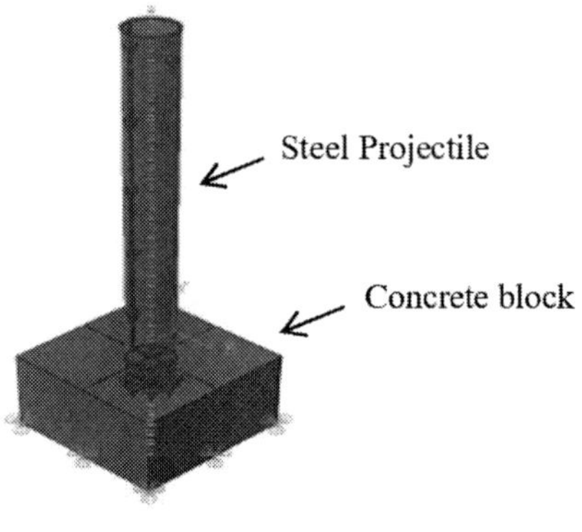

Figure 1. Loading and boundary condition of modelling.

2.4. Projectile simulation

All the steel projectile element nodes set an initial velocity of 4.0 m/s in a direction perpendicular to the surface rubberized concrete model. This value is given from experimental work carried out by other researchers in UTHM to compare the numerical and experimental results. In addition, the initial velocity of 5.0 m/s and 6.0m/s is also applied in this study as a projection to predict the energy absorption of modified rubberized concrete under different impact loads.

2.5. *Energy absorption*

The energy absorption was produced by the sudden collision between the rubberized concrete and projectile. The formula used to calculate the energy absorption is shown in equation (1). The energy absorption of FEA calculation by using formula due to the FEA cannot directly provide the output of energy absorption. Therefore, the absorption FEA and experiment computed using the same formula can provide a better validation.

$$E_a = m.g.X \tag{1}$$

Where E_a is the energy absorption of rubberized concrete under the impact load, m is the mass of the impactor, which is 80kg, g is the acceleration due to gravity at 9.81m/s^2, and X as the penetration depth of rubberized concrete after impact.

3. Results and discussion

3.1. *Mesh density sensitivity*

The mesh sensitivity test aims to ensure the result of the FEA is more sensibly accurate result. This mesh sensitivity test was only obtained for the concrete (0% RHA + 0% crumb rubber and simulate by using CDP model. The result of the FEA investigation utilizing various mesh densities for the rubberized concrete control sample is shown in table 4 and figure 2. It is indicated that the outcome of the FEA has been dramatically affected by the global size or the total number of elements. However, after the global size of 15, the penetration depth and calculated energy absorption do not change by an appreciable amount. Based on Kamal et al. [7], the processing time and storage required increased as the total element increased. Figure 2 shows that energy absorption of rubberized concrete increases as the total element increased and maintains consistently after the total element reaches 5200. As per Kamal et al. [7], the purpose of the mesh sensitivity study was to ensure the accuracy of the result and save the computational cost required at the same time. Therefore, the total element increased after 5200 was not required due to without much increase of the accuracy of prediction energy absorption. The global size of 15 is selected for an appropriate meshing size of all specimen models.

Table 4. Result of mesh refinement with energy absorption within CDP region.

Mesh density	Total elements	Penetration Depth of CDP model (mm)	E_a(J)
GB35	486	2.61	2.05
GB30	700	2.93	2.30
GB25	1152	2.83	2.22
GB20	2400	3.18	2.50
GB17	3888	3.35	2.63
GB15	5200	3.59	2.82
GB 13	8280	3.62	2.84

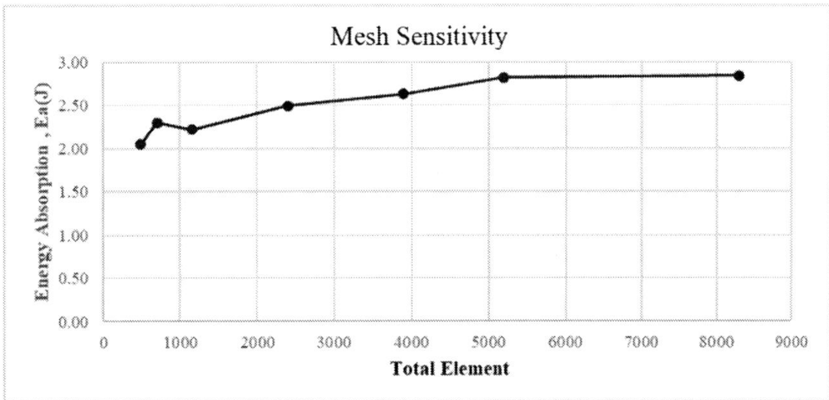

Figure 2. Result of mesh refinement for energy absorption within CDP region.

3.2. Comparison of three material descriptions

In this study, three different material description was used to simulate rubberized concrete. The energy absorption between the three models and experiment results, as well as its discrepancy, are shown in table 5. The average percentage discrepancies of all models are less than 20%, which is acceptable. The lowest average percentage discrepancy is CDP about 8.80%. The average percentage discrepancies of the DP and MDPC model were 13.71% and 19.94%, respectively. Figure 3 was presented the penetration depth of the three models compared to the experimental result. Although the DP model displays a good prediction of penetration depth, the penetration depth of CDP closer to an experimental result is more suitable for low-velocity impact [8]. Besides, MDPC obtained the highest percentage discrepancy for rubberized concrete block's energy absorption compared to the experimental result because the MDPC model has a better prediction interface between concrete and reinforced steel [9].

Table 5. Energy absorption of three model and experiment.

		Energy Absorption, E_a(J)				**Discrepancy %**		
RHA (%)	Crumb Rubber (%)	Experiment	CDP	DP	MDPC	CDP	DP	MDPC
0	0	2.68	2.82	2.50	3.53	5.28	6.45	31.96
10	5	5.40	4.84	4.54	5.05	10.32	15.84	6.40
10	10	6.95	6.28	5.64	8.44	9.60	18.82	21.47
					Average	8.80	13.71	19.94

Figure 4 displays the energy absorption of three models compared to experimental. The energy absorption for the CDP model is more nearly closed to the experimental result where has a better prediction. The difference in numerical accuracy of the three models is also contributed by the constitutive law of these models. The CDP modeling is based on the two main failure mechanisms as mentioned above, which means that the failure of the evolution of yield surface is controlled by two hardening variables. However, the DP model and MDPC model are constructed based on only one failure mechanism in tensile cracking and compressive crushing. The parameter of MDPC has an addition of a cap-shaped yield surface compared to the DP model. These reasons affect the overall performance of the three models and give the numerical accuracy differently. Further, concrete behaviour, including plastic behaviour, compressive behaviour, tensile behaviour, confinement, and damage mechanism, can well be replicated by using the CDP model. As a result, there was no surprise that the CDP has a better correlation with experimental than other models [10]. Since the CDP model presents the best prediction result to the experiment result, therefore is an appropriate material model to predict the energy absorption of modified rubberized concrete under different impact loads.

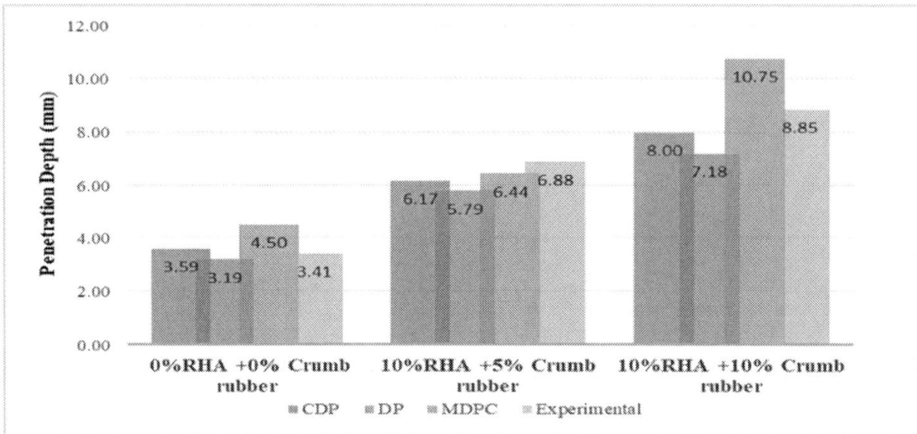

Figure 3. The penetration depth of rubberized concrete.

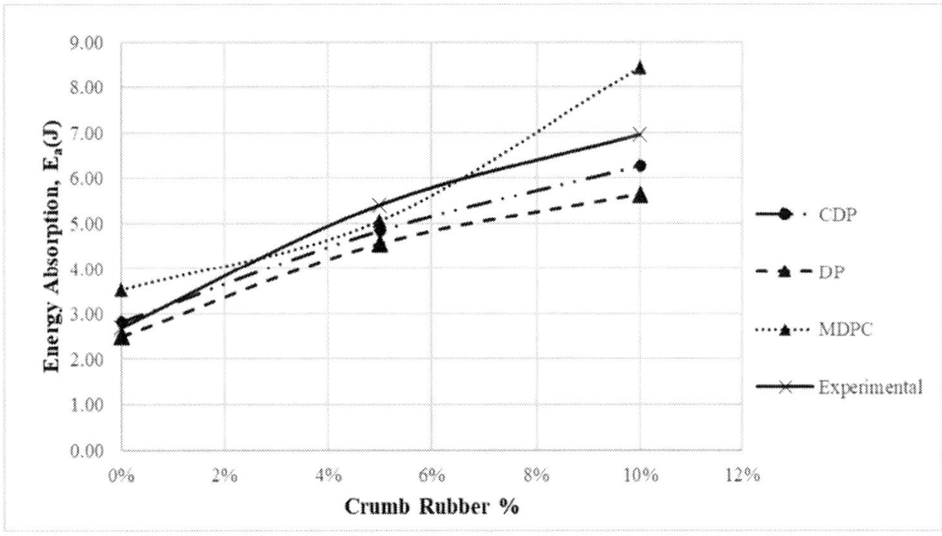

Figure 4. Energy absorption of rubberized concrete.

3.3. Failure of rubberized concrete

The failure mode of rubberized concrete from the FEM model and experimental are presented in figure 5. In addition, the numerical model exhibits reasonably close to that demonstrated by the experimental work [11]. It can be observed that the general outline of damage after the impact of the numerical model and experimental are in close agreement. Therefore, the FEM models have a good prediction of the failure mode of the rubberized concrete. On the other hand, the crater that produced under constant impact velocity was wider when used rubber as the sand replacement and RHA as cement replacement. The growth of the crater width as the percentage of rubber increases.

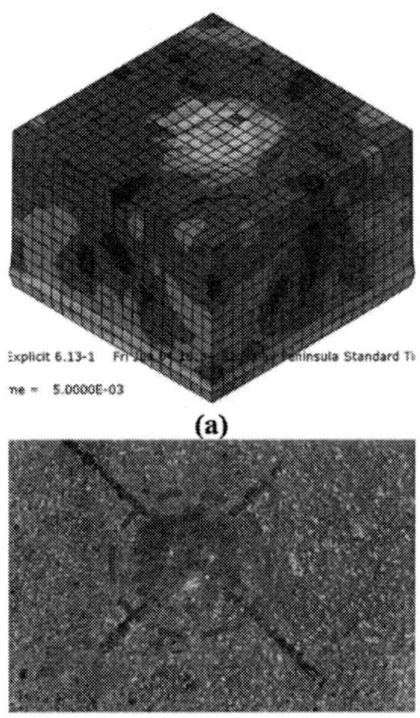

(a)

(b)

Figure 5. Result for rubberized concrete with 10% crumb rubber and 10% RHA after impact test (a) FEM model and (b) Experimental

3.4 Rubberized concrete under different impact loads

From the previous section, the CDP model has better accuracy under impact. Therefore, the CDP model is used in the rubberized concrete block under various impact velocities to predict the penetration depth and energy absorption. The result was summarized in table 6. From the table, the penetration depth of rubberized concrete was increased as the percentage of crumb rubber are also increase. This is due to the fracture of rubberized concrete under impact is softer than concrete without rubber. The rubber particles as the sand replacement can eliminate the stress concentration of void and delay cracks' occurrence and development [12]. Rubberized concrete can burden more deformation compared to normal concrete. As a result, the deformation of rubberized concrete increases as the percentage of crumb rubber increases.

On the other hand, the higher the percentage of crumb rubber, the higher the energy absorption of rubberized concrete, as shown in table 6. The rubberized concrete with 10% RHA and 20% rubber presented the highest energy absorption. The flexibility of concrete is increased by the addition of rubber and allow rubberized concrete to resist more impact, which means absorb more energy [13]. As mentions above, the rubber can fill the void of the concrete with the help of cementation to increase the density of rubberized concrete. The deceleration of impact increase as the plastic density increase [14]. Therefore, rubberized concrete can absorb or burden more energy compared to normal concrete, and the rubberized concrete with 20% rubber presented as optimum impact resistance and energy absorption [15].

However, the addition of rubber causes the compressive strength and brittleness of concrete to decrease. It can directly give an effect on the ability of energy absorption [16]. The addition of RHA plays a vital role to minimize the reduction of compressive strength and brittleness of rubberized concrete. The penetration depth of rubberized concrete increases as the rubber percentage increase, which the rubberized concrete does not lose too much strength under the help of RHA and has a good performance in energy absorption. Figure 6 was presented the energy absorption of rubberized concrete under different impact velocities. The energy absorption of rubberized concrete also increases as the percentage of crumb rubber increase under various velocities. The higher the impact velocity, the higher of energy absorption. The higher of velocity collide between rubberized concrete and projectiles steel, the higher kinetic energy produced. Therefore, more energy transfers to rubberized concrete for energy absorption.

Table 6. FEM result of rubberized concrete under different impact loads.

Velocity (m/s)	RHA (%)	Rubber (%)	Penetration Depth (mm)	E_a(J)
4	0	0	3.59	2.82
	10	5	6.17	4.84
	10	10	8.00	6.28
	10	15	8.05	6.32
	10	20	9.07	7.12
5	0	0	5.56	4.36
	10	5	9.52	7.47
	10	10	12.31	9.66
	10	15	12.33	9.68
	10	20	13.85	10.87
6	0	0	8.43	6.62
	10	5	13.65	10.71
	10	10	17.02	13.36
	10	15	17.03	13.37
	10	20	18.61	14.61

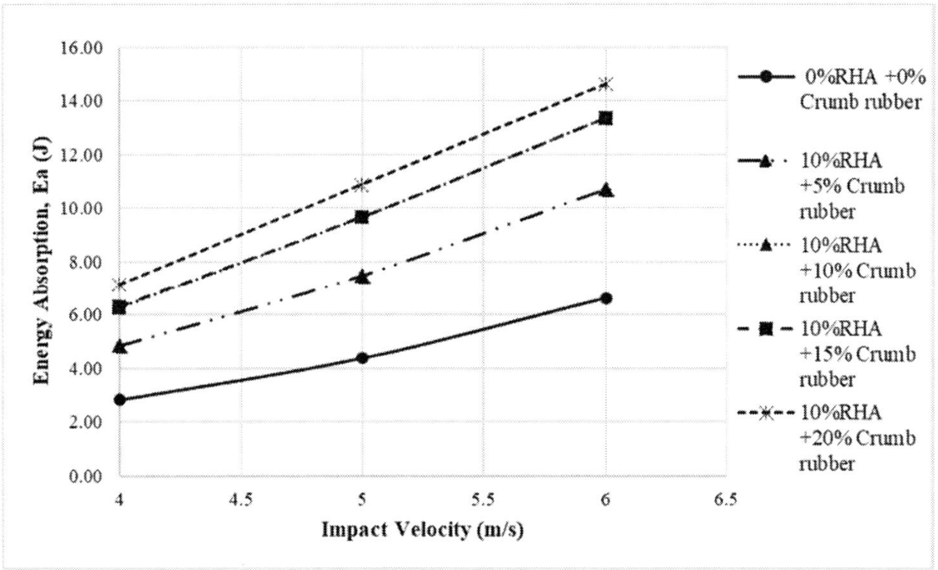

Figure 6. Energy absorption of rubberized concrete under different impact velocities.

4. Conclusion

An impact load test of rubberized concrete was successfully simulated by adopting three material descriptions, including DP, CDP, and MDPC. The numerical procedure using the FEA to effectively simulate modified rubberized concrete under impact loads was investigated accordingly. The CDP model produced a high accuracy result of FEA modelling compared with DP and MP. The average discrepancy percentage of the CDP model was 8.80% which is lower than DP and MDPC with 13.71% and 19.94%, respectively. The failure profile of rubberized concrete in the model is almost similar to the experiment's observation. Further, the growth of the crater widths is affected by the increment percentage of rubber. The CDP model has presented a better correlation with experimental compared to DP and MDPC models. Therefore, the CDP model was more suitable to replicate the rubberized concrete modelling under low-impact force. It can be concluded that crumb rubber in concrete generates a good result for energy absorption.

The elastic behaviour of concrete increased by the addition of rubber to resist more impact, which means absorb more energy. Besides, 10% of RHA for cement substitution was helpful for rubberized concrete to minimize the reduction of compressive strength and brittleness mechanism. More investigation of rubberized concrete under dynamic load is recommended to explore the dynamic response of rubberized concrete as a road barrier to reduce the injuries and deaths during road accident.

5. References

[1] Pham T M, Elchalakani M, Karrech A and Hao H 2019 Axial impact behavior and energy absorption of rubberized concrete with/without fiber-reinforced polymer confinement, *International Journal of Protective Structures.* **10** 154-173

[2] Senin M S, Shahidan S, Abdullah S R, Guntor N A and Leman A S 2017 A review on the suitability of rubberized concrete for concrete bridge decks *IOP Conf. Series: Materials Science and Engineering* **271** 012074

[3] Khatib Z K and Bayomy F M 1999 Rubberized Portland cement concrete, *Journal of materials in Civil Engineering* **11** 206-213

[4] Raveendra B R, Benipal G S and Singh A K 2005 Constitutive modelling of concrete: an overview, *Asian Journal of Civil Engineering* **6** 214-246

[5] Strukar K, Sipos T K, Milicevic I and Busic R 2019 Potential use of rubber as aggregate in structural reinforced concrete element-A Review, *Engineering Structures* **188** 452-468

[6] Mokhatar S N, Abdullah R, and Kueh A B 2013 Computational impact responses of reinforced concrete slabs, *Computers and concrete* **12** 37-51
[7] Kamal M, Rahman M M and Sani M S M 2013 Fatigue life prediction using simplified endurance function model, *Advances in Mechanical Engineering* **2013** 1-12
[8] Oucif C, Rama J K, Ram K S and Abed F 2011 Damage modelling of ballistic penetration and impact behavior of concrete panel under low and high velocities, Defence *Technology* **17** 202-211
[9] Mohamad M E, Ibrahim I S and Abdullah R 2015 Finite element modelling of interface shear strength at concrete-to-concrete bond *9th Asia Pacific Structural Engineering and Construction Conference (APSEC) & 8th Asean Civil Engineering Conference (ACEC)* (Kuala Lumpur, Malaysia)
[10] Raza A and Ahmad A 2019 Numerical investigation of load-carrying capacity of GFRP-reinforced rectangular concrete members using CDP model in ABAQUS *Advances in Civil Engineering* **2019** 1-21
[11] Tahmasebinia F 2008 *Numerical modelling of reinforced concrete slabs subject to impact loading* Master Dissertation (New South Wales: Wollongong University)
[12] Liu F, Chen G X and Li L J 2013 Dynamic behavior of crumb rubber concrete subjected to repeated impacts *The 2013 World Congress on Advance in Structural Engineering and Mechanics (ASEM13)* (Jeju) 2423-2434
[13] Khalil E, Abd-Elmohsen M, and Anwar A M 2015 Impact resistance of rubberized self-compacting concrete, *Water Science* **29** 45-53
[14] Jones M R and Zheng L 2013 Energy absorption of foamed concrete from low-velocity impacts, *Magazine of Concrete Research* **65** 209-219
[15] Ocholi A, Sanni M Y and Ejeh S P 2018 The impact resistance effect of partially replacing coarse aggregate with ground-rubber aggregate in concrete, *Nigerian Journal of Technology* **37** 330-337
[16] Sadrmomtazi A and Zanoosh R Z 2016 Concrete containing waste rubber particles under impact loading, *Bulletin de la Société Royale des Sciences de Liège* **85** 1328-1336

Acknowledgement
The authors would like to thank you University Tun Hussein Onn Malaysia (UTHM) for the financial support through GPPS Code H625.

Effects of Section Properties on Structural Behaviour and Failure Mode of Built-Up CFS Columns

W H Cheah[1], Z M Jaini[1], S N Mokhatar[1], S J Seyed Hakim[1], M Abu Seman[2] and M Kozlowski[2]

[1] Faculty of Civil Engineering and Built Environment, Universiti Tun Hussein Onn Malaysia, 86400 Johor, Malaysia
[2] Faculty of Civil Engineering Technology, Universiti Malaysia Pahang, 26300 Pahang, Malaysia
[3] Faculty of Civil Engineering, Silesian University of Technology, 44-100 Gliwice, Poland
Corresponding author: rizuan@uthm.edu.my

Abstract. Built-up CFS column is a type of structure that can be classified as an industrialized building system. This column has been widely used in the construction industry. It has relatively lightweight, easy to fabricate and provide efficient installation, thus suitable for the construction with difficulty inaccessibility. However, the main issue that arises from the built-up CFS column is its bearing capacity. The ultimate strength and displacement of the built-up CFS column are prominently governed by its section properties such as size and thickness. Therefore, this study intends to investigate the effects of section properties on the structural behaviour and failure mode of built-up CFS columns. The built-up CFS column was modelled in the three-dimensional using WELSIM, taking into account the nonlinearities of geometry, material criterion and contact surface. It was found that the built-up CFS column attains ultimate strength of 53.33 kN to 210.6 kN and displacement around 1.33 mm to 2.98 mm. When the size and thickness of square hollow sections are increased, the ultimate strength increases simultaneously but the displacement shows a decrement trend. Under compression force, it was observed that the built-up CFS column suffers distortional and flexural buckling as well as connector and stiffener failures.

1. Introduction

CFS members are made of cold-formed steel and commonly be applied to the structures with short span such as joists and roof trusses for commercial and industrial buildings [1]. CFS members that fabricated from more than one section and joined together to form a global structure is known as built-up members. The advantages of built-up members are that it creates symmetrical section properties, greater strength and resistance towards out-of-plane movement. Generally, built-up members with the desire size and thickness will be fastened by using stiffener (such as lacings and battens) and connector (such as bolts, screws and rivets) or welds to produce the new assembled shape [2]. Zhang et al. [3] stated that built-up members do offer one of the highest load-bearing capacity-to-weight ratios among several other structural components. The material itself is very workable, hence many shapes can be fabricated easily [4]. One of the famous built-up members is known as the built-up CFS column. This type of structure can be classified as an industrialised building system (IBS).

Built-up CFS column is relatively lightweight, easy to fabricated and provides efficient installation, thus suitable for the construction of building at the site with difficulty inaccessibility. Currently, the built-up CFS column has become popular in the construction industry due to its cost and sustainability. Despite that, bearing capacity still becomes the main issue that arises from the built-up CFS column. Moreover, CFS members are susceptible to buckling when applied as flexural and compression members [5]. The thin wall of CFS members will undergo local and distortional buckling, hence this issue must be taken seriously in the design stage. For the built-up CFS column, its ultimate strength and displacement are prominently governed by the section properties such as size and thickness. Other than that, fastener flexibility might affect stability. Although there are many previous studies have been done in inspecting the performance of built-up CFS column, the knowledge on the effects of section properties is still limited. Therefore, this study intends to investigate the structural behaviour and failure mode of built-up CFS columns concerning different section properties.

2. Built-up CFS column

Built-up CFS column consists of two or more parallel metallic sections and can be interconnected by lacings or battens. This type of column is widely used to accommodate its function as a primary structure. However, the built-up CFS column suffers from buckling and instability. It may be subjected to local and torsional buckling due to its thin-walled nature [6]. The buckling resistance of the built-up CFS column can be eliminated by the design modification. Young & Chen [7] suggested that web stiffeners can be added to enhance buckling resistance. Currently, there are two basic design methods for built-up CFS column, namely; the effective width method and the direct strength method. These methods are proposed by American Iron Steel Institute (AISI). However, only the effective width method is formally available for CFS members. The effective width method specifies the modified slenderness ratio to take the effect of shear force into account [8]. Meanwhile, the direct strength method incorporates many buckling phenomena into the thin-walled members where the buckling strength can be derived in a function of theoretical elastic buckling load [9]. The specifications for the design of CFS members such as provided by AISI S-100 and EC3 have become the main standard worldwide.

It was revealed that these codes of practices have a very good agreement regarding built-up members subjected to compression load [10]. However, EC3 does have limitations due to the inexistent interaction between the local and global buckling for the built-up CFS column. Besides, the section properties are vital in determining the structural behaviour. Section properties include the thickness, size and shape. Whittle & Ramseyer [11] conducted an experimental study on built-up CFS columns where three different thicknesses of cee channel were adopted. It was revealed that when the thickness of the cee channel is increased, the built-up CFS column can sustain an exceptional axial load and portray an improvement in strength [11]. Musa [12] investigated the structural behaviour of stiffened built-up CFS column that constructed using a combination of square hollow sections and steel plate. It was identified that the thickness of square hollow sections plays a vital role in governing the load-carrying capacity and displacement. In terms of failure mode, Kyvelou [13] found that section instability plays a paramount role. The increase in the slenderness ratio causes the ultimate strength to decrease uniformly [14]. Long and medium columns are prone to flexural buckling about the y-axis, where local buckling occurs when subjected to concentric axial compression for short columns [15].

3. Design and specifications

The design and specifications of the built-up CFS column are based on an experimental study conducted by Musa [12]. Built-up CFS column consists of three main components, namely square hollow sections as vertical members, steel plate as battens and rivets as connectors. These components are assembled to form the built-up CFS column. The schematic design of the built-up CFS column is shown in Figure 1. Since the section properties become the subject of interest, the detailed section properties of the built-up CFS column is provided in Table 1. Overall, the built-up CFS column has a height of 500 mm where the square hollow sections have the dimension of SHS 19x19, SHS 25x25 and SHS 32x32 with thicknesses ranging from 0.72 mm to 2.2 mm. Meanwhile, the battens have a size of 50 mm depth, 125 mm length and 1.0 mm thickness. The rivets are 4.8 mm in diameter and length of 7 mm.

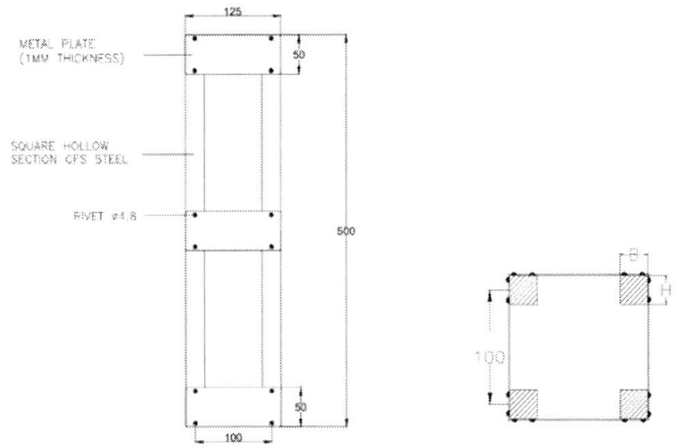

(a) (b)

Figure 1. Schematic design of built-up CFS column, (a) Elevation view, and (b) Plan view.

Table 1. Section properties of built-up CFS column.

Code	Section Properties	
	Size, *B* (mm) × *H* (mm)	Thickness, *t* (mm)
S1T1		0.72
S1T2		1.00
S1T3	19 × 19	1.20
S1T4		1.90
S1T5		2.20
S2T1		0.72
S2T2		1.00
S2T3	25 × 25	1.20
S2T4		1.90
S2T5		2.20
S3T1		0.72
S3T2		1.00
S3T3	32 × 32	1.20
S3T4		1.90
S3T5		2.20

4. Finite element modelling

The built-up CFS column was modelled and simulated using WELSIM, a finite element program. The numerical modelling is based on an explicit analysis considering the nonlinear plasticity. Therefore, the properties such as geometry, material criterion, constraint, loading and contact surface must be defined properly. Built-up CFS column was modelled in three-dimensional and discretized using tetrahedral elements with unstructured mesh. Figure 2 shows the built-up CFS column in the numerical modelling. The size of the mesh was automatically determined by WELSIM using an adaptive refinement method. It was identified that the size of mesh around 10 mm to 50 mm can produce convincing results. On the other hand, the material criterion of the built-up CFS column was derived from a combination of mechanical and hardening properties as tabulated in Table 2 and Table 3. This material criterion ensures that the plastic deformation can be achieved by the built-up CFS column and to avoid the fail-to-convergence condition. Hardening properties in term of multilinear kinematic hardening was used as proposed by Dewangan et al. [16] and Lu et al. [17].

The constraint is specifically used to represent the support and to control the translational. In the numerical modelling, the bottom part of the built-up CFS column was defined as fixed. Therefore, the constraint become true in all x, y and x directions. This restricts the horizontal and vertical movements as well as induces the reaction forces. On the other hand, the top part of the built-up CFS column was freed. The loading that shall be imposed to the built-up CFS column must be able to imitate the compression force. Therefore, a concentrated applied force in the downward direction was applied at the top part of the built-up CFS column. The pattern of compression force is a force-time history at the speed rate of 4.17 kN/minute with 15 sub-steps within a second. Meanwhile, the bond mechanism between battens to square hollow sections as well as rivets to battens and square hollow sections must be provided using a contact surface. In the numerical modelling, the perfect bonded was employed as a contact surface. It is requiring the definition of master and target geometries. Thus, the body of rivets was selected as the master geometries while the interface of battens to square hollow sections was defined as target geometries. In this condition, no sliding and separation between geometries are allowed.

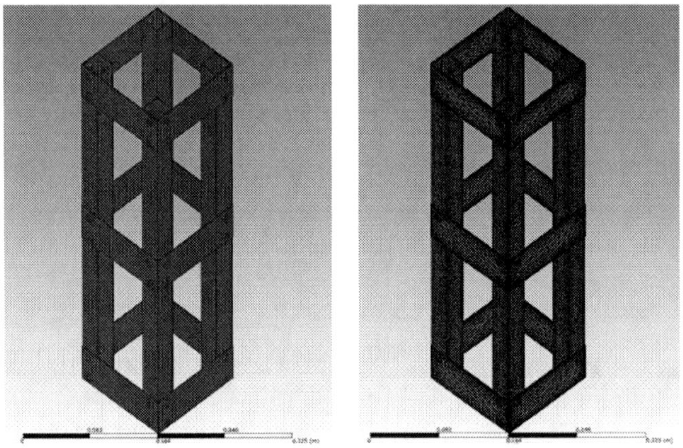
(a) (b)

Figure 2. The geometry of built-up CFS column, (a) Three-dimensional model, and (b) Discretization using tetrahedral elements.

Table 2. Mechanical properties.

Mechanical Properties	Value
Density, ρ (kg/m^3)	7850
Young's Modulus, E (GPa)	186
Poisson's Ratio, ν	0.29
Bulk Modulus, G (GPa)	148
Shear Modulus, S (GPa)	72
Yield Strength, σ_y (MPa)	350
Ultimate Strength, σ_u (MPa)	420

Table 3. Hardening properties.

Plastic Strain, ε_p	True Stress, σ_t (MPa)
0.0000	0
0.0029	350.46
0.0505	350.46
0.1078	395.37
0.1928	420.75
0.2208	420.75

5. Results and discussion

Figure 3 shows a comparison of strength-displacement profile and stress-strain curve between the numerical modelling and the experimental study. For the strength-displacement profile, the numerical modelling does provide a similar curve when compared with the experimental study. The ultimate strength obtained from the numerical modelling is lower than the experimental study. The percentage difference is around 7.75%. Likewise, the displacement obtained from the numerical modelling is slightly lower than in the experimental study. The percentage difference is about 2.64%. Based on this assessment, it can be upheld that numerical modelling can provide very convincing results. On the other hand, the stress-strain curve from the numerical modelling has a similar path as the multilinear kinematic hardening. The stress-strain curve from the numerical modelling reaches the elastic-plastic region and strain hardening, but unable to form softening phase. This indicates that the built-up CFS column undergoes failure when it reaches the yield stress. As the stress-strain curve has a negative gradient of tangent modulus, the numerical modelling is terminated.

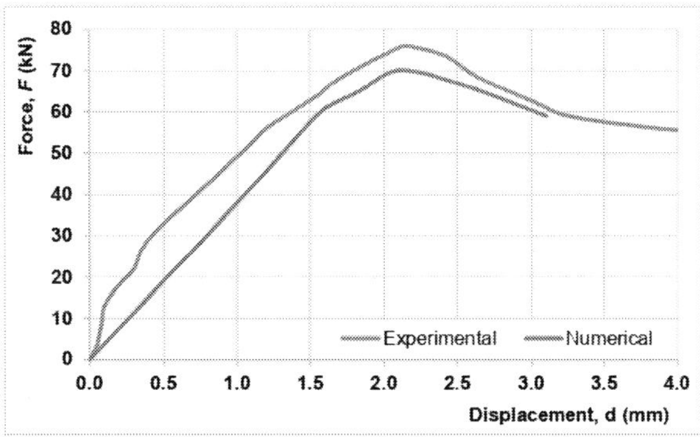

(a)

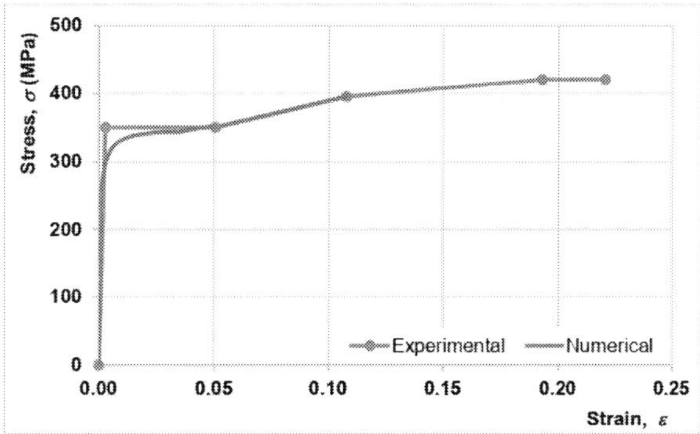

(b)

Figure 3. Comparison between numerical modelling and experimental study – (a) Strength-displacement profile, and (b) Stress-strain curve.

Figure 4 shows the effects of thickness on the ultimate strength of the built-up CFS column. It can be noticed that the thickness contributes significantly to the increment of ultimate strength. As the thickness increases, the ultimate strength of the built-up CFS column increases as well. This finding is similar to that observed by Muhammad Noor [18] and Lin [19], where the column's percentage of axial capacity become more conservative as the member thickness become thicker. It is no doubt that thickness is one of the section properties that will affect the structural behaviour and performance of built-up CFS columns. Although the ultimate strength of the built-up CFS column may be influenced by other factors such as buckling, the thickness gives a major impact on the moment of inertia and stiffness.

Figure 5 shows the displacement against the thickness of the built-up CFS column. The relation between the displacement and thickness is represented by a gradually negative gradient. Regardless of the size of square hollow sections, it can be observed that the increase in thickness will lead to the decrement of displacement. However, this trend is only true for the built-up CFS column with SHS 19x19 and SHS 32x32. For SHS 25x25, it can be seen that few thicknesses especially 1.2 mm and 2.2 mm have a bit higher values than their precedent and does not contribute to the smooth decrement of displacement. The finding for SHS 25x25 is likely contradicted to the previous experimental study by Kesawan et al. [20], which mentioned that the increment of web thickness will reduce the axial displacement of the hollow flange column.

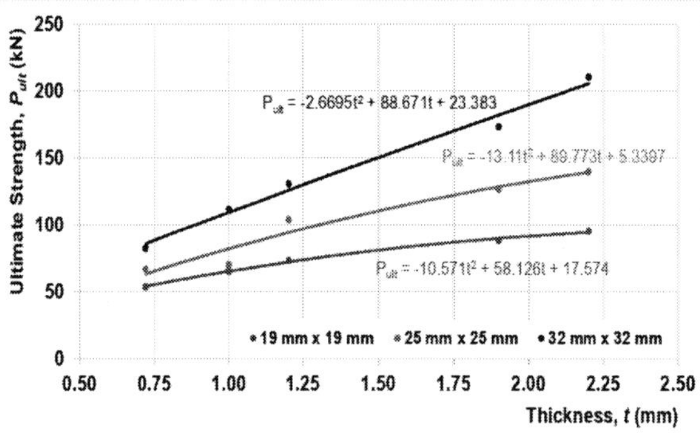

Figure 4. Ultimate strength versus to thickness.

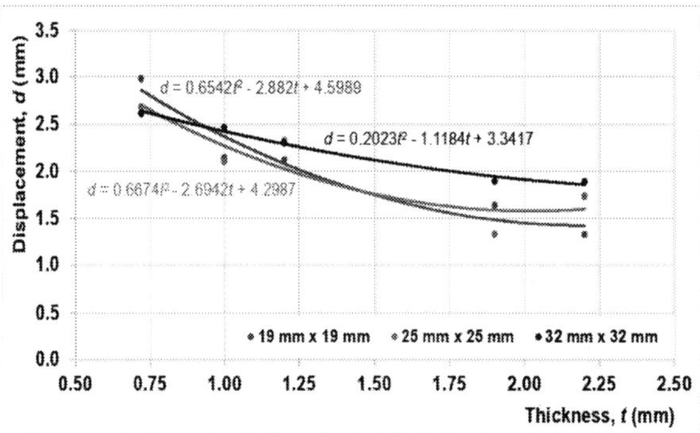

Figure 5. Displacement versus thickness.

The initial failure mode of the built-up CFS column is indicated by the yielding behaviour before it experiences a total collapse. It is noticeable that the built-up CFS column is less susceptible to buckling when the size getting larger. Similarly, the thickness tends to cause a lesser rate as it becomes thicker. According to Whittle & Ramseyer [11], the larger geometry of the structure has less buckling. Figure 6 shows the failure mode of the built-up CFS column. The failure mode may be localised or globalised across size and thickness. It can be observed that the built-up CFS column with SHS 19x19 suffers from distortional buckling throughout the whole length. Besides, there is stiffener failure occurs at the top and bottom parts of the built-up CFS column. Most built-up CFS columns with thickness 0.72 mm experience connector failure as the rivets stretched out from the battens and square hollow sections. On the other hand, SHS 25x25 and SHS 32x32 cause the built-up CFS column to experience the flexural buckling at the mid-height. Flexural buckling is always initiated at the member with a lower chord width-to-thick ratio [20]. In the numerical modelling, there is no appearance of lobe and bulging mechanisms as observed by Musa [12].

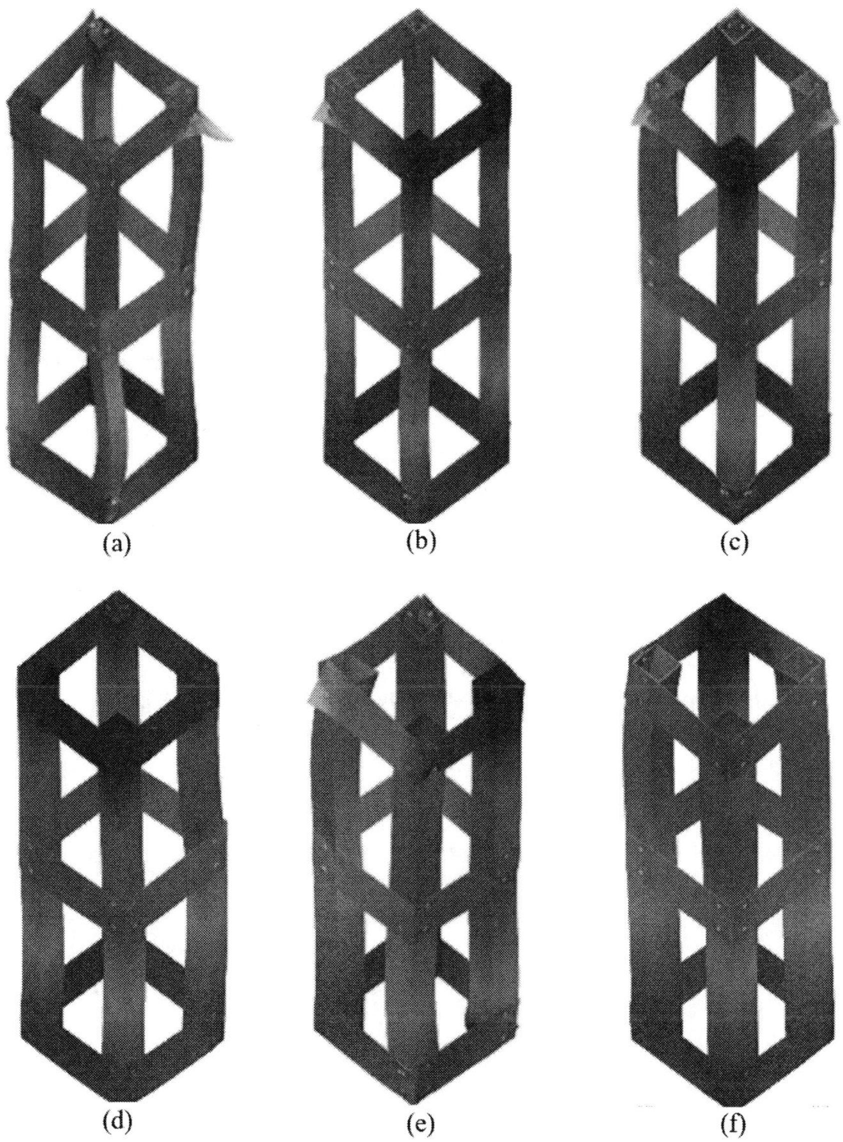

Figure 6. Failure mode of built-up CFS column, (a) SHS 19x19 (t = 0.72 mm), (b) SHS 19x19 (t = 2.2 mm), (c) SHS 25x25 (t = 0.72 mm), (d) SHS 25x25 (t = 2.2 mm), (e) SHS 32x32 (t = 0.72 mm) and (f) = SHS 32x32 (t = 2.2 mm).

6. Conclusion

Generally, this study is conducted to investigate the effect of section properties in terms of size and thickness. The numerical modelling was conducted using WELSIM, a finite element program. The effects of section properties in terms of the ultimate strength and displacement were quantitatively measured through the relationships of ultimate strength and displacement in corresponds to the thickness. Moreover, the failure mode was observed visually. It was found that the ultimate strength of the built-up CFS column getting stronger as the size becomes larger. Similarly, the thickness gave a favourable agreement. On the other hand, the displacement of the built-up CFS column shows a decrement trend when the size and thickness getting bigger and thicker. In terms of failure mode, it was found that the built-up CFS column suffers distortional and flexural buckling as well as connector and stiffener failures.

7. References

[1] Thiyagu K, Elanchezhiyan R, Kanaka P & Ganamanikarnika N 2019 Behaviour of cold formed built-up steel compression member *Int. J. Recent Technol. Eng.* **8** 8409–8413
[2] Muftah F, Mohd Sani M S H, Mohammad S & Tahir M M 2014 Ultimate load of built-up cold formed steel column *ARPN J. Eng. & Appl. Sci.* **9** 2095-2101
[3] Zhang K, Varma A H, Malushte S R & Gallocher S 2014 Effect of shear connectors on local buckling and composite action in steel concrete composite walls *Nucl. Eng. Des.* **269** 231-239
[4] Dandens A, Kreilis J & Andersons G 2011 Properties of cold-formed steel sections *Proc. 3rd Int. Sci. Conf. Civ. Eng.* (Latvia)
[5] Anbarasu M & Dar M A 2020 Axial capacity of CFS built-up columns comprising of lipped channels with spacers: Nonlinear response and design *Eng. Struct.* **213** 110559
[6] Wang L & Young B 2016 Behavior of cold-formed steel built-up sections with intermediate stiffeners under bending. I: Tests and numerical validation *J. Struct. Eng.* **142** 04015150
[7] Young B & Chen J 2008 Design of cold-formed steel built-up closed sections with intermediate stiffeners *J. Struct. Eng.* **134** 727-737
[8] Lu Yan, Zhou T, Li W & Wu H 2017 Experimental investigation and a novel direct strength method for cold-formed built-up i-section columns *Thin-Walled Struct.* **112** 125-139
[9] Georgieva I, Schueremans L, Pyl L & Vandewalle L 2012 Numerical study of built-up double-Z members in bending and compression *Thin-Walled Struct.* **60** 85-97
[10] Paiva F, Henriques J & Barros R C 2014 Comparative study of built-up members subjected to compression load, by the EC3 and AISC 360-05 *Eurosteel* **2014** 19-31
[11] Whittle J & Ramseyer C 2009 Buckling capacities of axially loaded, cold-formed, built-up C-channels *Thin-Walled Struct.* **47** 190-201
[12] Musa S 2019 *Effect of Thickness on Structural Behaviour of Stiffened SHS Cold Formed Steel Column* Undergraduate Thesis (University Tun Hussein Onn Malaysia)
[13] Kyvelou P 2017 *Structural Behaviour of Composite Cold-Formed Steel Systems* PhD Thesis (Imperial College London)
[14] Nie S F, Zhou T H, Zhang Y & Liu B 2020 Compressive behavior of built-up closed box section columns consisting of two cold-formed steel channels *Thin-Walled Struct.* **151** 106762
[15] Aruna G, Sukumar S & Karthika V 2019 Behaviour of cold-formed steel built-up closed columns composed by angle profiles *Asian J. Civ. Eng.* **20** 1037-1048
[16] Dewangan A, Bhatt G & Sonkar C 2020 Axial strength estimation of cold formed steel wall panels through numerical modelling *Emerging Trends in Civil Engineering* eds K Babu, H Rao & Y Amarnath (Springer) pp 91-100
[17] Lu L F, Huang G Y, Fang W Q & Yang D H 2012 Numerical simulation analysis on shear bearing capacity experiments of cold-formed steel self-drilling screws connection *Appl. Mech. Mater.* **166** 200-206
[18] Muhammad Noor Afandi A 2019 *Effect of Size on Structural Behaviour of Stiffened SHS Cold Formed Steel Column*. Undergraduate Thesis (University Tun Hussein Onn Malaysia)
[19] Lin C Y 1988 Axial capacity of concrete infilled cold-formed steel columns *Proc. 9th Int. Speciality Conf. Cold-Formed Steel Struct.* (Missouri)
[20] Kesawan S, Mahendran M, Dias Y & Zhao W B 2017 Compression tests of built-up cold-formed steel hollow flange sections *Thin-Walled Struct.* **116** 180-193

Acknowledgements
The authors would like to thanks the Civil and Structural Engineering Branch, Jabatan Kerja Raya Malaysia for their continuous assistance and support.

Sensor Modules for Enhancement of Safety Performance in Construction Safety Management

MAK Tuck Kiong[1], Loh Seng Yap[1], Eeydzah Aminudin[1,2], Rozana Binti Zakaria[1,2]

[1]School of Civil Engineering, Faculty of Engineering, Universiti Teknologi Malaysia, Johor Bahru, Malaysia
[2]UTM Construction Research Centre, University Teknologi Malaysia, 81310 UTM Skudai, Johor Bahru, Malaysia
Corresponding author: eeydzah@utm.my

Abstract. Safety management is very important for construction sector as it is the most hazardous working environments when it comes to occupational fatalities. The high cases such as injuries at the workplace, illnesses, and fatalities often caused harm to the workers and create the delay on the work progress. Therefore, this paper reviewed the whole operation process of the sensor modules that enhanced safety performance to prevent accidents occurred on construction site. By providing lab scale location-based safety management services to the workers, this technology able to allow a speedy response in the event of accident. From the benchmark conducted, there are three study that had been conducted previously using vibrational alert signal which identified as an accurate transmitted especially inside a danger zone, as the vibrational function of the sensor module able to function within the designated range. The second benchmark shows that previous study able to identify errors in the location information of the workers arising from obstacles, even though so the technology able to conveyed within the designated range as well. Thirdly, the previous study able to shows the information of a fall was conveyed quickly upon occurrence, and the previous study showed that fall able to identified using the context-aware information from the sensor module. The findings showed that the real-time location and context-aware information collected from the sensor module can be used to prevent accidents and respond quickly in the event of a fall. However, the vibration signal which generated by sensor module was insufficient to alert the workers when approaching the danger zone at construction site. Therefore to improve the whole development of prototype sensor module a further study is needed to incorporate the vibrational and alarm signal for recognising accidents upon occurred for enhancement safety management at construction sites.

1. Introduction

Construction risks at construction sites could be arisen from many sources. Due to the construction is tight scheduled, project-based, and decentralized occupation [1]. Second, construction risks also could generate by trait of trades of on-site workers that might affect construction safety [2].The implementation of safety management on construction site is either based on management-driven or technology-driven [3]. In general, the construction safety management could be enhanced by improve safety performance and minimise the rate of accidents on-site. From the study, it indicates that the management perspective which includes safety management processes such as safety education and training, and alter the individual/organizational characteristics such as workers' trade traits that could affect safety at site. However, these human safety errors could not be completely eliminated through safety education and training; thus, it leads to the implementation of innovative technologies that could assist and improve on existing management-driven safety management practices. These technical approaches that could enhance the management efforts [4]. The construction organisations are reluctant to implement innovative Information and Communication Technologies (ICT) mainly due to the project-based and fast-paced working environment [5]. The advanced development of ICT that made researchers gradually realised the vital of new technologies to many management practices, this including safety management [6]. Hence, by study the application of the beacon technology, which is gaining attention as a real-time location-tracking technology for the purpose of developing a location sensor module with integrated of vibrational and alarm signals that is appropriate for safety management of workers at sites.

2. Overview of construction industry

In comparing with other sectors, the amount of workers involved in construction site accidents has been accelerated yearly. Figure 1 shows the trends of construction accidents report based on the Korea ministry of employment and labour, and the annual statistics report between 2011 and 2016 [7]. One could say that the rise in construction site accidents are due to the increased in the volume of construction projects that made possible by advancement in construction plants and know-how that resulted in recruiting of a larger amount of construction workers at a construction project site, this issue could not be compensated by safety training and deployment of safety officers alone. The construction accidents are the most common types of accidents, and these accidents occur at a similar rate every year, because most of these safety-related accidents happen suddenly, there are certainly limits in safety management. Therefore, in order to minimise the accidents on construction sites, it is vital to implement more aggressive safety management at sites. However, the IoT is applied in a simplistic manner, and capable of gathering of various types of data extensively the actual utilization of such data has been minimal, although the radio-frequency identification (RFID) technology is deployed to construction sites, but it is not recognized as a technology for construction safety management from the aspect of construction site workers due to the presence of technical limitations pertaining to the RFID tag recognition rate, scope of recognition and mutual communication [8].

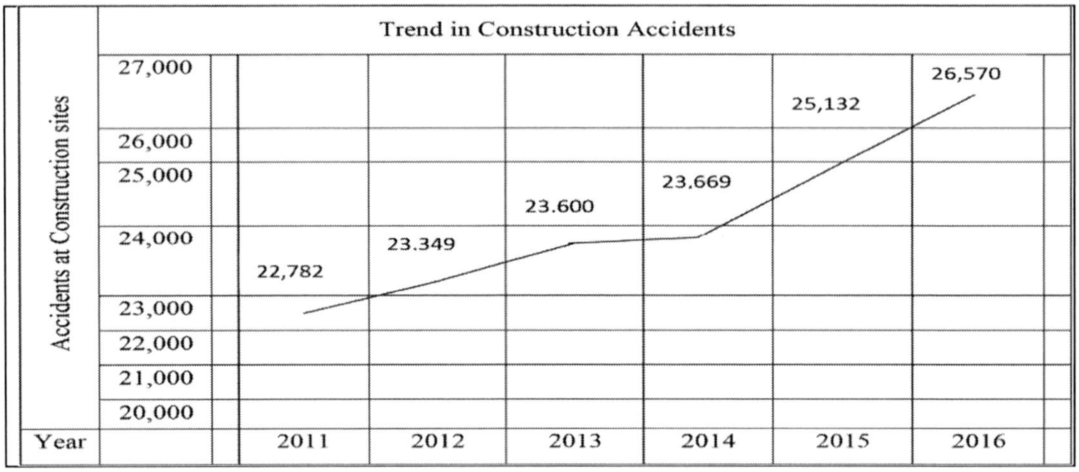

Figure 1. Trends in occupational accident victims in construction industry (Source: [7]).

3. Real-time location system

From the research, it indicates that through management based perspective which includes the safety management processes such as providing safety training and education, and also changing the workers' attitude and organisational characteristics towards construction safety. However, the workers' bad practices are difficult to change thoroughly through safety training and education. Hence, it is necessary for the construction industry to introduce various technologies that could help and upgrade the existing management driven safety management practices. These innovative technologies are for the enhancement rather to replace the management effort [4], however, the implementation in the construction site is slow due to the temporary project-based and fast-paced changing working site condition of the industry [5]. It is a matter of fact, with the gradually advances in the ICT, the researcher realised the significant of the new technology towards the construction and safety management practices [6]. Among the innovative technologies, the rear-time location using sensing devices have able to solve much of the construction safety related problems due to its ability to track and locate of workers, materials, and plants in real-time at site [7,9].

The aim of this study is to apply beacon technology for construction safety management, which has garner attention as a real-time location-tracking technology, and for the objective of developing an integrated sensor module which is appropriate for construction safety management of workers at construction sites. The research are expected to improve the construction safety of workers with the application of new ICT, and also lead to formation of ICT-applied system for labour and materials management. The application of FRID and RTLS will be proposed as a method for solving the construction safety at site. Table 1 showed some journals published on the impact of rear-time location towards construction safety at sites.

Table 1. Journals on the impact of rear-time location towards construction safety at sites.

Author	Journal Title	Technologies applied to RTLS	Performance of RTLS	Significant to site safety management
[10]	Control and application of accuracy positioning estimation based on real-time location system	The used of UWB antennas and IoT technologies in RTLS	UWB sensor module for RTLS	Development of ultra-wide band antennas for construction safety.
[11]	Applying Sensor-Based Technology to Improve Construction Safety Management	The various sensor-based technologies used in construction safety.	An effective way to collect, identify and process information	It provides real-time construction safety management
[12]	Accuracy Improvement of Real-Time Location Tracking for Construction Workers	To prevent work accidents and accuracy RTLS at construction sites,	Established a location tracking algorithm and concept of using assistant tags.	RTLS system for precision real-time positioning at sites.
[13]	Incorporating workers Awareness in the Generation of hazard proximity warnings.	Gather feedback from workers for creating an effective hazard proximity warning system for sites.	An improved hazard proximity warning system for field applications that utilize sensors	Implementation of advanced sensors by integrating the awareness of workers into the creation of hazard alarms.
[14]	An embedded Sensory System for Worker Safety: Prototype Development and Evaluation.	Using an embedded sensory system which enable workers to better perceive their surroundings.	Transmission of sensing signal is fast and long range.	Sensor is an effective device for data transmission.
[15]	A Framework of On-Site Construction Safety Management Using Computer Vision and Real-Time Location System.	Construction safety management system using Fast R-Convolution Neural Network (CNN)–based computer vison and Bluetooth Low Energy (BLE)- RTLS	Enable moving objects to be tracked, and the next location can be predicted.	To warn workers approaching hazardous zone through a smartphones, via loud sounds and vibrations.

[16]	Tracking System using LoRaWan Technology	Long Range transmitter enabled tracking device to perform as data accumulation server	LoRaWAN platform replace the use of IoT infrastructure on the existing Wireless Standard.	LoRa is a low power, wide area network, and with long-range communication link.
[17]	Implementation of Safety Management System for Improving Construction safety Performance.	Risk identifying and monitoring safety at construction projects.	To minimise risk at construction site.	To enhance safety at construction site.
[18]	Building information modelling in combination with real-time location systems and sensors for safety performance enhancement.	BIM could analyse whether the PPEs are worn through integrating pressure sensors and positioning technologies.	A wireless network architecture consisted of end nodes, repeaters/checkpoints and coordinators is considered.	RTLS and virtual construction are created to judge whether the worker needs to wear a helmet and give a warning.
[19]	Design and Implementation of Location and Activity Monitoring System Based on LoRa	Sensor for real-time data gathering and transmitting it to server	It could gather data precisely and with long distant reception	User could trace the activity and location data via smartphone
[20]	Automatic Fall Detection Using Smartphone Acceleration Sensor	Using smartphone and a detection algorithm to improve fall detection	Using smartphone for locating fall.	To improve fall detection and minimise false alarm rate.
[21]	Technology's Role in Safety Management	Virtual reality and Augmented reality	Safety context and real-time data of workplace condition	It can send location details to managers in field office for monitoring

3.1. The advantages of technology in construction safety management

Some of these technologies could be applied jointly to generate safety benefits on the workplace. For instance, the laser scanners or UAVs are often used to scanning at site and to capture a detailed set of data before creating a point cloud for the deployment of a BIM model. Table 2 showed some of the potential safety benefits of technology in construction sector.

Table 2. Potential Safety Benefits of Technology in Construction (Source: [21]).

Technology	Potential safety benefit	Example
Building Information Modelling (BIM)	•Enable risks elimination or substitution (design for safety). •Improve risks recognition and identification. •Improve safety inspections. •Enhance safety planning, awareness or communication.	•Automated safety-rule checking platform to identify potential risks before commencing work and suggest risk prevention measures [13] •Enable site personnel report near-hits to managers and supervisors to detect and mitigate potential workplace risks [22].
Laser Scanning	•Benefit for risks recognition and identification. •Improve safety planning, awareness or communication. •Enhance safety conditions of construction site.	•Identify blind spots of a heavy construction equipment in the job site using a 3-D laser scanner [23]. •Take accurate measurements of heavy equipment and the surrounding environment prior to start of work [10].

Radio Frequency Identification (RFID)	•Facilitate risks recognition and identification. •Assist in safety planning, awareness or communication. • Benefit for site safety compliance [8].	•RFID tags embedded in PPE such as a hard hat to identify real-time. •RFID tags attached to safety gear to support and check on-site compliance with safety procedure [8]
Robotics and automation	•Enable risks prevention •For effective safety planning, awareness or communication. •For preventing safety or health problems.	•Perform welding tasks in high-risk situations using articulated robots •Use a pipe manipulator for concrete pipe laying in trenches [24].
Unmanned aerial vehicles (UAVs)	•Improve risks identification. • Enable safety planning, awareness or communication. •For effective safety inspections.	•Perform effective safety inspections for high-rise structural steel work. •Perform reality capture and take accurate measurements in high-risk or hard-to-reach areas such as bottom of a bridge [25].
Sensors	• For enhancement of construction safety management. •For effective safety planning, awareness or communication.	•Utilize oxygen and temperature sensors to facilitate intelligent monitoring of confined spaces on site and send warning signals in case of an emergency [26].
Warning systems	•Enable risks identification and alerting. •Enhance safety planning, awareness or communication.	•Use sensing technologies to warn workers against potential hazards at construction site [27].

3.2. Utilised of sensing technology in construction safety

Sensing technologies are used for construction safety, for instance, in accident forewarning system, safety route prediction and planning, integrated safety management, structural health monitoring, safety training, and education and highly dangerous operations management. Some of applications of sensing technology are showed in table 3.

Table 3. The adoption of sensing technology in construction safety [11].

Adoption in Construction safety		Sensing Technology	Significance of usage
	Risk Avoidance	Radio frequency identification , ultra-wide band, Zigbee, sensing devices and wireless sensor network	For plants and workers tracking and locating
Safety Design	1. For construction site plants and workers planning.	Radio frequency identification, ultra-wide band, Vision sensing	For planning & scheduling of plants and manpower
Hazard identification	1. To identify & categorise of hazardous areas.	Radio frequency identification, ultra-wide band, Ultrasound, Vision sensing, sensors and wireless sensor network	Barring workers from entering hazardous area, and find out workers' risk behaviour.
Integrated safety management	1. QA & QC management of building materials and resources. 2. Health and safety management of workers.	Global positioning system, radio frequency identification, ultra-wide band, ultrasound, Vision sensing, sensing devices and wireless sensor network	Checking building materials and resources' quality. Introduce new ideas for workers' safety at site.
QA & QC for Structural & building health		Sensing devices and wireless sensor network	Monitor structural & building health

Workers' Safety training and education	Global positioning system, radio frequency identification , Vision sensing	Provide safety briefing and lesson
Accident Forewarning system	Global positioning system, radio frequency identification, ultra-wide band. Zigbee, ultrasound, sensing devices and wireless sensor network	Finding effective forewarning methods for preventing plant collision at site.
High risk operations management	RFID, UWB, Zigbee, WLAN	Danger zone safety control

3.3. Sensor module

This sensor module consisted of a device which could collects beacon signals for RTLS, a 2nd device for collecting data on acceleration and angular velocity that caused by fall, and a 3rd wireless networking device that provides the information collected earlier.

The beacon signals is able for real-time locating, this sensor module system with a built-in communication device of Bluetooth 4.0 or higher. Similarly, to be able to measure the context-aware data of workers at site, the MPU-6050 (Inven Sense Inc.) is used as a chip to measure the acceleration and angular velocity of fall. The MPU- 6050 is built-in with a 3-axis gyro sensor and a 3-axis acceleration sensor. It also used nRF52832 which manufactured by Nordic for micro controller unit (MCU) and SPI and GPIO interfaces. Thus, enable the sensor chip and LoRa transmission and reception chip control, and equipped the Bluetooth function, that is characterised by low power consumption when sending and receiving wireless beacon signals. The sensor module is shown in figure 2.

Sensor Module (MPU-6050)	
micro controller unit MCU(nRF52832)	
Battery	SWITCH
Bluetooth Low Energy (BLE) Module	Semtech SX1276 (LoRa) Long Range communication link

Figure 2. Sensor module architecture [28].

3.4. The long range communication and Gateway wireless network

The sensor module's Long Range communication and Gateway are using wireless network technology. The collected data of indoor location, tilt, and acceleration are sent through Long Range communication. The Long Range communication used chip for sending and receiving of data. The chip required frequency and bandwidth's range is set at between137MHz and1020MHz for communication (ttps://www.lora-alliance.org/).The benefits of Long Range communication are suitable for low power sensor device and no need to install many gateway due to it has long- distant transmission ability. However, the disadvantages of LoRa communication is low transmission speed if the volume of data transmitted is large, to compensate this weakness, it is required to reduce the amount of transmission data.

3.5 The server device

The deployment of Server device for filtering and processes the data received by the sensor module. The server device also perform the real-time processing of position data, and context data of workers that received by the sensor device, and also providing the information to the site manager for action. Similarly, it also transmits alert signals to the workers that approaching the hazardous area for avoiding accidents to occur.

3.6. The creation of advanced sensor for construction safety

The development of new sensor that incorporating vibration and siren signal into the sensor module. This new sensor with enhanced safety feature will alert the worker who wearing it when approaching danger zone. The sensing signal received by sensor receiver which worn by the worker who approaching the danger zone will calculate the distance and sends the data to the server and to the manager, and at the same time , it will triggered the vibration signal and siren signal of sensor module. Figure 3 shown the safety management system architecture of RTLS.

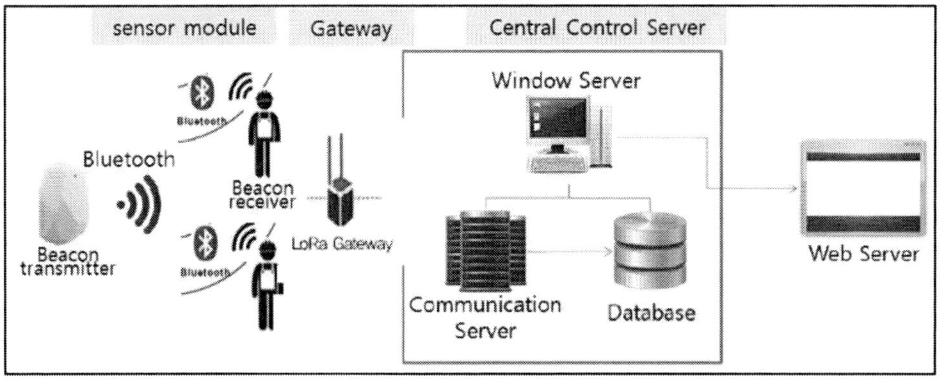

Figure 3. Safety management system architecture of worker Monitoring System for Activity Recognition and construction site RTLS [12].

3.7. Potential targeted experiment using the integrated sensor with vibration and siren signal feature.
First experiment was performed at the Chow Kit Trade Centre construction project site, and at the Lower roof level with potential high risk of accident. When the worker who worn a prototype integrated sensor module receiver with vibration and alarm signal approaching the danger zone that with three sensing transmitters installed at appropriate location of the danger zone with a designated radius of between 5m and 10m, the sensing receiver device will produced vibration and siren signal to alert the worker. The vibration intensity and interval of the sensor module was programmed to the requirements and adjusting accordingly to the circumstances of the construction site. For instance, a strong vibration intensity is required initially when a worker is entered a danger zone, and then weak vibration intensity could be set at certain intervals thereafter. The result of the experiment showed the outcome performance of prototype was effective to prevent potential accidents at construction sites.

Second experiment was performed at the similar area as first experiment to find out the effect of the signals weakened with obstacles positioned between the signal transmitters and receiver, for instance, the presence of a concrete wall of certain thickness and to find out for errors in the location information. At most of the construction site, there are usually have many construction plants and materials placed at different location. In the previous studies have shown that a limitation to indoor locating has been described as the numerous errors in location positioning measurements due to the present of concrete walls of certain thickness on site. To verify whether this problem could be resolved by using the prototype integrated sensor module, a worker who warn a prototype integrated sensing receiver was approaching a 200mm thick concrete wall on site as an obstacles, and establishing an danger zone with a radius of 10m .the experiment results indicated that accuracy of distance measurement was accurate, the alert signals were properly generated and transmitted within the 10m radius of the designated danger zone.

Third experiment was performed at same location as above to see the aware of context of fall. This experiment was to find out whether in the event of a fall accident, the context-aware system could become aware of the context of fall worker, and provide the related information to the field manager quickly in order to take necessary emergency measure to assist the victim. The study result shown that worker who worn the prototype integrated sensor module with acceleration and gyro sensor could receive help due to the data of fall detection could be inform by the system to the field manager during the fall event through using the real-time location-tracking technology and response swiftly to the situation.

3.8. Summary of the research gap

The current construction safety management could be improved with further enhance on the information technology, with a wide range of systems and technologies, for instance, the improvement on the current modules, however, it need to tab the technical limitation that might prevents the substantial improvements. For further improvement in safety management of workers at construction sites, it is necessary to apply IoT-based information technology, there has recently vital for persons and things to manage their own safety. Hence, in order to limit the accidents at construction sites, it is necessary to develop a prototype with the application of real-time location-tracking technology that based on beacon signal technology and a context-aware technology incorporated into sensor technology to form an enhanced sensor module with vibration and siren feature for speedy, effective and efficient response was fabricated in this study. From the outcomes of the experiments, it has shown that it would be possible to implementing the enhanced sensor module system to the construction sites.

4. Conclusion

Currently, the construction safety management that using the sensor modules with the integration of a vibration motor could give an alert signal to the worker that wearing the device, however, it could further improve by incorporating an alarm system within the sensor module. Thus, in order to enhance the safety management system at site, a prototype sensor module with real-time location tracking technology that based on beacon technology and context-aware technology incorporated with vibration and alarm system in sensor technology was fabricated for experiments, and the results showed it is possible to apply to the construction site. This chapter has presented the methodology to study sensor modules with beacon-based location tracking technology and context-aware technology. The experiments were carried out for fall of context-aware, wireless indoor positioning technique of real-time location tracking, and the degree of signal strength weakened with the present of obstacles at site that caused errors in the location measurements. The designed experiments and its procedure were for compliment to the objectives of this study.

5. References

[1] Wei T C, Hew C M, Huang Y H, Bria T A and Lin Y H 2021 Exploring the relationship between safety climate and worker safety behavior on building construction sites in Taiwan, *Sustainability* **13** 3326

[2] Othman I, Shafiq N and Nuruddin M F 2017 Effective Safety Management in Construction Project *IOP Conf. Series: Materials Science and Engineering* **291** 012018

[3] Zhao Z, Goh Y M and Li Q 2015 Overview and analysis of safety management studies in the construction industry, *Safety Science* **75** 337-350

[4] Calvetti D, Meda P, Goncalves M C and Sousa H 2020 Worker 4.0: The Future of Sensored Construction Sites, *Buildings* **10** 169

[5] Lee S and Moon J 2015 Case Studies on Glazing Robot Technology on Construction Sites *Proceedings of the 32nd ISARC* (Oulu)

[6] Zhou Z, Irizarry J and Li Q 2013 Applying advanced technology to improve safety management in the construction industry: a literature review, *Construction Management and Economics* **31** 606-622

[7] Oh J W and Kang J K 2018 A Study of RTLS-based Construction Worker Safety Management System Model, *International Journal of Engineering &Technology* **7** 365-369

[8] Zhaozong M and Zhen Li 2016 RFID Tag as a Sensor - A Review on the Innovative Designs and Applications, *Measurement Science Review* **16** 305-315

[9] Li H, Chan G, Wong J K W and Skitmore M 2016 Real-time locating system applications in construction, *Automation in Construction* **63** 37-47

[10] Aktas A 2019 Control and application of accuracy positioning estimation based real-time location system, *An International Journal of Optimization and Control: Theories & Applications* **9** 82-88

[11] Zhang M, Cao T, Zhao X 2017 Applying sensor-based technology to improve construction safety management, *Sensors* **17** 1841
[12] Kim H and Han S 2018 Accuracy improvement of real-time location tracking for construction workers, *Sustainability* **10** 1488
[13] Can K, Louis J and Albert A 2020 Incorporating worker awareness in the generation of hazard proximity warnings, Sensors **20** 806
[14] Cho C and Park J 2018 An embedded sensory system for worker safety: prototype development and evaluation, *Sensors* **18** 1200
[15] Zhang J, Zhang D, Liu X, Liu R and Zhong G 2019 A Framework of on-site Construction Safety Management Using Computer Vision and Real-time Location System *International Conference on Smart Infrastructure and Construction* (Cambridge)
[16] Zade S G, Ambatkar A A, Bhagat P J, Ambatkar V B, Korde K A and Nagne K B 2020 Tracking system using LoRaWAN technology, *International Research Journal of Engineering and Technology* **7** 1448-1551
[17] Yiu N S N, Chand D W M, Sze N N, Ming Shan and Chan A P C 2019 Implementation of safety management system for improving construction safety performance: a structural equation modelling approach, *Buildings* **9** 89
[18] Dong S, Li H, Yin Q 2018 Building Information Modelling in combination with real-time location systems and sensors for safety performance Enhancement, *Safety Science* **102** 226–237
[19] Lin S, Ying Z and Zheng K 2019 Design and implementation of location and activity monitoring system based on LoRa, *Computer Science* 1-15
[20] Dong T T, Truong H and Dang T K 2016 Automatic fall detection using smartphone acceleration sensor, *International Journal of Advanced Computer Science and Application* **7** 123-129
[21] Karakhan A and Alsaffar O 2019 Technology's role in safety management, *Professional Safety (PS) Journal* **64** 43-45
[22] Shen X and Marks E 2015 Near-Miss Information Visualization Tool for BIM, *Journal of Construction Engineering and Management* **142** 04015100
[23] Kivrak S and Kia F R 2018 A virtual blind spot identification system for construction project, *Scientia Iranica Transactions A: Civil Engineering* **25** 109-117
[24] Skibniewski M J 2014 Research Trends in Information Technology Applications in Construction Safety Engineering and Management, *Frontiers of Engineering Management* **1** 246-259
[25] Albeaino G, Gheisari M and Franz B W 2019 A systematic review of unmanned aerial vehicle application areas and technologies in the AEC domain, *Journal of Information Technology in Construction* **24** 381-405
[26] Cheung W F, Lin T H and Lin Y C 2018 A real-time construction safety monitoring system for hazardous gas integrating wireless sensor network and Building Information Modelling technologies, *Sensors* **18** 1-24
[27] Gambatese J A, Lee H W and Nnaji C A 2017 *Work Zone Intrusion Alert Technologies: Assessment and Practical Guidance* Final Report (Salem: Oregon Department of Transportation)
[28] InvenSense Inc 2013 *MPU-6000 and MPU-6050 Product Specification Revision 3.4* (Sunnyvale: InvenSense Inc)

Acknowledgements
This research would also like to acknowledge Q.J130000.2451.09G77for the contribution.

IOP Conf. Series: Materials Science and Engineering 1200 (2021) 012025 doi:10.1088/1757-899X/1200/1/012025

Talent as a Spearhead of Construction 4.0 Transformation: Analysis of Their Challenges

Santi Edra Nisa Lau[1,5], Eeydzah Aminudin[1,2,5], Rozana Zakaria[1,5], Chai Chang Saar[3], Ahmad Farhan Roslan[4], Zuhairi Abd Hamid[4], Maria Zura Mohd Zain[4], Zafira Nadia Maaz[6] and Abdul Hadi Ahamad[1,5]

[1] School of Civil Engineering, Faculty of Engineering, Universiti Teknologi Malaysia, Malaysia
[2]UTM Construction Research Center, Universiti Teknologi Malaysia, 81310 UTM Skudai, Johor Bahru, Malaysia
[3] Faculty of Engineering, Computing and Science, Swinburne University of Technology Sarawak, Malaysia
[4]Construction Research Institute of Malaysia, Malaysia
[5] Department of Structure and Materials, School of Civil Engineering, Faculty of Engineering, Universiti Teknologi Malaysia, Malaysia
[6] Department Quantity Surveying, Faculty of Built Environment and Surveying, Universiti Teknologi Malaysia, Malaysia
Corresponding author: santiedranisalau@gmail.com

Abstract. Global industries are investing in technology to accelerate digital transformation. Construction is also most likely to be digitalised based on current technology trends. However, technology adoption is not the only ingredient to successfully transform the construction industry towards the fourth industrial revolution (4IR). This transformation requires additional changes for the employees. It is expected to significantly impact the talent landscape, ranging from job categories to skill sets. While this transformation holds excellent benefits, it also poses many challenges. This paper discusses the challenges that individuals, construction companies and governments face from a talent perspective. The data is obtained from literature review results and content analysis through focus group discussion. A focus group discussion was conducted among experts with high knowledge in both the construction industry and 4IR. Information obtained from the discussion was used to identify and categorise the determining challenges. The study revealed nine (9) major talent challenges that the construction industry is currently facing, such as inadequate high skilled talent, lack of education and training to widen talent readiness, talent job security, lack of awareness or clarity of 4IR, dependency on outside talent, employer's readiness, negative attitude of future talent towards changes, the potential of emigration of highly trained or qualified talent, and strong resistance towards new changes and technologies. 4IR can be implemented effectively in the Malaysian construction industry if key challenges that hold the talent are overcome. In conclusion, an active role from quadruple helix collaboration positively assist the transformation.

1. Introduction

Shifting to the fourth industrial revolution (4IR) can improve the work condition where dangerous and strenuous tasks are replaced with safer and more efficient technology innovations and inventions. For example, autonomous construction implementation such as drones can help track current project progress [1], especially in hazardous conditions or large areas, without putting workers at risk.

The previous three revolutions should be defined first to concisely understand the term "Fourth Industrial Revolution". The first revolution brought changes through the creation of steam engine; the second revolution characterised by electricity for mass production, and the third revolution began at the end of the 20th century with information technology as a backbone, culminating in the significant advances made in the Information and Communications Technologies (ICT) [2]. Then, digital integration and intelligent engineering led to the next level of revolution, named the fourth industrial revolution. This revolution represents a fusion of technologies that allow smart, autonomous, and decentralised environments to happen [3]. The fourth industrial revolution (4IR) term has been used interchangeably with industry 4.0. To simplify, the Industrial Revolution represents the development of industries over the years in terms from as a result of technology maturity. Thus, 4IR represent the latest technology development. Nevertheless, industry 4.0 is a term that focuses on the manufacturing industry and has a similar concept to "Integrated Industry", "Smart Industry", or "Smart Manufacturing" [4]. From the construction industry perspective, similar ideas can be found under the terms Construction 4.0 [5][6] or digital construction [7].

Construction 4.0 brings tremendous positive impacts towards society, economic, and governance aspects of the construction industry. It heightens the expectations while enriching opportunities within this industry. For instance, Unmanned Aerial Vehicles (UAVs) helps in monitoring the progress of construction projects [5]; advanced material construction produces lightweight structural components that help to reduce waste CO_2 emission and energy consumption [5]; and BIM enhances the current communication process, provides a collaborative platform, and supports interoperability [8]. Overall, there is a modestly positive impact not only in the construction industry but, potentially create ripple effect across industries.

However, this transformation requires more than mere technology implementation. It is impossible to transform successfully without uplifting current talent. The talent is profoundly essential but often forgotten during the transformation process. Talent is defined as individuals who can make valuable contributions to the company and its operational performance [9]. Today's technological advancement will change how people work and lead to new job creation. According to World Economic Forum [10], 65% of children will ultimately work in entirely new job categories that do not exist yet. For instance, artificial intelligence is seen as a technology that allows robots to replace human tasks. While this transformation holds excellent benefits in enhancing the working process, it comes with challenges. As mentioned in World Economic Forum [11], the construction industry will face challenges in recruiting talent in the future. Sirotek and Firlus [12] noted that the biggest trigger to the 4IR adoption is the availability of skilled labour. As the Construction 4.0 continues to grow, it is crucial for the Malaysian construction industry to address knowledge expansion on talent issues niche. Discussion on aspects of how Construction 4.0 continues to change how and where people work, new challenges are generated are precursors towards the Construction 4.0 success.

Hence, this paper aims to address the following research questions: 1) What challenges are posed by Construction 4.0 from a talent perspective? and 2) How to solve the talent gap in 4IR in the context of the construction industry? This paper will be presented in five sections. This paper begins with an assessment of the construction industry's state towards 4IR, followed by building up the research question and theoretical background on 4IR in section 2. Section 3 discuss on the methodology of the study. Section 4 explores the finding and results of the study which followed by a discussion of the theoretical. Conclusion and future research are addressed in section 5.

2. Literature Review

The section involves a review of current literature and answering research questions concluding with talent gap identification.

2.1. Construction 4.0

Construction 4.0 is a term adopted from industry 4.0 and widely used to represent the 4IR in the construction industry. Several documents have mentioned this term, such as the Malaysian

Construction 4.0 Strategic Plan (2021-2025) and the report by Roland Berger [13]. The European construction industry federation (FIEC, 2015) mentioned: *"Fourth industrial revolution", known for all EU industry as "Industry 4.0", is known in our sector as Construction 4.0."* One of the difficulties in defining construction 4.0 is that different people have different interpretations. To clarify this term, this paper presents the following definition. Construction 4.0 is characterised by integrating two pillars: digitisation of the construction industry and industrialisation of the construction process [14]. As noted by Osunsanmi et al. [15], it represents digitalisation where a set of technologies work together to enhance project performance and increase client satisfaction. Another definition proposed by Chen et al. [16], Construction 4.0, is reusing technologies younger than 20 years in the design and construction processes to help increase productivity.

The construction industry is falling behind other sectors, such as services, manufacturing, and agriculture, regarding their contribution to Malaysia's economy. Their contribution is relatively small as compared to other sectors. In 2020, the construction sector contributed 0.4 per cent to the growth of Malaysia's economy, contracted 19.4 per cent from the preceding year due to restrictions on the economy imposed in response to the COVID-19 pandemic [17]. Despite the COVID-19 situation, digital adoption in all Malaysian sectors, including construction, has been accelerating. Thus, their impact should be highlighted, especially in 4IR, as it is related to other sectors' success.

2.2. Challenges faced by construction industry talent

COVID-19 situation has helped in accelerating digital adoption in all Malaysian sectors, including the construction industry. Due to Movement Control Order (MCO) restriction enforced by the Malaysian government, all construction activities have stopped working unwillingly. Through the advent of technologies, project monitoring, status tracking, and scheduling tasks can still be done online instead of face-to-face meetings [18]. Therefore, embracing construction 4.0 will equip the industry with the right technologies, especially during these difficult days.

While the impending change holds great promise, it also posed many challenges. The main challenges are mainly associated with human resources [19][20]. Malaysian construction industry faced problems such as shortages of human resources and a dearth of productivity [21]. According to World Economic Forum [11], the new revolution will struggle to recruit the "digital" talent plus the image of low job security posed by the construction industry. Utting [22] explained that skills shortage usually occurs when employers cannot fill in the specialised skill needs within that occupation. Labours' qualification holds a vital key towards the project productivity enhancement as lack of skilled labour influences the decline in construction project productivity [23]. Tariq & Sonntag [24] attempt to study the benefits and challenges through a survey study. Three significant challenges were identified: knowledge limitation, financial resource limitation, and technology awareness limitation. Even though it shows challenges from different perspectives, knowledge limitation can be considered under the talent concept. More barriers or obstacles in Construction 4.0 implementation are identified in table 1.

Table 1. Barriers to Construction 4.0 transformation from talent perspective.

Barriers	References
Skill shortage	[25], [26], [27], [28], [29], [30], [31]
Lack of understanding of disruptive changes	[10], [24], [26], [27], [29] [32], [31]
Workforce strategy not aligned to innovation strategy	[10], [26], [29]
Lack of worker acceptance	[26], [27]
Talent off-boarding	[33]
Job disruptions	[30], [31]
Resistance to change	[30], [32]

To prevent a worst-case scenario, the individual can take a proactive approach by upgrading their skill through self-learning or attending training. Relying too much on the company's initiative can hinder the progress of development. Therefore, there is a need to identify talent barriers and factors that could support strategies, inducing a smoother transformation of construction 4.0 technologies.

2.3. Solving the talent gap in 4IR in the context of the construction industry

Implementing new technology and jump on board with Construction 4.0 is not an easy journey. There are always challenges when it comes to transformation. The transformation is possible provided concerted efforts from quadruple helix (industry, government, academia, and users/civil society). Table 2 introduces the strategies developed from the literature review to solve the talent gap in Construction 4.0.

Table 2. Strategies to solve to Construction 4.0 transformation from a talent perspective.

Strategies	References
Human resource competencies assessment	[34], [35]
Training programmes	[10], [24], [34], [36]
Review of different human resource development initiatives	[34], 35], [36]
Attract multi-skilled or new talent recruitment through attractive strategies	[34], [36]
Promote transparent career paths to avoid employees' uncertainty about career options	[36]
Collaboration with educational institutions	[10], [36]
Promote initiatives that spur innovation and improve industry attractiveness	[36]
Apprenticeships and academic programmes	[10], [36]
Collaboration in and across industry	[10]

To support the above strategies, there must be strong collaboration from the quadruple helix. For the government, strong policies or attractive initiatives are needed to support the process. For industry, they need to put talent development and skills requirements at the centre of their growth. For academia, they need to introduce and increase the awareness of future talent towards 4IR. For users/civil society, they require a new mindset to meet the expectation of the future.

3. Methodology

This paper involved input from many leading companies along the construction value chain, such as contractors, project owners and developers, academics, and government and civil society leaders.

3.1. Data Collection

The data were collected from the focus group discussion session among construction experts across the Malaysian construction industry stakeholders. The participants of the focus group was chosen due to their ability providing trustworthy insight on human behaviour aspect [37]. Participants for this study were selected according to the criteria of: 1) position, 2) directors and senior leaders within engineering-led organisations, 3) knowledge, and 4) implementation of at least one of Construction 4.0 enabling technology in their organisation. The reason directors and senior leaders were selected because they have a deep understanding on organisational structures and problems faced in the process of introducing 4IR.

Few studies reported that the appropriate size ranges from as few as four and as many as fifteen [38,39]. In this study, a total of 32 participants were involved the session. The participants were further divided into five groups, A, B, C, D, and E, and led by a moderator in a loosely structured discussion style. Each group must consist of different domains of the construction industry. A large group may be beneficial in generating more opinions. In contrast, a smaller group is favourable to decipher sensitive issues as 4IR is still embryonic and more opinion matters. Figure 1 and Figure 2 illustrated the tabulation of respondents involved during the discussion session.

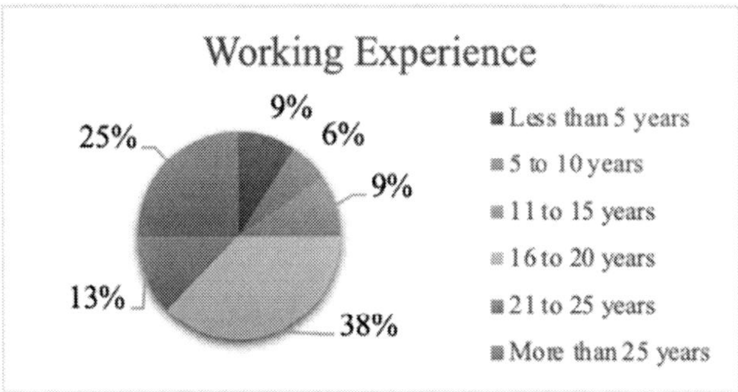

Figure 1. Participants' working experience.

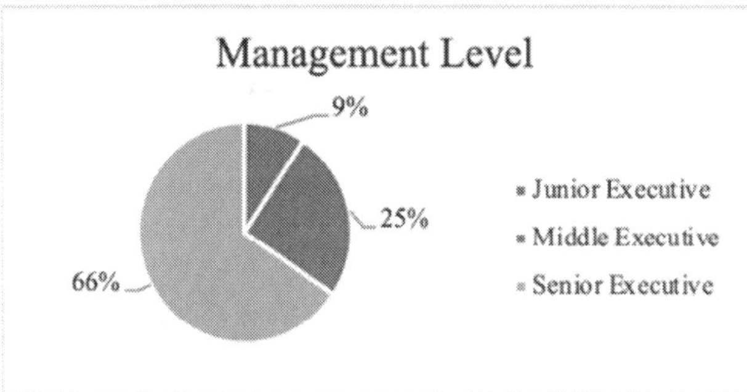

Figure 2. Participants' level of management.

Figure 1 shows that majority of the respondents have 16-20 years of working experience in the construction industry. According to their work experience, it can be assumed most of the participants who attended the discussion has a vast knowledge of the company from project, financial, and staff management. They manage the construction project at the site and delivers the information of all construction activities to associates. Figure 2 describes participants' management level, and the result shows that more than half of the participants were senior executive level. They are individuals at the highest level of control and able to give the information needed.

3.2. Data Analysis

A blank butcher paper was presented during the discussion session, and participants need to discuss and brainstorm all challenges of 4IR in construction in a group assigned at the early stage. Their responses will be recorded both on butcher paper and laptop. The content analysis method will be used to analyse the result. Content analysis provides a systematic coding of data by categorising the information to discover patterns undetectable by merely listening to the conversation recorded [39]. In the first place, the challenges listed were documented. The study identified 20 elementary challenges. They were next, sorted, classified, and grouped by their central element. Similar challenges will be grouped into one prominent obstacle. Then they were commented and, in some parts, transformed into a proper form. The final list consisted of 9 talent challenges and problems in 4IR adoption.

4. Result and discussion

The results of the survey are discussed in this section. The analysis revealed important findings of challenges in the construction industry to 4IR from a talent perspective.

4.1. Challenges of talent for 4IR in construction

In response to the first research question, this paper identified the same challenges as the literature and also identified an additional challenge focused on talent among construction workers.

This paper identify nine (9) major challenges that can harness the transformation and adoption process; (1) inadequate high skilled talent, (2) lack of education and training to widen talent readiness, (3) job security, (4) lack of awareness or clarity of 4IR, (5) dependency on outside talent, (6) employer's readiness, (7) negative attitude of future talent towards changes, (8) the potential of emigration of highly trained or qualified talent, and (9) strong resistance towards new changes and technologies.

Based on table 3, the result analysis shows that the participants highly agree that inadequate skill talent is the most challenging part of digital construction. It contributed to 80% of level agreement (four out of five groups agreed). Introducing 4IR means individuals need to prepare themselves with high skill and have a high literality. Lack of qualified personnel is frequently mentioned in 4IR related literature papers [29]. The most demanding skills are information and data literacy, communication and collaboration, digital content creation, safety and security, and problem-solving [40]. Therefore, attending training or any related 4IR courses is one of the best ways to upgrade talent skills.

Table 3. Challenges relating to talent in adopting 4IR in construction.

Representative challenges quotes	Central challenges	Group A	B	C	D	E	Frequency	Agreement level
Lack of skilled labour	Inadequate high skilled talent	/					4/5	80%
Insufficient talent pool supply			/					
Talent availability				/				
Skilled resources availability					/			
Education and training readiness	Lack of education and training to widen talent readiness	/					3/5	60%
Education readiness			/					
Lack of skilled student skill						/		
Lack of demand (project)	Talent job security		/				3/5	60%
Job security				/				
Not enough jobs of the future to cater to industry					/			
Awareness on 4IR	Lack of understanding or clarity of 4IR			/			3/5	60%
No benchmarking of the status					/			
Lack of awareness						/		
Enforcement of foreign workers	Dependency on outside talent			/			2/5	40%
Dependency on outside talent						/		
Employer's readiness	Employer's readiness			/			2/5	40%
No clear vision, objective						/		
Wrong attitude of students	The negative attitude of future talents towards changes					/	1/5	20%
Emigration of highly trained or qualified people	The potential of emigration of highly trained or qualified talent					/	1/5	20%
The existing workforce is in the comfort zone	Strong resistance towards new changes and technologies					/	1/5	20%

Next, digital transformation to Construction 4.0 will bring a challenge to job security. Three out of 5 groups agreed on the impact of job security, which contributed to 60% of the agreement level. The advent of 4IR will increase the dominance of technology cause fears of individuals over future job security. Then, the challenges followed by lack of education and training to widen talent readiness with an agreement level of 60% (three out of five groups agreed). A financial problem can be noted as one of the main problems leading to a lack of education and training readiness. As mentioned early, the high cost of training, education, and consultant fees [27] demotivated construction employers towards transformation. In this context, the development of cost-benefit analysis costs can help estimate implementation costs and the profit obtained in the future [27]. Even though a high price is needed at the earlier implementation stage, significant benefits can be found later. Thus, appropriate education and training programmes must be included in the strategic development.

In this study, the lack of understanding or clarity of 4IR is highlighted as a barrier with an agreement level of 60% (three out of five groups agreed). It can be supported with a survey from Future of Jobs [10]; 51% of the construction players lack understanding of the disruptive changes. This result can be related to timid actions being taken by companies to adopt 4IR due to insufficient understanding and awareness. However, in few cases, the awareness of digitalisation is well understood, but the definition of construction revolution and industrial revolution are not well-defined. It causes another problem of unclear status either from people or from the organisation dimension. In this context, a guideline for successful implementation concerning the specific characteristics and requirements of the construction environment can be provided for companies to consider before implementing new technologies [27] to increase the level of understanding and clarity among individuals or organisations.

The other challenges that construction 4.0 faces are dependency on outside talent and employer's readiness, which contributed to 60 % of the agreement level. It is followed by the wrong attitude of students (20%), emigration of highly trained or qualified talent (20%) and the existing workforce is in the comfort zone (20%). Identifying this readiness is significantly needed as it will enable companies to measure shortages and future steps towards the digital transformation process [41]. Gürdür et al. [42] stated in their study that companies need a roadmap for organisational readiness assessment purposes. If this problem is not well addressed, this will eventually create a digital divide on the organisation and industry level. The organisations or industries with inadequate focus on digitalisation will be wiped out and left behind from the market. As mentioned in the table above, some construction players show a strong resistance towards the transformation. Most of the reason is due to their reposeful feeling with the conventional method. Changing to Construction 4.0 will force them to move from their comfort zone. In conclusion, moving forward to 4IR, the construction industry talents need to enhance and speed up their adaptability in technology advancement to avoid being left behind by other industries.

5. Concluding remarks

The construction industry is one of the least digitised industry in the country. Indystry4WRD, initiated by the manufacturing industry, was introduced back in 2018, while Construction 4.0 strategic plan has just been announced earlier in 2021. Even though it is a bit late compared to the manufacturing industry, Malaysian construction's positive effort should be praised. The output of this research shows that he construction industry positively catching up on information digitalization, communication, and collaboration. At the end of the process, the adoption will pave the way for more advanced vital technologies and increase the efficiency and productivity of the industry.

This paper has shed light on the aspects of 4IR adoption and its talent gap. Therefore, a holistic shift of talent in the context of awareness, education, skill, and competencies should exhibit the era of 4IR, especially in the construction industry. The future talent needs to be highly adaptable, resourceful, resilient, and interdisciplinary for interaction and collaboration in the industrial market [40]. Whether this emerging technology will influence the job security in a company is also a question worth pondering. At this moment, it is important for talents to prepare themselves with training, and taking active role to improve their lifelong learning. In a nutshell, efforts to close the talent gaps among construction players need to be grounded based on future skill requirements due to rapid technological changes.

6. References

[1] Rane S B, Potdar P R and Rane S 2019 Development of project risk management framework based on industry 4.0 technologies: Benchmarking, *An International Journal* **28** 1451-1481
[2] Sima V, Gheorghe I G, Subić J and Nancu D 2020. Influences of the industry 4.0 revolution on the human capital development and consumer behavior: A systematic review, *Sustainability* **12** 4035
[3] Santos C, Mehrsai A, Barros A C, Araújo M and Ares E 2017 Towards Industry 4.0: an overview of European strategic roadmaps, *Procedia Manufacturing* **13** 972-979
[4] Hermann M, Pentek T and Otto B 2016 Design principles for industrie 4.0 scenarios *In 2016 49th Hawaii international conference on system sciences (HICSS)* 3928-3937
[5] Craveiroa F, Duartec J P, Bartoloa H and Bartolod P J 2019 Additive manufacturing as an enabling technology for digital construction: A perspective on Construction 4.0, *Automation in Construction Journal* **4** 6
[6] Maskuriy R, Selamat A, Maresova P, Krejcar O and David O O 2019 Industry 4.0 for the construction industry: Review of management perspective, *Economies* **7** 68
[7] Woodhead R, Stephenson P and Morrey D 2018 Digital construction: From point solutions to IoT ecosystem, *Automation in Construction* **93** 35-46
[8] Al-Ashmori Y Y, Othman I, Rahmawati Y, Amran Y M, Sabah S A, Rafindadi A D U and Mikić M 2020 BIM benefits and its influence on the BIM implementation in Malaysia ,*Ain Shams Engineering Journal* **11** 1013-1019
[9] Wiblen S and McDonnell A 2020 Connecting 'talent'meanings and multi-level context: A discursive approach, *The International Journal of Human Resource Management* **31** 474-510
[10] World Economic Forum 2016 The future of jobs: Employment, skills and workforce strategy for the fourth industrial revolution Global Challenge Insight Report (Cologny/Geneva: World Economic Forum)
[11] World Economic Forum 2016 Shaping the future of construction: A breakthrough in mindset and technology In World Economic Forum(Cologny/Geneva: World Economic Forum)
[12] Sirotek S and Firlus B H 2016 *Does organizational learning pay off? A case study of Norwegian and German firms regarding the link between organizational learning and the maturity of Industry 4.0 implementation* Master's thesis (Norway: Norwegian University of Science and Technology) pp 1-198
[13] Berger R 2016 Digitization in the construction industry *Think Act* 116
[14] Forcael E, Ferrari I, Opazo-Vega A and Pulido-Arcas J A 2020 Construction 4.0: A literature review, *Sustainability* **12** 9755.
[15] Osunsanmi T O, Aigbavboa C and Oke A 2018 Construction 4.0: the future of the construction industry in South Africa, *International Journal of Civil and Environmental Engineering* **12** 206-212
[16] Chen Q, de Soto B G and Adey B T 2018 Construction automation: Research areas, industry concerns and suggestions for advancement, *Automation in Construction* **94** 22-38.
[17] Mahidin M U 2021 *Malaysia Economic Performance Fourth Quarter 2020*. Retrieved on June 20, 2021 from https://www.dosm.gov.my/v1/index.php?r=column/pdfPrev&id=cnBscXhyMENra2JzUVhWRTZ4NU44QT09
[18] CIDB 2020 Embracing Construction Revolution CIDB Malaysia (Kuala Lumpur: CIDB) pp 1–60
[19] Leong W D, Teng S Y, How B S, Ngan S L, Abd Rahman A, Tan C P, Ponnambalam S G and Lam H L 2020 Enhancing the adaptability: Lean and green strategy towards the Industry Revolution 4.0, *Journal of Cleaner Production* **273** 122870
[20] Ariffin K A Z and Ahmad F H 2021 Indicators for maturity and readiness for digital forensic investigation in era of industrial revolution 4.0, *Computers & Security* **105** 102237

[21] Alaloul W S, Musarat M A, Rabbani M B A, Iqbal Q, Maqsoom A and Farooq W 2021 Construction Sector Contribution to Economic Stability: Malaysian GDP Distribution, *Sustainability* **13** 5012

[22] Utting P 2010 The risks of skills shortage in construction *Proceedings 5th Built Environment Conference* (Durban)

[23] Hussain S, Xuetong W and Hussain T 2020 Impact of Skilled and Unskilled Labor on Project Performance Using Structural Equation Modeling Approach, *SAGE Open* **10** 2158244020914590

[24] Masood T and Sonntag P 2020 Industry 4.0: Adoption challenges and benefits for SMEs, *Computers in Industry* **121** 103261

[25] Koch V, Kuge S, Geissbauer R and Schrauf S 2014 Industry 4.0: Opportunities and challenges of the industrial internet, *Strategy & PwC* 5-50

[26] Whysall Z, Owtram M and Brittain S 2019 The new talent management challenges of Industry 4.0 *Journal of Management Development*

[27] Oesterreich T D and Teuteberg F 2016 Understanding the implications of digitisation and automation in the context of Industry 4.0: A triangulation approach and elements of a research agenda for the construction industry, *Computers in industry* **83** 121-139

[28] Wankhede V A and Vinodh S 2021 Analysis of Industry 4.0 Challenges using Best Worst Method: A case study, *Computers & Industrial Engineering* 107487

[29] Turkyilmaz A Dikhanbayeva D, Suleiman Z, Shaikholla S and Shehab E 2021 Industry 4.0: Challenges and opportunities for Kazakhstan, *SMEs Procedia* CIRP **96** 213-218

[30] Kumar P, Bhamu J and Sangwan K S 2021 Analysis of Barriers to Industry 4.0 adoption in Manufacturing Organizations: an ISM Approach, *Procedia CIRP* **98** 85-90

[31] Kamble S S, Gunasekaran A and Sharma R 2018 Analysis of the driving and dependence power of barriers to adopt industry 4.0 in Indian manufacturing industry,,y *Computers in Industry* **101** 107-119

[32] Suresh N, Hemamala K and Ashok N 2018 Challenges in implementing industry revolution 4.0 in INDIAN manufacturing SMES: insights from five case studies *International Journal of Engineering & Technology* 7136-139

[33] Sivathanu B and Pillai R 2018 Smart HR 4.0–how industry 4.0 is disrupting HR, *Human Resource Management International Digest* **26** 7-11

[34] Ghobakhloo M 2018 The future of manufacturing industry: a strategic roadmap toward Industry 4.0, *Journal of Manufacturing Technology Management* **29** 910-936

[35] HM Government 2013 *Construction 2025: Industrial Strategy: government and industry in partnership* (London: Crown)

[36] World Economic Forum 2018 An Action Plan to solve the Industry's Talent Gap, *World Econ. Forum* **2** 1–17

[37] Orzes G, Poklemba R and Towner W T 2020 Implementing industry 4.0 in SMEs: a focus group study on organizational requirements *Industry 4.0 for SMEs* eds D Matt, V Modrak, H Zsifkovits (Cham: Palgrave Mcmillan) pp 251-277

[38] Muijeen K, Kongvattananon P and Somprasert C 2020 The key success factors in focus group discussions with the elderly for novice researchers: a review, *Journal of Health Research* **34** 359-371

[39] O Nyumba T, Wilson K, Derrick C J and Mukherjee N 2018 The use of focus group discussion methodology: Insights from two decades of application in conservation, *Methods in Ecology and evolution* **9** 20-32

[40] Flores E, Xu X and Lu Y 2020 Human Capital 4.0: a workforce competence typology for Industry 4.0, *Journal of Manufacturing Technology Management* **31** 687-703

[41] Hizam-Hanafiah M, Soomro M A and Abdullah N L 2020 Industry 4.0 readiness models: a systematic literature review of model dimensions, *Information* **11** 364

[42] Gürdür D, El-khoury J and Törngren M 2019 Digitalizing Swedish industry: What is next?: Data analytics readiness assessment of Swedish industry, according to survey results, *Computers in Industry* **105** 153-163

Acknowledgements
The authors would like to thank Malaysian Ministry of Higher Education for funding this study under the grant FGRS, Registration Proposal No: FRGS/1/2020/TK01/UTM/02/3 & UTM (cost center no: R.J130000.7851.5F391), GreenPROMPT research team members and Department of Construction Management, UTM CRC, together with the opportunity to conduct the research.

Masonry Dust Risk Control Practices and the Barriers to the Adoption of Engineering Controls in Reducing Risk due to Dust Exposure in Masonry Work

N H Abas[1], M Mad Ali[1] and N A Abas[2]

[1] Jamilus Research Centre, Faculty of Civil Engineering and Built Environment, Universiti Tun Hussein Onn Malaysia, 86400 Parit Raja, Johor, Malaysia
[2] Kolej Kemahiran Tinggi Mara Sri Gading, Jalan Kluang, 86400 Batu Pahat, Johor, Malaysia
Corresponding author: nhaslin@uthm.edu.my

Abstract. The construction industry is known to be among the most vulnerable industries in terms of occupational safety and health. One of the riskiest activities in construction involves masonry, which reports several cases involving occupational health diseases due to the exposure to silica dust. Thus, this study aims to investigate the perception toward health risks due to exposure to dust in masonry work. A survey research method was adopted for this study, which involved the participation of 25 active construction contractors in Selangor and Kuala Lumpur. The findings reveal that respiratory problems and health diseases are the most perceived risks associated to dust exposure in masonry work. There are various barriers to the adoption of engineering control in masonry work such as high costs, lack of awareness of dust hazards, and workers' attitudes. The current practical intervention methods used by contractor firms are respiratory protection, wet method, and sweeping compound. This study provides information on the current masonry work environment and the barriers of the adoption of engineering control technologies. Additionally, this study suggests that the key players in the construction industry should take an active part to increase the implementation of engineering controls in a construction project, and not rely solely on the use of PPE.

1. Introduction

The construction industry is known to be the most vulnerable in terms of personal safety and health [1,2]. This has been proven through the high number of accidents and fatality rates reported every year [3]. However, statistics emphasize on safety hazards, whilst the injuries due to exposure to health hazards are loosely reported. Managing the exposure to health hazards is equally important, as it is part of the employer's compliance with existing occupational safety and health (OSH) legislations. For example, the manual published by the International Labour Office in Geneva [4] states that "The concerned work should be safe, and conditions on the construction site should not cause damage to life, health and professional skills". The importance of ensuring a worker's health is also highlighted in the Occupational Safety and Health Act (OSHA) 1994 – Section 15. The health and injuries reported indicate that the construction industry is a dangerous occupation at a high-risk environment, whereby employees are at a higher risk of injury and even fatality due to poor safety and health implementations [5]. With that, the occupational health hazards in the construction industry can be broken down into six categories: asbestos-related diseases, silica-related diseases, noise-induced hearing loss, hand-arm vibration syndrome, musculoskeletal disorders, and dermatitis [6].

Construction works such as masonry, drilling, sandblasting, crushing, tunnelling, and grinding increase personal exposure to silica [7]. Silica dust, also known as respirable crystalline silica (RCS), is created when working with silica materials like concrete, mortar and sandstone [8]. The effects of dust exposure in construction workers' health are asthma, chronic obstructive pulmonary disease (COPD), silicosis, and lung cancer [8]. In the United Kingdom (UK), less than 100 cases were reported every year between 1996 and 2009, and among the cases reported were death from silicosis [9]. In Australia, about 587,000 workers were exposed to silica dust while working and estimated that 5758 of these workers will develop lung cancer over their lifespan [10]. While in Malaysia, in 2019, about 101 cases of occupational lung diseases have been reported, and 61 cases were confirmed as occupational lung diseases by the Occupational Health Division [11]. However, these record cases are general and not specifically caused by masonry dust.

There are several control measures commonly implemented at the construction site to minimize the effect of exposure to silica dust. Examples of dust exposure controls are a sweeping compound for clean-up activities, wet method, use of local-exhaust ventilation (LEV) and area ventilation, training, and PPE [12,13]. This includes controls such as substituting with less hazardous materials or processes such as, replacing concrete or brick to steel that can be determined during the design stage. Nevertheless, several researchers agreed that total substitution or elimination of silica from the construction process may not be a practical alternative [14].

The present study proposes the use of engineering controls to reduce the hazards and risk of exposure to silica dust as it is practical, more economical, and can be implemented at any stage of the construction progress. However, some studies reported that the implementation of engineering/design control is irregular and lack systematic adoption in the masonry sector [11]. Nij et al. [12] in their study revealed that even though engineering controls such as, wet dust suppression or local exhaust ventilation, can effectively reduce the respirable quartz exposure value (at more than 70% reduction factors), the use of a respiratory protection is the most widely used preventive measure in the construction industry. This is worrying because respiratory protection might not always reduce exposure sufficiently and should be combined with other control measure(s) to reduce exposures to an acceptable level [12].

Therefore, this study aims to investigate barriers to the adoption of engineering controls in masonry work, using the same survey research method by Ghosh et al. [11]. The survey research method that was adopted for this study involved the participation of 25 active construction contractors in Selangor and Kuala Lumpur. These participants were selected from the G6 and G7 contractor firms, who are registered under the Construction Industry Development Board (CIDB).

1.1. Intervention method for exposure to dust

Masonry workers are exposed to silica dusts while performing their work [15]. The risks came from operating on concrete and stone materials such as mortar, brick and asphalt that contain concentrated amounts of crystalline silica. The silica-containing dust is emitted and enters the workers' respiratory system. Silica is a natural material found in most clay, rocks and sand and even in concrete and bricks. According to HSE [16], when masonry activities such as cutting, drilling, grinding, etc. are carried out, the silica produces fine dust called respirable crystalline silica (RCS). Workers may develop various lung diseases by breathing in RCS such as silicosis, lung cancer, chronic obstructive pulmonary disease (COPD) [16][17]. RCS can travel deep into workers' lungs and cause silicosis, an incurable and sometimes deadly lung disease. Usually, construction workers who are affected, are not aware of the symptoms because of the risk triggered by long-time of silica exposure.

Several intervention methods have been introduced and commonly used in the construction industry to reduce the risks of exposure. Intervention may be designed to avoid hazards at any of the three locations, namely, at the source, in the path between the source, and at the receiver as shown in figure 1.

Figure 1. Intervention methods to prevent hazard exposure [12].

Ghosh et al. [12] stated that adopting an increased number of engineering controls can minimize the risk of dust exposure to workers. Common ways of reducing dust exposure in the construction industry are by using personal protective equipment (PPE), use of local exhaust ventilation system (LEV), removing wet dust by using (cooling water), and influencing worker behaviour through education and training [13]. Other than that, an old and reliable dust control technique is to wet or mist a surface that is to be cut or drilled [18]. For example, when using a jackhammer, wetting the surface with mist or spray of water helps to reduce the amount of airborne dust [19]. Several research have reported the reductions in respirable dust by either wetting a blade or sparing a surface [18,20].

Moreover, another tool to lower the masonry dust hazard is local exhaust ventilation (LEV). Based on WHS [17], LEV is a more efficient way to control highly toxic contaminants such as RCS dust, from the point or specific sources. A review article authored by Croteau et al. [21] has shown that the use of LEV can significantly decrease airborne particulate exposures caused by various construction-related activities. Based on short-term sampling, it has been shown that local exhaust ventilation (LEV) and wet techniques can reduce silica and respirable dust exposure by >90% [13]. Whereas, full-shift measurements, however, showed lower exposure reductions (<50%) when dust collection equipment was used [13].

The use of PPE as intervention at the receiver is considered the least desirable approach to controlling construction health hazards [12]. However, respiratory protection is primary. If the respirator is not used properly, the functions will be decreased. Research have shown that respiratory protection equipment is often not available or not used sufficiently and that it may not offer adequate protection from respirable crystalline silica for many tasks [13]. For instance, respirator protections are inadequate as it limits dust exposure only for the individual, while co-workers and nearby personnel are still exposed to it [18].

2. Methods

For this study, a survey research method was adopted. The overall research process of the study included the following four steps: (i) selecting the respondents; (ii) developing the survey instrument; (iii) conducting the survey and collecting data; and (iv) analyzing the data.

2.1. Respondent's selection

Twenty-five (25) construction employees from the G6/G7 contractor firms in Selangor and Kuala Lumpur were selected as the number of interests. The reason for choosing construction projects in Selangor and Kuala Lumpur is because both the states record the highest number of domestic project constructions. A total of 21 G7 contractors and 4 G6 contractors participated in this study, in which 72% of the firms had more than 30 years of experience in construction activity. The participating firms are currently undertaking construction projects such as residential, office, educational, highway, and railway projects. Residential projects comprise the majority of participating firms' current construction projects, contributing to about 16 firms (64%). Meanwhile, office building, educational building, highways, and railways are the minority with 3 firms (12%), 4 firms (16%), and 1 firm (4%), respectively.

This study involved collecting data through a survey method distributed to each participant from each firm. Among the 25 participants who participated in the survey, 6 participants hold the Safety and Health Officer (SHO) position, 3 were project managers (PM), and 19 were engineers. Furthermore, the frequency distribution of the participants based on their working experience shows that 15 out of 25 of the participants have less than 5 years of working experience, 9 participants whose working experience range 6-10 years, and only 1 participant has working experience within the range of 11-15 years.

2.2. Survey instrument development

A survey method was applied in order to achieve the objective of this study. The individual survey form was composed of a combination of two types of question divided into three (3) parts as follows:

(i) Part A: Demographic information of the respondent. This part is concerned with the contractors' firm and the representatives' demographic information. It was a checklist question which comprised of information of contractors' firm such as, firms' CIDB registration grade; firms' year of operation; number of employees in the firm; and types of current projects. On the representatives' information, questions such as respondents' position in the organization and working experience in the construction industry were asked.

(ii) Part B: Evaluating of the respondents' perception of risk associated with masonry dust and types of dust control that firm applied in masonry work. In part B, the respondents were asked to indicate their perception of the dust risk and health impact of exposure to dust in masonry works. This is to evaluate more about respondents' understanding of the dust risk. Then, respondents were also asked about the intervention method for controlling masonry dust implemented at their construction sites.

(iii) Part C: Identifying respondents' perceptions of the barriers to the adoption of engineering control technologies in the masonry work. The purpose of this part is to gather information from the respondents' perspective about barriers to the adoption of engineering control technologies in the masonry work.

2.3. Data collection

The surveys were conducted between August and October of 2019. The researcher contacted and asked the participants' permission to participate in the research after explaining the objective and scope of study, before distributing the survey form. The surveys were held by a site visit to the construction projects and interviewing with different participants at different locations and times. The method took approximately 30 minutes for each participant. The survey began with a brief summary of the project objective, followed by a review of the consent information and assurance that respondent identities would never be linked to the responses and that only the aggregate data would be published.

3. Results and discussion

3.1. Respondents' perception on dust risk and intervention method associated with exposure to dust hazards in masonry works

3.1.1. Awareness of dust risk in masonry activity. Most of the participants (88%) were aware about the dust risk in masonry activities, whilst the remaining 3 participants (form engineer group) were not aware about the risk. This might be due to their lack of working experience and their scope of work, which did not expose them to the hazards and risks. For example, an engineer may just carry out their duty such as monitoring the construction activity and are not directly involved with the masonry dust. Overall, it is indicated that majority of the respondents are knowledgeable regarding the risk.

3.1.2. Health impact from exposure to dust in masonry work. The respondents were asked to indicate their opinion about the health impact from exposure to dust in masonry work, as shown in figure 2. A total of twenty-five (25) responses were identified, such as respiratory problem, health diseases, skin and eyes irritation, and fever. The respondents identified mainly of potential risk with dust exposure are 59% of respiratory problems such as asthma, in which 19 out of 25 respondents stated this. The next mainly health impact from dust exposure is health diseases such as lung cancer and silicosis (25%). This is in agreement with the finding by Akbar-Khanzadeh [22], which states that exposure to crystalline silica can result in both respiratory and non-respiratory health effects. The main dust-related diseases affecting construction workers are lung cancer, silicosis, chronic obstructive pulmonary disease (COPD), and asthma [8]. Other than that, 10% of the respondents answered that exposure to dust could also cause skin irritation and sore eyes.

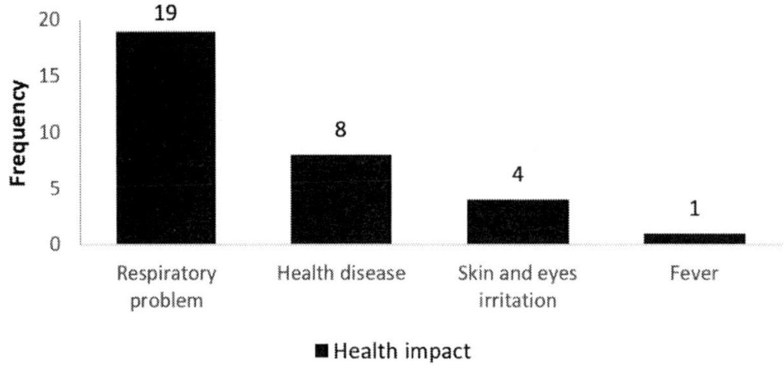

Figure 2. Health impacts from exposure to dust in masonry works.

3.1.3. Intervention method that your firm implemented to prevent dust hazards. Figure 3 shows the respondents' answers on the type of dust control that are implemented at their construction sites. Several methods were identified such as respiratory protection, wet method, and sweeping compound, in which most of the respondent's answers are the combination of more than one method. It can be clearly seen that most of respiratory protection intervention method is being implemented at all construction sites. Meanwhile, 84% of the construction sites used the wet method, followed by sweeping compound (64%). It is identified that 14 out of 25 sites implemented the combination of respiratory protection, wet method, and sweeping compound.

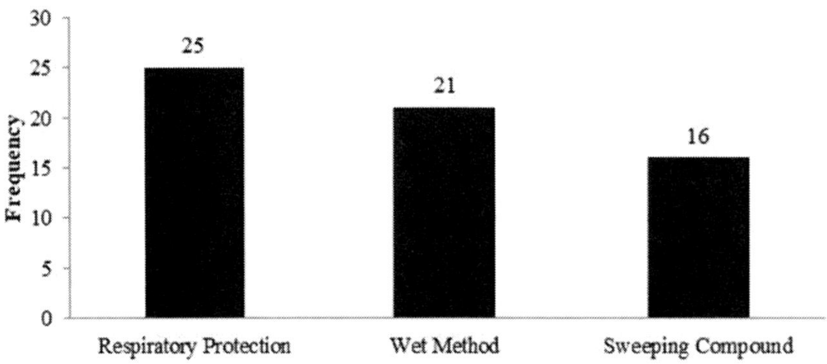

Figure 3. Types of dust control in construction site.

According to NIOSH [7], engineering control and work practices provide the best protection for workers and must be implemented first, before respiratory protection is used. Answers pertaining to respiratory protection equipment (RPE) that was most frequently used by the respondents was to prevent worker exposure to dust hazard. However, majority of the respondents provided inappropriate RPE to their workers. For example, when the workers were exposed to masonry dust, they only wore dust masks, which are not suitable respiratory protection. Ghosh et al. [12] agreed that RPE is not used properly or not available, and that this may not provide sufficient protection from RCS for any tasks.

Respondents also answered that they applied engineering control such as wet method to minimize the dust generation in masonry work. For example, a worker used a jackhammer to chip concrete by using a water spray attachment to control dust emissions. Previous studies agreed that wet method is very useful in reducing dust level [14, 22, 20]. However, mist or spray of water must be in the right angle and applied constantly at the point where the jackhammer hits the surface. This method is to wet the dust before it can spread away and effectively reduce airborne dust. Moreover, employers should train workers on the proper use of wet methods to reduce visible dust [23].

3.1.4. Best controls for preventing exposure to dust hazard. With regards to the respondents' perception of the best control for preventing exposure to dust effect in the workplace, most respondents opined that using appropriate PPE as the way to control workers exposure to dust hazard. This contradicts the principle of hierarchy of risk controls that state PPE should be is used when other control measures are not feasible and where additional protection is needed [24]. The PPE is to be used by the workers wearing the appropriate filters such as, P1 filters, that are for mechanically generated particulates like silica [25]. Meanwhile, NIOSH recommends the use of half-face piece particulate respirators with N95 or better filters for airborne exposures to crystalline silica at concentrations less than or equal to 0.5 mg/m3 [26]. Despite that, employers need to make sure that the RPE is suitable for work and fits the user's face.

Additionally, respondents also answered that job rotation or short working hours in administrative control is the best control for preventing exposure to dust hazard. This is because job rotation can reduce the time that workers are exposed to a hazard [24]. For example, respondents recommend masonry workers can be rotated in 4 hours to limit exposure to dust risk in a long time. HSE [27] stated that exposure should never normally exceed the silica maximum exposure limit (MEL) of 0.3 mg/m3 when measured or estimated for any 8-hour period, called a time-weighted average (TWA). Therefore, employers must reduce exposure to RCS dust as far as it is reasonably practicable below the MEL to make sure workers worked in a safe environment.

Other administrative measures mentioned by the respondents were to have a toolbox meeting or safety meeting before work commences. Based on Nawi et al. [28], a toolbox meeting is to discuss the health and safety matters which involve all the construction workers. Employers also need to make sure employees understand and do what they need to do when they are exposed to dust hazards. Hence, workers would be aware of the dust risk in masonry work if employers held daily health and safety briefings before the work commences.

In addition, answers pertaining to hazard identification, risk assessment and risk control (HIRARC) was identified as the way to control the exposure to dust hazard. DOSH [24] stated that the purpose of HIRARC is to enable employers to plan, introduce, and monitor preventive measures to ensure that the risks are adequately controlled at all times.

3.2. Respondents' perception of the barriers to the adoption of engineering control in masonry work

Analysis of the open-ended questions revealed that the key barriers to the adoption of engineering control technologies in the masonry work that can be categorized under 3 emergent themes, namely due to expensive cost, lack of awareness to dust hazard, and worker's attitude. The majority of respondents (47%) answered that to provide engineering controls such as LEV is expensive and not practical for construction activities. Expensive cost was also highlighted for providing appropriate type of PPE (such as RPE) compared to dust masks. Wong and Soo [29] reported that several construction companies in West Malaysia are reluctant to provide their workers with PPE and invest in safety equipment. Hence, the protection at work is inadequate, allowing the workers to work with risk at their worksites [5]. Therefore, the contractor must take the responsibility to put allocation in providing appropriate PPE to the workers even though PPE is lease feasible in the hierarchy of risk control.

Other than that, 36% of respondents answered that the employers' and employees' lack of knowledge and awareness of the impact from exposure to dust hazards is one of the hindrances for adopting engineering control technologies in masonry activities. The effect of exposure to dust is latent/delayed, in which the worker can only experience the effect after several years. Therefore, it is a responsibility for employers to provide appropriate intervention methods to reduce the risk of exposure to dust. For the effective management of work-related safety and health risks and hazards, it is crucial to the ability to recognize and identify hazards [25]. When workers lack adequate safety and health education, hazardous risk on construction sites may not be identified by the workers. Safety practices at work are vital in promoting safety amongst themselves on the construction site [30,31,32].

In addition, 17% of the respondents also added that workers' attitude is one of the obstacles in implementing Engineering control technologies in masonry work. Due to workers' stubbornness, they refuse to wear respiratory protection as it is uncomfortable to work in. Feeling uncomfortable with the equipment when doing their work on-site and seeing it as a barrier to their work output are the reasons most employees refuse to wear PPE.

4. Conclusion

Based on the results, the majority of respondents identified and indicated the masonry risks such as respiratory problems and lung diseases such as silicosis and lung cancer. The findings of this objective revealed that the main hindrance for implementing engineering control technologies in masonry work are cost issues. Other than that, the respondents also stated that the lack of awareness from employers and employees about masonry dust risk is an obstacle to perform engineering control technologies. With high costs, and the lack of awareness and proper attitude has caused low adoption rates of engineering control masonry work. The most common intervention methods applied by respondents in construction sites are respiratory protection equipment, wet method, and sweeping compound.

This study also contributes to an understanding of the risk related to exposure to dust hazards in masonry work, which will increase the safety performance in the construction site. It is particularly important for the current masonry work environment that the barriers to the adoption of Engineering control technologies can be identified. Hence, the key players in the construction industry may take an active part to increase the implementation of engineering control in construction projects.

This study has several limitations. Firstly, it is only focused within 2 states in Malaysia, Selangor and Kuala Lumpur. The concentration of the study within Selangor and Kuala Lumpur may not be adequate. Next, this study only involved 25 respondents such as the project manager, the safety and health officers, and the engineers. Future study should be carried out by adding more respondents, for example, designers and masonry workers, to obtain more understanding of intervention methods implemented for dust risk and the barriers for adopting engineering control technologies in masonry work accurately. The different populations and working environments will also obtain a different outcome.

5. References

[1] Sarkam S F, Shaharuddin L S, Zaki B M, Masdek N R N M, Yaacob N J A and Mustapha M 2018 Factors Influencing Safety Performance at the Construction Site, *International Journal of Academic Research in Business and Social Sciences* **8** 1057–1068.

[2] Nawi M N M, Ibrahim S H, Affandi R, Rosli A and Basri F M 2016 Factor Affecting Safety Performance Construction Industry, *International Review of Management and Marketing* **6** 280–285.

[3] DOSH 2020 *Occupational Accidents Statistics by Sector* Retrieved on March 1, 2021 from www.dosh.gov.my

[4] International Labour office 2012 *Safety, Health and Welfare on Construction Sites* Technical Report (Geneva: International Labour Office Geneva).

[5] Yakubu D M and Bakri M I 2013 Evaluation of Safety and Health Performance on Construction Sites (Kuala Lumpur), *Journal of Management and Sustainability* **3** 100–109.
[6] World Health Organisazation (WHO) 2001 *Occupational health - A manual for primary health care workers*,(Cairo: World Health Organization).
[7] Centers for Disease Control and Prevention, National Institute for Occupational Safety and Health 2002 *Health Effects of Occupational Exposure to Respirable Crystalline Silica* Retrieved on May 30, 2020 from https://www.cdc.gov/niosh/docs/2002-129/
[8] HSE 2013a *Construction dust. Health and Safety Executive* Retrieved on October 10, 2020 from http://www.hse.gov.uk/pubns/cis36.pdf
[9] Sen S, Mitra R, Mukherjee S K, Das P and Moitra S 2015 Silicosis in Current Scenario: A Review of Literature, *Current Respiratory Medicine Reviews* **12** 56–64.
[10] Cancer Council 2014 *Occupational Cancer Risk Series Silica Dust* (Australia: Cancer Council Australia).
[11] DOSH 2021 *Occupational Poisoning and Diseases Statistics 2019* Retrieved on August 8, 2021 from https://www.dosh.gov.my/index.php/statistic-v/occupational-diseases-statistic/3869-2019/file
[12] Ghosh S, Langar S and Bhattacharjee S 2015 A Cross Sectional Study of the Perceptions of Large Contractors towards Prevention through Design *Proceeding of the 51st ASC Annual International Conference* (Texas)
[13] Nij E T, Hilhorst S, Spee T, Spierings J, Steffens F, Lumens M and Heederik D 2003 Dust Control Measures in the Construction Industry, *British Occupational Hygiene Society* **47** 211–218.
[14] Muianga C V, Rice C H and Succop P 2009 Silica dust control in small-scale building/structure demolition operations using good work practice guidance *J. Phys.: Conf. Ser.* **151** 012055.
[15] Ertas A 2010 *Prevention through Design: Transdisciplinary Process* (Texas: The Academy of Transdisciplinary Learning and Advanced Studies. Texas) pp 1–52.
[16] HSE 2013b Control of exposure to silica dust- A guide to employees (United Kingdom: Health and Safety Executive).
[17] WHS 2013 Silica: Technical guide to managing exposure in the workplace, *2*, 1–22 Retrieved on October 1, 2020 from https://www.worksafe.qld.gov.au/__data/assets/pdf_file/0008/83186/silica_managing_workplace.pdf
[18] Lahiri S, Levenstein C, Nelson D I. and Rosenberg B J 2005 The cost effectiveness of occupational health interventions: Prevention of silicosis, *American Journal of Industrial Medicine* **48** 503–514.
[19] USDoL 2017 Control of silica dust in construction- Jackhammers or Handheld Powered Chipping Tools.
[20] Flynn M R and Susi P 2003 Engineering controls for selected silica and dust exposures in the construction industry- A review, *Applied Occupational and Environmental Hygiene* **18** 268–277.
[21] Croteau G A, Flanagan M E, Camp J E and Seixas N S 2004 The efficacy of local exhaust ventilation for controlling dust exposures during concrete surface grinding, *Annals of Occupational Hygiene* **48** 509–518.
[22] Akbar-Khanzadeh F, Milz, Ames A, Susi P P, Bisesi M, Khuder S A and Akbar-Khanzadeh M 2007 Crystalline silica dust and respirable particulate matter during indoor concrete grinding - Wet grinding and ventilated grinding compared with uncontrolled conventional grinding, *Journal of Occupational and Environmental Hygiene* **4** 770–779.
[23] US Department of Labour 2013 *Controlling Silica Exposure in Construction While Operating Jackhammers* Retrived on June 30, 2020 from http://www.ehsdb.com/resources/Images-4/Silica_images/Jack%20hammers.pdf.
[24] DOSH 2008 *Guidelines for Hazard Identification, Risk Assessment and Risk Control (HIRARC)* (Putrajaya: DOSH).

[25] Lingard H, Leifels K, Rahnama S, Fletche H and Harley J 2018 *Applying the hierarchy of control to occupational health risks in construction: Barriers to effective decision-making* (Melbourne: Centre for Construction Work Health and Safety Research, RMIT University)
[26] NIOSH 2008 *Respiratory Protection Recommendations for Airborne Exposures to Crystalline Silica* (Washington DC: Department of Health and Human Services)
[27] HSE 2001 *Controlling Exposure to Stonemasonry Dust* (United Kingdom: Health and Safety Executive).
[28] Nawi M N M, Ibrahim S H, Affandi R, Rosli A and Basri F M 2016 Factor Affecting Safety Performance Construction Industry, *International Review of Management and Marketing* **6** 280–285.
[29] Wong S S and Soo A L 2019 Factors influencing safety performance in the construction industry, *Journal of Social Science and Humanities* **16** 1–9.
[30] Mohammed Y ., Shamsul B M T and Bakri M I 2017 Assessing Workers Safety Management Knowledge on Construction Site, *International Journal of Engineering Research and Science* **3** 20–26.
[31] Kerry T V, Abas N H, Mohd Affandi H and Md. Amin S 2021 Stakeholder's perceptions on the significant factors affecting safety management implementation at construction sites, *Malaysian Construction Research Journal Special Issue* **13** 68-80
[32] Abas N H, Yusuf N, Rahmat M H and Tong Y G 2021 Safety Personnel's Perceptions on the Significant Factors thatAffect Construction Projects Safety Performance, *International Journal of Integrated Engineering* **13** 1-8

IOP Conf. Series: Materials Science and Engineering 1200 (2021) 012027 doi:10.1088/1757-899X/1200/1/012027

Readiness of Malaysian Small and Medium Enterprises Construction Companies for Building Information Modelling Implementation

N K Arif [1], M F Hasmori [1], R Deraman [1], M N Yasin [1] and M A Mohd Yassin [2]

[1] Department of Civil Engineering, Faculty of Civil Engineering and Built Environment, Universiti Tun Hussein Onn Malaysia, 86400, Parit Raja, Batu Pahat, Johor, MALAYSIA
[2] Department of Quality, KSK Group Berhad, KSK Building, 32, Jalan Yap Ah Shak, 50300 Kuala Lumpur, MALAYSIA
Corresponding author: mfikri@uthm.edu.my

Abstract. Building Information Modelling (BIM) has shown its efficiency in helping the construction industry players. The BIM helped to achieve better integration of project information and to enhance collaboration between stakeholders and the industry players from the early phase of projects until the end. Small and medium-sized companies (SMEs) have been under-represented in studies on BIM adoption. Although the technology to implement BIM is readily available and rapidly maturing, the adoption of BIM is still slow. Malaysian government support the adoption of BIM-oriented Integrated Design (ID) approaches in their national markets, however, the implementation of BIM software support systems among Small and Medium Enterprises (SMEs) in Malaysia is considered inadequate. In terms of the implementation of information technology, the construction sector in Malaysia still lags behind other sectors. To have a better knowledge of BIM in SMEs, it is important to first identify the major barriers to BIM adoption in SMEs and then examine potential solutions. Hence, the main motivation of this study is to investigate the readiness of Malaysian SMEs construction companies for BIM implementation, to determine the challenges of BIM implementation, and to develop a strategy to overcome the challenges. An online questionnaire survey was conducted and as many as 115 responses were collected. A quantitative approach was used for this research. This study uses a descriptive analysis method which comprised of structured questionnaires developed from reviews of past researches. The finding of this study indicate that BIM adoption in Malaysia is still relatively low. Many actions are needed to promote BIM adoption among construction SMEs in Malaysia.

1. Introduction

The existence of modern technologies in this era is seen to be essential for the growth of construction industries sectors in Malaysia. Building Information Modelling (BIM) has shown its efficiency in helping the industry players to analyze and develop projects design. The BIM has thus helped to achieve better integration of project information and to enhance collaboration between stakeholders and the industry players from the early phase of projects. Even though the technology to implement BIM is readily available and rapidly maturing, the adoption of BIM is still slow [1][2][3]. The ideas of Building Information Modelling (BIM) had been introduced in the Architecture, Engineering and Construction (AEC) industry to overcome problems in construction projects. Each construction player uses BIM for different purposes [4][5].

The implementation of building information modeling (BIM) in construction projects includes the digital representation of the physical and functional characteristics of the components that represent a construction project [6, 7 & 8]. In the construction industry, BIM has been able to solve several key problems. The BIM concept has embedded information and communication technology (ICT) into the construction process. Until ventures start, BIM incorporates knowledge from the outset of the building phase [9][10][11]. Therefore, by creating a digital model before the completion of the project, BIM incorporates multiple construction processes involved in handling building construction. BIM itself is not a technology, it is an information-rich, model-centric business process with the potential to transform project execution and add value over the entire life cycle of the infrastructure assets – plan, design, build and BIM is a system of knowledge about how to build things [12][13][14]. Technology is what enables the transformative potential of BIM for the construction industry. Technology allows intelligent 3D models to be developed and used and links both project parties and stakeholders.

2. Building Information Modelling definition and concept

Building Information Modelling (BIM) is a tool that converts the process of construction from conventional fragmentary processes to an integrated digital process [15][16]. It is assisted by numerous instruments, technology, and contracts for collecting, producing, evaluating, and managing the digital representation of the project model generated. BIM is well-known as an efficient connectivity and coordination n-D modeling platform and its use has expanded from the 3D geometric model to include planning, costing, and facilities management [17]. Throughout the project life cycle, BIM implementation has brought important benefits to the construction industry. BIM has demonstrated its ability to increase productivity and performance in the design stage. BIM will integrate auto-scheduling and cost analysis in the construction phase and facilitate better project delivery management and efficiency. Compared with the traditional working model, using BIM methods will help us achieve coordination, cooperation, and integration while enhancing the flow and processing of information.

Recognizing the above benefits, most AEC companies have begun to use BIM in their projects and have again tried not to introduce conventional approaches, as BIM improves their efficiency and dramatically reduces the requests for information and rework, and also applies to the handling of demolition waste. Although the possible advantages of the technologies can appear obvious, BIM's industry acceptance and adoption rate varies [18]. Ghaffarianhoseini et al. (2016) found that although the new software solutions are readily accessible, major hidden technology restrictions in the construction industry prevent their widespread acceptance. In other words, the BIM adoption in the construction industry is at early level.

3. Methodology

This study aims to explore and discuss the suitable methods for the current study as well as research design and process flows. The importance of any research method, methodology, or design lies in its ability to help the researcher accomplish the research purpose and goals most efficiently and conveniently possible. The significance of this chapter lies in the ability to establish methodological approaches and process flows, as it explains the steps and all aspects necessary to be taken as data and outcomes are gathered, evaluated, and addressed. A literature review-based questionnaire survey was done to accomplish these aims. A semi-structured interview to evaluate the hypotheses was then performed. The Internet was the source for a literature review to analyze the implementation of BIM technologies from a wider perspective.

In this research, materials such as books, journals, international conferences, and scientific papers were used. To evaluate BIM implementation in the global sense, McGraw Hill Construction 2014 reports on BIM were introduced and used to define variables for assessing BIM level based on the understanding and implementation of BIM technology, the usage of BIM, and the extent of use of BIM. The questionnaire is therefore structured to examine the participants' knowledge and awareness of BIM in this work, the current level of readiness for BIM adoption in Malaysian construction industry SMEs, the challenges to implementing BIM, and the strategy to overcome the challenge. To accomplish particular aims in this research, that is evaluating the current level of readiness for BIM adoption in Malaysian construction industry SMEs, there are many different methods of investigating and a variety of methods to consider. In this research, the method selected as the most fitting for this study is a quantitative method which is by sending structured questionnaires to construction industry professionals within SMEs as targeted respondents. This research is called targeted sampling. It is a purposeful, systematic method by which a controlled list of specified populations within geographical districts is developed and detailed plans are designed to recruit adequate numbers of cases with each of the targets. Statistical Package for the Social Science (SPSS) was used to analyse data gathered in this study. After collecting data and information via online questionnaires, the analysis process begins. The frequency distribution, reliability analysis, and descriptive analysis provided by SPSS software enable researchers in understanding the results better. Quantitative data intake was carried out using an online questionnaire.

4. Results and discussion

The data obtained by the methods specified in the previously were analyzed and discussed. The objectives of the study have been achieved by systematic analysis of the collected data. Hence, this research has collected data through online questionnaire surveys within Malaysian SME construction companies and has been analyzed and discussed. All obtained results are recorded for in-depth analysis purposes. The data is presented in tables.

The questionnaire was developed based on the awareness of respondents to BIM adoption level, BIM usage, and frequency of using BIM. The questionnaire was distributed to the construction workers online who live in Malaysia to achieve the objectives of this study. As much as 115 sets of questionnaires obtained from construction industry workers in Johor, Malaysia. The scaling method was also used to analyze each response from the respondent's choices. In this section, the questionnaire results were analyzed based on the research objectives using the findings from SPSS Analysis. The result of the readiness of BIM adoption and the level of awareness in Malaysian construction industry SMEs have been obtained and compared with the other country.

4.1. Frequency distribution analysis

Frequency distribution was conducted to know in-depth about the findings of the demographic data among respondents that have been collected. It is used to analyze data from section A from the questionnaire which contained the demographic profile of the respondents. In this section, there were 4 questions to identify the background of the respondents who participated in the survey and 9 questions to indicate the readiness of BIM adoption in a Malaysian construction company in Johor. Demographic insights are used to validate more specific information to have a deeper understanding of the attributes of the respondent. Validation of respondents is the first few questions that were asked in the online questionnaire to validate the respondent's background and their involvement in Small to Medium Enterprise (SME) Construction Companies. Respondents have to choose two (2) choices of the answer "yes or no". At the same time, they were asked to select their profession, type of organization they work for and other question that indicates the readiness of BIM adoption in SMEs Construction Company. This study manages to collect 115 valid views of the respondents. The analysis begins with a description of the demographic profile of the respondents as indicated in Table 4.1 below. For frequencies analysis, all 115 respondents are included in the said analysis to know in-depth about the respondents participate in the survey.

Table 1 shows the frequency and percentage of the demographic profiles of the respondents. A total of 115 respondents, 100% are from Small to Medium Enterprise (SME) construction companies at the same time-based in Johor. It showcased that all the respondents are working in Johor SMEs Construction Company. The second demographic characteristic is the respondent's profession. The highest group who answered the questionnaire are a civil engineer (42.6%), followed by a building surveyor (36.5) and the least are those who work as an architect (24%). These results indicated among all the respondents involve in this survey, the highest age group are mixed of various construction industry players. Their willingness to participate in this survey is mainly because of their involvement in the BIM area. On the year of service of the respondents in the construction company, only 25 respondents have experience working in the company for between 10 to 14 years which is 21.7%. The highest services year is 39.1% or 45 employees already been working in a construction company that is between 1 to 4 years and 5 to 9 years respectively.

Table 1. Frequency and percentage of respondents' profile.

No	Description	Frequency	Percentage (%)
1	SME Company		
	Yes	115	100
2	Johor Construction Company		
	Yes	115	100
3	Profession		
	Civil engineer	49	42.6
	Building surveyor	42	36.5
	Architect	24	20.9
4	Experience in construction industry		
	1-4 years	45	39.1
	5-9 years	45	39.1
	10-14 years	25	21.7

The next demographic characteristics from table 2 are the awareness of BIM and experience of using BIM in the construction project. Based on the table in 4.2, 115 respondents indicating 100% are aware of the BIM. Next, only 11 respondents indicating 9.6% have experience using BIM, meanwhile 104 respondents or 90.4% have not experienced it. Nevertheless, this shows that although the awareness of the BIM is widely promoted the level of adoption and implementation are still low to be compared to the other country.

On the usage of BIM software tools, only 17 respondents or 14.8 % used Revit (Autodesk) whereas 98 respondents, or 85.2% have not used any of the BIM software. This indicates that Malaysian construction company needs to be updated on the latest version of software of BIM and works on reaching the full adoption of BIM. Next, there are 88.7% of respondents have rated as a beginner in proficiency level in applying BIM tools and 11.3% have no experience. This indicates the level of proficiency in applying BIM tools is still relatively low to be compared with another country that has advanced.

Next, the study has shown that the majority of the respondents which is 80.9% have chosen poor adoption to describe their adoption level of BIM in the organization. Meanwhile, only 22% have no adoption at all. However, this indicates that the state of BIM adoption among Malaysian SMEs construction companies is still at a low level. Those rates give a drawback indicator of the BIM adoption in the Malaysian construction industry. Next, on the year of experience in implementing BIM for construction projects, only 17.4% or 20 employees are implementing it for 2 to 5 years. The majority which is 82.6% or 95 of the employees has less than 2 years of experience. This shows that not many of the industry players have much experience in implementing BIM in a construction project.

Furthermore, the survey has also shown that only 9.6% have attended BIM workshops or talks and 90.4% have not. To overcome this challenge, practical application of the technology must be implemented. Therefore, it is recommended that construction businesses begin implementing BIM and urge their teams to practice it. This will make it easier for businesses to incorporate BIM into their operations. The practical BIM application's adaptation and familiarity will have a major influence on professionals' understanding of BIM implementation. This indicates that BIM has not been fully adopted by the Malaysian SMEs construction company.

Table 2. Frequency and percentage of BIM adoption readiness in Malaysian construction industry SMEs.

No	Description	Frequency	Percentage (%)
1	BIM awareness		
	Yes	115	100
2	BIM experience		
	Yes	11	9.6
	No	104	90.4
3	BIM software tools usage		
	Revit (Autodesk)	17	14.8
	None	98	85.2
4	BIM tools proficiency level		
	Beginner	102	88.7
	No experience	13	11.3
5	BIM adoption level		
	Poor adoption	93	80.9
	No adoption at all	22	19.1
6	Years of implementing BIM		
	Less than 2 years	95	82.6
	2 to 5 years	20	17.4
7	Attended BIM workshop		
	Yes	11	9.6
	No	104	90.4
8	Numbers of BIM project		
	1 project	11	9.6
	2-5 projects	104	104
9	BIM exposure		
	Industry led training	7	6.1
	None/Self-taught	108	93.9

In addition, the result has shown that 90.4% of the respondents have carried out 2 to 5 projects on BIM, meanwhile, only 9.6% have worked on only 1 project of BIM. This still shows that the adoption of BIM is still low in the construction project to be compared to the other country that has used BIM for more than 10 projects. Lastly, the result from the survey has shown that the majority of the respondent which is 93.9% are exposed or learned about BIM by self-taught or none and only 6.1% are by industry Led training. This data is important to make an assumption later based on the descriptive and regression analysis on the readiness of BIM adoption in Malaysian SMEs construction companies.

4.2. Descriptive analysis
The descriptive analysis allows the researcher to summarize the analysis of data in a more meaningful way. Descriptive analysis for this study was run on 115 respondents who are working in a Malaysian SME construction company in Johor.

Table 3. Mean score range.

Range	Level
1.00 – 2.33	Low
2.34 – 3.67	Medium
3.68 – 5.00	High

Table 4. Response for challenges in BIM implementation.

Number	Description (Qs)	Mean	Standard Deviation
CH4	The high initial cost of software	4.12	1.240
CH6	Reluctance to adopt BIM from the management level	4.12	0.880
CH9	Lack of BIM standards, guidelines, and protocols	3.97	0.968
CH3	High cost of implementation process	3.84	0.790
CH5	Lack of demands from clients	3.84	0.790
CH1	Lack of information sharing in BIM	3.51	0.852
CH7	Lack of enforcement for implementing BIM by the government in construction projects	3.51	0.852
CH10	No legal or contractual agreement on BIM	3.07	0.856
CH2	Lack of professionals	2.97	0.837
CH8	Lack of best practice and guidance within the industry	2.97	0.837

From the table 3 and table 4 above, it shows that the mean for CH4 and CH6 is 4.12 which are the highest and these indicate that most of the respondent cannot afford the cost of the BIM software and the reluctance from management level to adopt BIM is one of the top challenges. The table also shows the second-highest mean is CH9 which is 3.97 as this indicates most respondents are also lacking BIM standards, guidelines, and protocols. They need clear guidelines and protocols for them to achieve total BIM adoption. Next, it is followed by CH3 and CH5 which have the same value of mean 3.84. This proves that the high cost of the implementation process of BIM and lack of demands from clients are as well listed as the top 5 challenges. Next, most respondents also agreed that they are lacking information sharing in BIM and enforcement for implementing BIM in construction projects by the government which shows the mean of 3.51. In addition, some of the respondents also agreed that they have no legal or contractual agreement on BIM as the mean is 3.07. Lastly, the lowest mean in this variable with 2.97 indicates that respondents are lacking professionals and best practices and guidance within the industry.

Table 5 shows the mean for IM1 and IM5 is 4.18 which indicates a majority of respondents agreed that providing information of BIM from the start for the top management in an organization and promoting awareness by industry's contributors are the best strategy to overcome the challenges of BIM implementation. The above table also shows the second-highest mean are IM3 and IM10 with the same value 4.12 which indicates that the respondents need to be provided BIM education and certifications to overcome these top three challenges. The table also shows that some of the respondents completely agreed that a strategic approach model is required to assist construction players to implement BIM in construction projects as the mean for IM4 is 4.08. Next, most of the respondents are also agreed that they need encouragement from top managers in the organization to implement BIM as the result shows the mean of 3.93 and followed by IM2 and IM9 with the value of mean 3.84 respectively. Lastly, the lowest mean is from IM8 with a mean value of 3.11. This also shows that BIM organizational structure needs to be established to overcome the challenge.

Table 5. Implementation strategy to overcome the challenges.

Number	Description (Qs)	Mean	Standard Deviation
IM1	Provide information on BIM from the start for the top management in an organization	4.18	0.874
IM5	Promotion and awareness by industry's contributors	4.18	0.874
IM3	Provide BIM education and certifications	4.12	0.880
IM10	Development of BIM standard legal or contractual agreement	4.12	0.880
IM4	A strategic approach model is required to assist construction players to implement BIM in construction projects.	4.08	0.785
IM6	Encouragement from top managerial in the organization to implement BIM	3.93	1.041
IM2	Employing BIM specialists	3.84	0.790
IM9	Government should provide BIM guideline	3.84	0.790
IM7	Enforcement for implementing BIM by the government in construction projects	3.83	0.973
IM8	Establish BIM organizational structure	3.11	0.915

5. Conclusion

The recent state of BIM adoption and readiness in Malaysian construction industry SMEs has been determined, the challenges to implementing BIM in Malaysian construction industry SMEs have been identified and the strategy to overcome the challenges has been recognized. The construction sector collaborates with CIDB to promote BIM adoption in Malaysia, raising awareness among professionals and organizations and encouraging them to adopt and implement the process at their workplace. Questionnaire survey was distributed through online platform to all the construction industry players.
The adoption and use of the BIM technology concept within the company has a major influence on professionals and management abilities, according to the findings of this study. If a company implement BIM in its everyday operations, the degree of BIM implementation and team awareness rises, and training and promotion seminars become ineffective if it is not applied. Willingness and dedication always lead to successful adaption participation and teamwork to improve BIM implementation awareness. The perspectives of experts on BIM adoption and advantages were explored in this study, and suggestions were made appropriately.

The most significant finding of this study is that BIM adoption in Malaysia is still relatively low in the Malaysian construction industry sectors. As a result, the adoption of BIM is delayed. The construction industry is struggling to comprehend how BIM may be implemented across parties. According to the research, BIM was implemented in a fragmented environment to create models, but BIM implementation in a full BIM project management is inefficient. As a result, intensive workshops to promote BIM should show the process of BIM implementation rather than only advocating the advantages and benefits of implementation, according to this study. To generate actual confidence for adoption, a case study demonstrating the entire system process and its effectiveness in improving the work environment is required.

6. References

[1] Ezcan V, Goulding J S, Kuruoglu M and Rahimian F P 2013 Perceptions and Reality: Revealing the BIM Gap Between the UK and Turkey, *International Journal of 3-D Information Modeling* **2** 1-15

[2] Latiffi A A, Brahim J and Fathi M S 2015 Roles and responsibilities of construction players in projects using building information modeling (BIM) *12th IFIP International Conference on Product Lifecycle Management* (Doha) pp 173-182

[3] Othman I, Al-Ashmori Y, Rahmawati Y, Mugahed Amran Y and Al-Bared M 2021 The level of Building Information Modelling (BIM) Implementation in Malaysia, *Ain Shams Engineering Journal* **12** 455-463

[4] Li Y, Deng S and Zhang Y 2019 Research on the motivation to contribution and influencing factors of university students a semi-structured interview based on qualitative research *IOP Conf. Ser.: Mater. Sci. Eng.* **563** 052095

[5] Liu S, Xie B, Tivendal L and Liu C 2015 Critical Barriers to BIM Implementation in the AEC Industry, *International Journal of Marketing Studies* **7** 162-171

[6] Construction Industry Development Board (CIDB) 2017 *Malaysia Building Information Modelling Report 2016* (Kuala Lumpur: Construction Industry Development Board)

[7] Azhar S, Khalfan M and Maqsood T 2012 Building information modelling (BIM): Now and Beyond, *Australasian Journal of Construction Economics and Building* **12** 15-28

[8] Al-Ashmori Y Y, Othman I, Rahmawati Y, Amran Y M, Sabah S A, Rafindadi A D U and Mikić M 2020 BIM benefits and its influence on the BIM implementation in Malaysia, *Ain Shams Engineering Journal* **11** 1013-1019

[9] Collis J and Hussey R 2003 *Business Research: A Practical Guide for Undergraduate and Postgraduate Students* (Houndmills, NY: Palgrave Macmillan)

[10] Zahrizan Z, Ali N, Haron A, Marshall-Ponting A and Hamid Z 2013 Exploring The Adoption Of Building Information Modelling (BIM) in the Malaysian Construction Industry: A Qualitative Approach, *International Journal of Research in Engineering and Technology* **2** 384-395

[11] Haron A T 2013 *Organisational Readiness to Implement Building Information Modelling: A Framework for Design Consultants in Malaysia* PhD Thesis (Salford: University of Salford Manchester)

[12] Memon A H, Abdul Rahman I, Memon I and Azman N I A 2014 BIM in Malaysian Construction Industry: Status, Advantages, Barriers and Strategies to Enhance the Implementation Level, *Research Journal of Applied Sciences, Engineering and Technology* **8** 606–614

[13] Latiffi A, Mohd S, Kasim N and Fathi M 2013 Building Information Modeling (BIM) Application in Malaysian Construction Industry Project management reference model View project Building Information Modeling (BIM) Application in Malaysian Construction Industry, *International Journal of Construction Engineering and Management* **2013** 1–6

[14] Khan R A, Liew M S and Ghazali Z B 2014 Malaysian Construction Sector and Malaysia Vision 2020: Developed Nation Status, *Procedia - Social and Behavioral Sciences* **109** 507–513

[15] Joblot L, Paviot T, Deneux D and Lamouri S 2017 Literature review of Building Information Modeling (BIM) intended for renovation projects, *IFAC-PapersOnLine* **50** 10518–10525

[16] Li P, Zheng S, Si H and Xu K 2019 Critical Challenges for BIM Adoption in Small and Medium-Sized Enterprises: Evidence from China, *Advances in Civil Engineering* **2019** 1–14

[17] Kong S W R, Lau L T, Wong S Y and Phan D T 2020 A study on effectiveness of Building Information Modelling (BIM) on the Malaysian construction industry *IOP Conference Series: Materials Science and Engineering* **713** 012035

[18] Ghaffarianhoseini A, Doan D T, Zhang T, Ghaffarianhoseini A, Naismith N and Tookey J 2016 A BIM readiness & implementation strategy for SME construction companies in the UK *Proc. of the 33rd CIB W78 Conference* (Brisbance)

IOP Conf. Series: Materials Science and Engineering 1200 (2021) 012028 doi:10.1088/1757-899X/1200/1/012028

Key Design Issues in the Construction Project: Conceptual and Detailed Design Review Phases

N N Sarbini[1], A N Abdul Aziz[1], A N Mazlan[1], N H Abdul Shukor Lim[1] and Mohd Fairuz Ab Rahman[2]

[1] Department of Structure & Materials, School of Civil Engineering, Faculty of Engineering, Universiti Teknologi Malaysia 81310 Johor Bahru, Johor
[2] Department of Occupational Safety and Health, Ministry of Human Resources 62530 Wilayah Persekutuan Putrajaya, Malaysia
Corresponding author: noornabilah@utm.my

Abstract. The Malaysian construction industry is among the highest ranked industries that contributes to accident and fatality rates. The Guidelines on Occupational Safety and Health in Construction Industry (Management) (OSHCIM) has introduced guidelines for design risk reduction in the construction industry. It involves determining root causes and mitigating the risk by control of the causation of accidents. The objective of this study is to identify the key design issues during the conceptual and detailed design review phases in the construction project. Although some of the design considerations have already been proposed by the OSCHIM guidelines, however, further implementation need to be diverse and versatile by considering the current design practices that are already being practiced by the industry practitioners. The methodology of this research involves document reviews, and a questionnaire survey that was used to collect answers from respondents in the construction industry. According to the results, the key design issues in the conceptual phase involve the site condition, environmental influences, ground or ground water condition, existing structure, demolition, services, adjacent to major infrastructure, traffic disruption, access for works, adjacent to other properties and projects, and site restriction. Moreover, those found in the detailed phase involve the mechanized construction system, installation of prefabrication components, ease of process, structural opening, edge line, fall hazard, accident prevention, layout optimization, ease of activities, permanent safety features, provision of access, and fall hazard. However, it is the duty of the construction industry practitioners to make sure that all other design issues are also included into their construction project risk analysis. As proposed by the OSHCIM guidelines, the review of design risks in the construction project should consider safety in all the stages: during construction, operation, maintenance, modification, and demolition.

1. Introduction

Due to the diverse and competitive nature of the construction industry, the implementation of occupational safety and health continues to be challenging [1]. Besides, safety concerns at sites are becoming the priority for any construction project planned, as the construction industry involves difficult and complex activities [2]. The authority and other key players have employed many campaigns and approaches in the construction industry, which include enhancing safety knowledge, improving work procedures, and surveying the effectiveness of the current safety practices [3]. The frequent spate of accidents related to the construction industry is a matter of concern to all. Steps must be taken to address this issue of public confidence over occupational safety and health issues at worksites.

Statistics from the Social Security Organization (SOCSO) indicates that 8,841 accidents were reported in the construction industry in 2019, as compared to 4,330 cases in 2011 [4]. In line with the SOCSO report, the Department of Occupational Safety and Health (DOSH) reported 144 deaths in the year 2019, as compared to 183 deaths in 2017, as shown in figure 1 [5]. In particular, fatal accidents in the construction industry decreased by 16% (11.28 per 100,000 workers), as compared to 13.44 per 100,000 workers in 2018. This made the construction industry come second in place after the manufacturing industry, which contributes to the highest fatality rates. However, the total number of accidents in the construction industry has increased quite significantly, by 18% (3.81 per 1,000 workers, with the total number of accidents reported as 4,863), as compared to 3.11 per 1,000 workers, with a total number of accidents reported as 3,911 in the previous year, 2018, as shown in figure 2.

While numerous approaches have been employed to reduce accidents, the number of accidents in the construction industry in Malaysia has kept increasing. Concerned with the trend, DOSH, in collaboration with the Construction Industry Development Board (CIDB), has issued a Guideline on Occupational Safety and Health in Construction Industry (Management) (OSHCIM) in 2017. The guideline had proposed three important stages to reduce risks related to design issues in construction projects. The stages include conceptual, detailed, and pre-construction design reviews. The idea of OSCHIM is to use the higher hierarchy of control methods, such as risk elimination, substitution, and engineering control, during project design stages, so that only residual risks will need to be controlled during and after construction stages. This article focuses on key design issues that the designers need to give priority upon during the conceptual and detailed design stages.

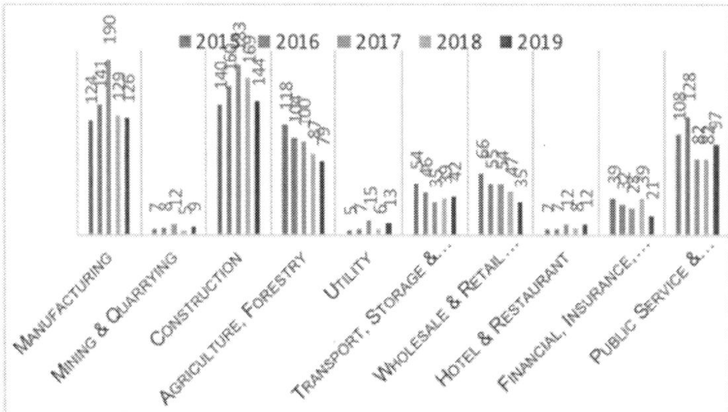

Figure 1. Statistics on fatalities in Malaysia [5].

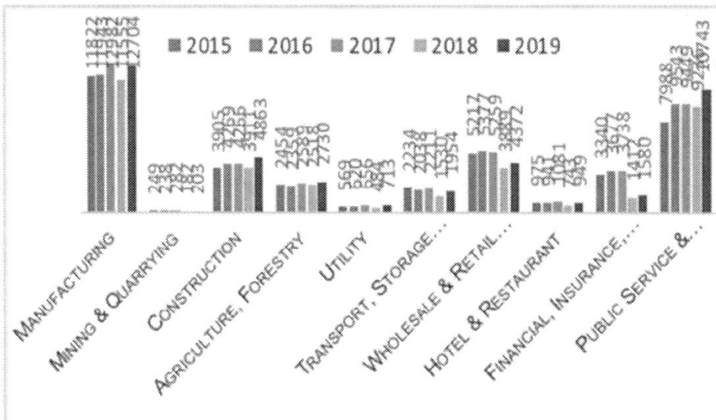

Figure 2. Statistics on accidents in Malaysia [5].

2. Literature Review

Accidents in the job are common in the construction industry, and they can result in permanent disability and a high rate of death [6]. This is evident from comparable estimates of fatal and non-fatal accidents that occur in various industries, including manufacturing. Accident statistics in the construction industry show that there is still a high rate of accidents in the Malaysian construction industry, implying that the industry is one of the key sectors that require a significant and rapid overhaul of current site safety practices. However, when basic information for mapping effective interventions, such as statistics on incidents, source of injury or death, occurrences of different types of accidents, possible causative factors, and mitigation strategy are only insufficiently available, ensuring safety becomes extremely difficult [7].

To avoid incidents at construction sites, the root causes of accidents must be investigated. Efficient and appropriate preventive action against injuries can only be taken by recognizing and determining the root causes of accidents. Many accidents are preventable and most, if not all construction accidents at the site, should be treated as a failure of management [8]. Awareness of the causes of accidents would also promote the development of a better, healthier, and more favourable working condition at construction sites. In general, there are several effective accident prevention strategies that may be used by employers and employees to reduce injury rates in the construction site, as reported by Abdul Rahim et al. [9]. These include engineering control and safety training. However, they are still lacking in suitability and practicality. In order to propose preventive approaches to mitigate the accident rate at construction sites, the preventive measure must be suitable and in line with the cause of the accident. This was solved by the Guidelines on Occupational Safety and Health in Construction Industry (Management), which proposed five key elements, as shown in figure 3.

There are three main risk review stages involved in OSHCIM. They are related to design issues that may occur before the design is being realized into the construction phase. During every phase, all types of personnel involved (e.g. consultants, clients, and contractors) need to review design issues that will involve the proposed project. The conceptual stage involves the risks that are associated with geotechnical, existing infrastructure, traffic management, site surroundings, security, and hazardous material. Meanwhile, the detailed design stage need to involve design risks related to prefabrication, heavy lifting, confined space, falling from heights, temporary works, layout, access for maintenance, health hazard, weather, emergency route, and others. Lastly, the last stage of the risk review is related to minimizing the design risk, based on approaches such as temporary work, specialist design, weather, and others. Undoubtedly, the overall design risk review focuses on the life cycle of the structure, which includes construction, operation, maintenance, modification, and demolition.

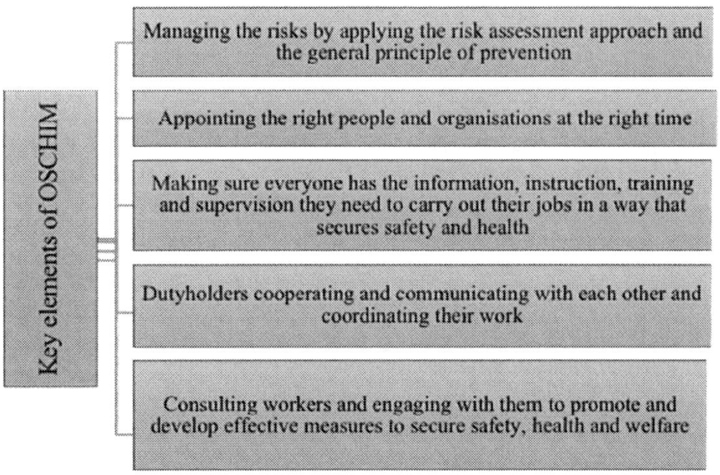

Figure 3. Key elements for risk reduction [10].

Practitioners have long ago already considered the design risks highlighted by OSHCIM. Civil and structural engineers play an important role in designing the structure so that it is stable, robust and strong. Meanwhile, mechanical and electrical engineers play their role in ensuring that services and utilities are designed suitably and safely. Moreover, the contractor and client also play their role in supporting the design proposed by the designers. All other parties such as architect, quantity surveyor and land surveyor also play their role to ensure the success of a project. OSHCIM was introduced to guide the practitioners in a more systematic and feasible design risks analysis. It will help to ensure that all possible risks are mitigated effectively, so that only minimum residual risks exist in the future. However, since this guideline is still new, it is lacking in terms of its implementation and exemplars. Therefore, this study, by considering the key design issues, intended to assist the practitioners to start considering to apply the OSHCIM guideline into their construction project. By considering key design issues, it will give them an idea to more possible design risks that may be existing in the project. Furthermore, they can also look into other foreseeable design risks that may be existing due to their proposed design.

3. Methodology

The research methodology of this study is outlined in figure 4. A document search was conducted, based on the Malaysian Department of Occupational Safety and Health, Malaysian Social Security Organization, published reports and journal papers. Based on a critical review of the literature, a questionnaire was designed in accordance with the design issues. A validation process was conducted to the survey questionnaire before distributing it to the respondents in the pilot survey phase and the actual survey program. The data that were collected were then analysed and reported in this article.

The questionnaire was designed to capture the main factors that have contributed to the increasing number of accidents that have occurred in the Malaysian construction industry. The questions were organized into two thematic parts. The first part included the respondent's demographic information. The second part involved design issues that have caused accidents. The details on the design issues that have caused accidents have been focused according to the scopes.

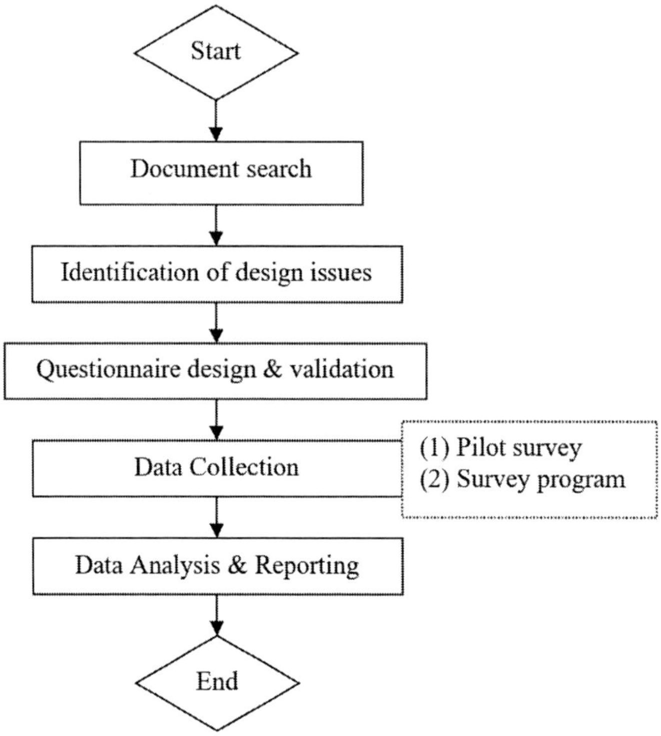

Figure 4. Flow chart of research methodology.

Table 1 shows the details of the design issues with respect to the review stages which was considered as a scope in this study. The questionnaire included multiple-choice questions with predefined answers, offering respondents the possibility to choose and rank from several options or the possibility to grade on a "1" to "5" scale. In this scale, 1 represented "no effect", 2 represented "slightly effect", 3 represented "significantly effect", 4 represented "very significant effect" and 5 represented "extremely significant effect".

Table 1. Scope of design issues.

No.	Conceptual stage	Detailed design stage
1	Geotechnical	Prefabrication
2	Existing infrastructure	Falling from heights
3	Traffic management	Layout
4	Site surroundings	Access for maintenance

After a verification of the survey questions, a pilot-test was conducted on a small sample of respondents, to ensure the questionnaire's clarity and comprehensibility, as well as its ease of completion. The pilot test included eight construction practitioners with extensive experience; five designers and three site engineers were among the respondents. Meanwhile, a total of 100 professionals were selected for the actual survey program, including safety officers, designers, project managers, site engineers and quantity surveyors. After the completion of the questionnaire survey, Microsoft Excel software was used to record the data by category. An analysis of questionnaire data was then finally carried out to meet the research objectives.

4. Results and Discussion

4.1. Design issues related to construction industry accidents

In general, there are many design issues in the construction industry since the structure itself is a complex model made from a design sketch. The results of the present review showed that there are a few standard risks that can be related to design considerations. According to the analysis, the causes of construction accidents were primarily related to the method of implementation (23.1%), which included issues such as incorrect work procedures, lack of knowledge, and failure to follow work procedures. The second most important factor was management (19.1%), which was due to a lack of work supervision, lack of technique guides, lack of communication, lack of teamwork, lack of information flow, no significant knowledge provided to the workers, no warning system, noncompliance with safety regulations, and an inadequate safety policy.

Design complexity came third in place (18.4%), because of its complexity, not being constructible, incomplete and inconsistency of design, safety not considered during design, and lack of safety by eliminating or avoiding design (lack of safety control in design). The unique nature of construction (16.3 %) came next, which includes issues related to working operation, working at height, limited working space, working in variable hazards, and transient workforce, indicating that this industry is more dangerous than others. Moreover, factors such as site condition (6.8%) and issues that are material related (5.4%), equipment related (7.5%), and external related (3.4%) all play a role in the cause of accidents in construction sites. Figure 5 and figure 6 mapped the issues in relation to its respective design considerations. The figures take account of the guidelines proposed by OSHCIM, and are also aligned with the previous literatures and statistical data from DOSH and SOCSO.

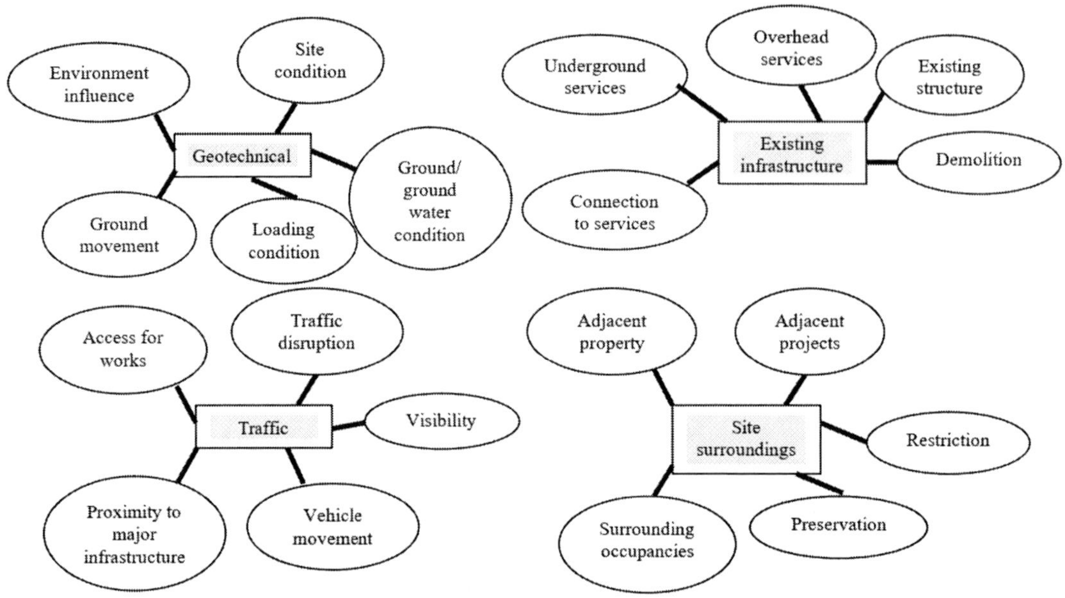

Figure 5. Design issues in the conceptual design review stage.

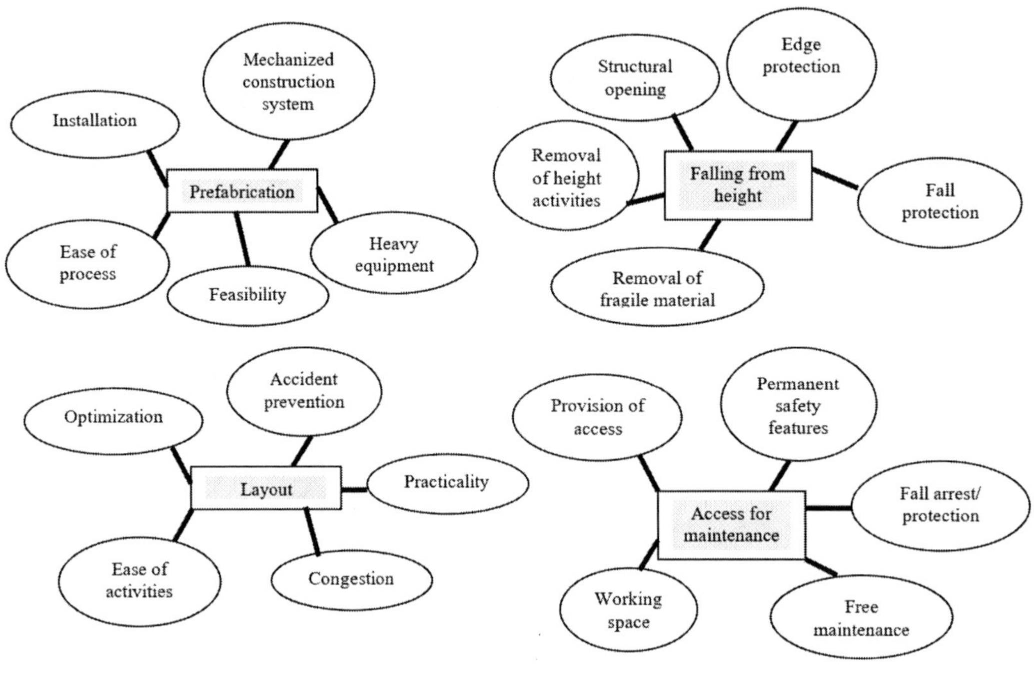

Figure 6. Design issues in the detailed design review stage.

4.2. Demographic information

The survey responses involved 58 male respondents and 42 female respondents. The respondents were from various age categories: 18 – 25 years (1%), 26 – 35 years (48%), 36 – 45 years (34%) and 46 – 55 years (17%). Meanwhile, the greatest portion of the respondents were designers or consultants at 36%, followed by safety officers at 27%, site engineers at 22%, project managers at 11% and others at 4%. Lastly, the largest category of respondents had a working experience of less than 5 years (45%), followed by 40% of the respondents having 5 – 10 years of working experience, and 15% of the respondents having working experience of more than 5 years.

4.2. Key design issues

4.3.1. Conceptual stage

The contribution of all three factors from the geotechnical aspect (stability of soil, soil condition or classification, and susceptibility to flooding due to ground water condition) on design related accidents in the construction site were ranked by a majority of respondents (54-67%) at four (very significant effect). Meanwhile, 40-44% ranked them at five (extremely significant effect), and only 2-5% ranked them at three (significant effect). The ranked results show that most of the problems encountered in construction projects are due to unforeseen ground conditions that could have been avoided if suitable precautions had been carried out. The results further indicate that all these factors should therefore be taken seriously during design risk analysis.

The existing infrastructure aspect is also as important as the geotechnical aspect. This is shown by the results of the analysis, which indicated that all the factors (existing structure, demolition process and services) were ranked at five by a large number of respondents (39-46%), at four by the majority (47-53%), and at three by a small percentage of respondents (3-9%). Designers thus need to identify and consider all these factors before the beginning of the construction phase. This is to ensure that they can provide a safe environment during the construction stage. Designers need to eliminate all the foreseeable risks, and if that is not possible, they need to modify the designs or provide a less risky option. In addition, they must ensure that appropriate information, on the reasonably practicable steps they have taken to reduce or control those risks, is included in the safety and health file.

According to the results obtained for the traffic management aspect, all the factors (adjacent to major infrastructure, traffic disruption and access for works) were ranked at five by 41-56% of the respondents, at four by 43-52% and at three by a small portion of 1-7%. Practitioners in the construction industry generally already know the massive impact of neglecting or ignoring this aspect. Current practice, a designer needing to provide a Traffic Management Plan (TMP) to ensure effective traffic management implementation at work zones. But, considering the issue early during the design phase will helps to promotes better planning and design for the on-site implementation teams, ensuring proper traffic management.

The results related to the site surroundings aspect (adjacent to other property, restriction and adjacent to other projects) show that the majority of the respondents ranked it at five (43-56%) while the others ranked it at four (42-48%). This shows that all the factors of this aspect have such a huge influence in the contribution to an accident or failure on the construction site. The client thus need to provide a brief on the necessary prior information (relevant to the planning, management and monitoring of the construction phase and the coordination of health and safety matters during the construction phase) to the principal designer and contractor. Table 2 summarize the priority for the associated key design issues.

Table 2. Ranking of key design issues for conceptual stage.

Issues	Priority	Key design issues	Remarks
Geotechnical	1	Site condition	Soil is among the hardest properties to predict and therefore a systematic technical consideration should be carried out during the design stage.
	2	Environmental influence	A construction project can caused impact to the environments and therefore risks involved need to be minimized even as early as during the design stage.

	3	Ground/ ground water condition	There are issues related to the ground water including water penetration through capillary which caused dampness and water ponding at ground floor level. Besides, a ground water issue if not properly catered will caused blockage. All of these should be avoided as early as during the design stage.
Existing infrastructure	1	Existing assets/infra-structure	An existing asset onsite should be identified as early as during the design stage so that any important decision such as a protection can be planned accordingly.
	2	Demolition process	When carrying out a new development it may involves a demolition process to the existing structure, this process need to be identified with its associated risks as it will affect the infrastructure that may be existed adjacent to the site.
	3	Services/ utilities	The existing services or utilities (especially those overhead and buried underground) need to be identified even as early as during the design stage to avoid disruption and failure if it is found later during the construction stage.
Traffic	1	Adjacent to major infrastructure	Design should consider avoiding a traffic jam from happen and the machineries road is diverted from any adjacent major infrastructure ways such as airport, school and religion buildings.
	2	Traffic disruption	When a new development is planned especially at busy and compact area, it will cause a traffic disruption especially due to the massive construction activities. It is the duty of the designers to make sure the disruption should be minimized if not to be avoided.
	3	Access for works	Access for works are important especially for a safety and security purposes. The access should be safe and secure so that any unauthorized vehicles or persons will no mistakenly entering the area.
Site surroundings	1	Adjacent to other property	Design should consider risks involves when the new development is to be constructed adjacent to other property. An example is the construction activities involves from the design project may cause failure or defect to the adjacent property.
	2	Site restriction	Restrictions from the site surrounding area need to identified in early design stage, so that it can be minimized and if not, a better management will be planned for the future.
	3	Adjacent to other projects	Design should consider risks involves when the new development is to be constructed adjacent to other project. The examples of risks are impact to the environment due to an intolerable projects.

4.3.2. Detailed design stage

The use of a prefabrication system in a construction project can benefit the practitioners in several ways. These include project quality, better monitoring, and speed of construction. However, the use of the system, if not properly monitored and planned, will initiate design risk issues. The risks can come during the process and installation. Besides, since it involves the use of a heavy machine and mechanized system, further risks will be introduced. Most of the respondents agreed that all the proposed considerations related to prefabrication in this study (mechanized construction system, installation of component and ease of process), are indeed key design issues related to prefabrication.

According to results on the falling from heights aspect, all the factors (structural opening, edge protection and fall protection) were ranked at five by more than half of the respondents (56-59%). Meanwhile, 40-43% of the respondents ranked it at four. These results show the high level of importance that need to be put under consideration during the pre-construction and construction phase. The designers need to reduce floor openings in their structures, to prevent falls from heights. The project need a safe access designed and planned prior to the construction, to avoid falls when working at a height. When this is discovered early, any accidents can be avoided and prevented.

The layout considerations for material, equipment, pedestrian and traffic flow (related to accident prevention, site arrangement optimization and ease of construction activities) during the construction stage was ranked at four by the majority (maximum of 56%). Meanwhile, they were ranked at five by maximum of 33% of the respondents, whilst they were ranked at three by maximum of 11%. This indicates that the factors are very important to ensure the smoothness of workflow in the construction phase, while providing a safe environment in the construction site.

Finally, for the access for maintenance (permanent safety features, provision of access and fall arrest), the results show that most of the respondents rank it at four and five. It is known as the maintenance activities include inspection, testing, measurement, replacement, adjustment, repair, upkeep, fault detection, replacement of parts, servicing, lubrication and cleaning. All these activities need safe access and sufficiency for the maintenance worker to work on it, in order to avoid accidents from occurring. Table 3 summarize the priority for the associated key design issues.

Table 3. Ranking of key design issues for detailed design stage.

Issues	Priority	Key design issues	Remarks
Prefabrication	1	Mechanized construction system	When prefabrication components to be used, risks will exist due to its mechanized system which used heavy equipment and machineries. The risks involves should be identified and measured.
	2	Installation	The use of prefabrication components will need an installation at their prescribed locations. The risk such as those related to drop material can occurred and therefore should be identified and evaluated.
	3	Ease of process	The process of prefabricated components installation can be identified as early as at design stage, so that the easiest and less risks method will be selected.
Falling from height	1	Structural opening	Any high risk opening within a structure for a new development should be minimized if cannot be avoided. This is especially apply for a floor opening to prevent falling from height.
	2	Edge protection	Any structural edge should be protect to prevent falling from height.
	3	Fall protection	If any structural opening or structural edge cannot be avoided from being constructed, a possibility of designing a fall protection should be considered during the design stage.
Layout	1	Accident prevention	The site and structural layout should be designed with accident prevention.
	2	Optimization	The site and structural layout should be designed to promote smooth traffic and assist the asset life cycle activities.
	3	Ease of activities	The site and structural layout should be designed so that the asset life cycle activities can be optimized and facilitated.
Access for maintenance	1	Permanent safety features	Safety features can be designed as a permanent component to facilitate the maintenance activities such as air-conditioner ledge and roof system.
	2	Provision of access	The access for maintenance works should be safe enough to prevent accidents.
	3	Fall arrest/protection	Fall protection can be designed as a part of building feature or permanent components to facilitate the maintenance works.

5. Conclusion

Growing at a rapid pace, Malaysia's construction industry continues to contribute to the country's economy. Due to the nature of the complex activities involved in the construction industry, this industry is regarded as among the dangerous in terms of its risks and needs for occupational safety and health. The Guideline of Occupational Safety and Health in Construction Industry (Management) has been issued by DOSH to decrease the risks involved in the construction industry. The approach of the guideline is to mitigate design risks involved during the conceptual, design and pre-construction stages to ensure that only residual risks exist during the life cycle of the building or structure. The guideline has highlighted several design considerations. However, more specific design considerations should also be taken from the perspectives of the practitioners (such as designers and contractors). In order to start with the risk analysis for the respective design considerations, this study has reported the key design criteria. Nevertheless, the risk analysis should not stop at the key design criteria only, as the practitioners should also include other design risks that will potentially be foreseeable along the life cycle of the building or structure. In this way, it is believed that accidents in the construction site can be prevented, or if not, then at least minimized. This will promote safety and security priority, and will benefit the practitioners directly, and the community indirectly.

6. References

[1] Hosseinian S S and Torghabeh Z 2012 Major Theories of Construction Accident Causation Models: A Literature Review, *Int. J. of Advances in Engineering & Technology* **4** 2231-2263

[2] Gursharan S 2012 Management and Administration, Monitoring and Audit of Construction Projects, *Masters Builders Journal* **2** 77-82

[3] New Straits Times 2020 *Construction Related Deaths and Injuries Alarming* Retrieved on August 5, 2021 from https://www.nst.com.my/news/nation/2020/02/565830/construction-related-deaths-and-injuries-alarming

[4] SOCSO 2018 *Annual Report* Retrieved on June 1, 2021 from https://www.perkeso.gov.my

[5] DOSH 2019 *Fatal Accident Case* Retrieved on June 1, 2021 from https://www.dosh.gov.my

[6] Chan A, Chan D, Wong F and Yam M 2008 Work at Height Fatalities in the Repair, Maintenance, Alteration, and Addition Works, *J. of Construction Eng. and Management* **134** 527-535

[7] Rafindadi A, Othman I, Napiah M and Mikic M 2020 Rate of Occurrence of Fatal Accidents in Malaysian Construction Industry after BIM Implementation, *Int. J. of Eng. and Management Research* **10(2)** 49–76

[8] Asan A and Akasah Z 2015 Developing an Accident Causation Model for Accident Prevention at Building Construction Sites *International Civil & Infrastructure Engineering Conference 2014* (Kota Kinabalu)

[9] Abdul Hamid A R, Abd. Majis M Z and Singh B 2005 Accident Investigation at Construction Sites *Civil Engineering Research Seminar (SEPKA)* (Johor Bahru) pp 306–321

[10] OSHSIM 2017 *Occupational Safety and Health in Construction Industry (Management) Guideline* 2017 (Putrajaya: Department of Occupational Safety and Health, Ministry of Human Resources) 1-74

Acknowledgments

The authors wish to acknowledge financial assistance from Ministry of Higher Education and Universiti Teknologi Malaysia through the research university grant (RUG) of UTM and collaborative research grant (CRG) of R.J130000.7351.4B527 and R.J130000.7351.4B528.

IOP Conf. Series: Materials Science and Engineering 1200 (2021) 012029 doi:10.1088/1757-899X/1200/1/012029

A Study of The Practical Use of Green Engineering-Based Technology for Building Sustainability: Users' Perspective

Amirul Hakim Hasan[1], **Sushilawati Ismail**[2]

[1,2] Department of Civil Engineering, Faculty of Civil Engineering and Built Environment, Universiti Tun Hussein Onn Malaysia, 86400 Parit Raja, Batu Pahat, Johor, MALAYSIA
Corresponding author: sushila@uthm.edu.my

Abstract. Sustainable construction has been the main priority in the global construction industry. For that reason, Malaysia is moving towards green building approach to promote built environment sustainability. Therefore, it is essential to raise awareness about the environmental-responsible practice among construction players particularly on the implementation of green engineering-based technology. This study aims to explore the existing green engineering-based technology and to investigate the preferred green engineering-based technology to upgrade the sustainability of existing residential buildings. The study focuses on the existing technology to be incorporated in the operation of buildings. A comprehensive literature review was carried out and 384 feedback was collected from questionnaire survey distribution among respondents in Johor Bahru. Various green engineering-based technology includes solar energy, wastewater treatment, rainwater harvesting, window shading, housing landscape, natural ventilation, smart PDLC film glasses, roof thermal insulation, and smart home control panel are discussed in this paper. From this study, it was found that that the housing landscape is the most preferred with the highest mean index value followed by daylighting system and solar panel. The results indicate the preferred green engineering-based technology is because of low cost and familiarity among users. The future study should consider relating the awareness of individuals on sustainability with their preferred green technology. Perhaps, the coverage of this study should be extended by involving respondents from all over Malaysia.

1. Introduction

Sustainable development is an important aspect of building construction in modern times. The development that incorporated a sustainability approach required sustainable principles in construction to be practiced. Thus, one of the solutions to support sustainable practice in construction is green engineering-based technology. It is a method to promote sustainable techniques like incorporating smart and advanced equipment in buildings to ensure more economical, high efficiency, and environmental-friendly. The adoption of green engineering-based technology in construction activities could promote sustainable buildings in terms of energy usage and minimization of waste. It enhances the use of wastes to be reused and recycled into useful products. This promotes less usage of natural resources and saves money. Sustainability is achieved by re-evaluating the issue that causes environmental pollution to design a more constructive way of green engineering to be more sensible to be applied in building [1].

In order to promote building sustainability, every phase of the building life cycle has to be guided by sustainable principles. Energy efficiency and application of renewable energy are among the important element to attend to. Many contractors recently implementing a sustainable approach to preserve the environment at the same time conducting their construction activities [2]. Materials should be selected based on their significance of impacts to the environment as well as to the well-being of humans throughout the construction lifecycle. Besides, sustainable building features perhaps could maximise operational efficiency which contributing to the energy consumption efficiencies. However unfortunately, green practice in construction has been neglected among contractors due to unfamiliar exposure to this approach [3]. This situation needed further attention to make sure the construction practitioners aware of the technology that exists for protecting the environment, while simultaneously improving human health and well-being, and conserving valuable resources like water and energy. Therefore, the exploration of green engineering-based technology is necessary. The objectives of this study are to discover green engineering-based technology that contributing to building sustainability as well as preferred choice of users incorporating such technologies into their residence.

1.1. Green engineering-based technology

Green engineering-based technology is the incorporation of sustainable features in a product by modifying the existing technology to reduce implications towards ecosystem [4]. The equipment or technology based on sustainability purposes need to be designed to be practical to install and show benefits towards consumers. Commonly known as eco designs, which has endless conveniences that serve towards sustainable practices among users [5]. Eco design equipment able to reduce usage of energy and reduce emission of greenhouse gases towards environment [6]. This means eco design need to be environmentally efficient when in operation. The execution of eco design in retrofitting buildings need to focus towards achieving green building criteria. Thus, the green based technology is appropriate to replace the conventional technology in order to operate better towards sustainability focused.

Green engineering-based technology must ensure to not neglect the detrimental effects towards environment during production and operation [7][8]. The green technology needs to be durable and user friendly. The high durability of technology avoid frequent replacement as this will save resources in production [9]. The technology which is sustainability driven need to consume minimal amount of energy as this will provide great energy saving. For instance, power utilization can be reduced with installation of green engineering for the operation of buildings [7]. Addition to this, the ecoefficiency growth increases in time when existing buildings are been retrofitted with green engineering.

Green engineering-based technology need to be profitable with the inclusion of eco-friendly requirements [10]. Underdeveloped nations faced barriers in practicing green engineering for buildings such as shortage of venture capital and exposure of its application [7]. Insufficient of proper execution of green engineering-based technology among stakeholders is caused by their ignorance about the applications and practicability [11]. This study perhaps could provide collective insight and idea in incorporating green engineering-based technology into building project development or even in redevelopment of existing building.

2. Materials and methods

There were two parts involved in this study which are comprehensive literature review and questionnaire survey.

2.1. Literature review

Literature review is important to gain information about related scope of research from past studies. In this study, comprehensive literature review was carried out to achieve the main objective of the study which is to explore the existing green engineering-based technology in promoting sustainable building. The green engineering-based technology explored in literature review are based on existing technology. The most used databases for literature review were ScienceDirect, Scopus, ResearchGate and Springer Link as they provide interdisciplinary research databases. The Boolean operators consist of 'AND' or 'OR' were used with keywords in identifying chosen journals. The keywords and Boolean operators were used together which consist of "solar energy" AND "sustainability" OR "energy saving", "wastewater treatment" AND "sustainability", "rainwater harvesting" AND "sustainability, "window shading" OR "window tint" AND "sustainability", "landscape" AND "sustainability", "natural ventilation" AND "sustainability", "PDLC" OR "smart window" AND "sustainability", "roof insulation" AND "sustainability", "smart home" OR "building automation system" AND "sustainability". Next, the results were filtered according to publication years which were between 2010 to 2021.

2.2. Questionnaire survey
A questionnaire survey was used to investigate the preferred green engineering-based equipment to upgrade the sustainability of existing residential building. The questionnaire was distributed to respondents which are residents in Johor Bahru. A total of 384 feedbacks were collected and considered sufficient for analysis based on Krejcie and Morgan's sample size table [12]. A brief information with photo of each green engineering-based technology were included in the questionnaire form. This is to give an idea to respondents regarding the green engineering-based technology in helping them to answer the survey. The first section of questions was using 5-point Likert scale to identify their preferable on the listed technologies, meanwhile the second section asking on price ranges that they willing to spend for each of the item. The IBM SPSS Statistics software is used to analyse the data obtained based on the mean index as shown in equation (1).

$$Mean\ Index = \frac{\Sigma(1X_1+2X_2+3X_3+4X_4+5X_5)}{\Sigma(X_1+X_2+X_3+X_4+X_5)} \quad (1)$$

where; X_1=Number of respondents on scale 1
X_2=Number of respondents on scale 2
X_3=Number of respondents on scale 3
X_4=Number of respondents on scale 4
X_5=Number of respondents on scale 5

3. Results and discussion
This section presents in details the findings of the study from comprehensive literature review and analysis of the questionnaire survey.

3.1. Exploration of the existing green engineering-based technology in promoting sustainable building
From the literature study, there are nine types of green engineering-based technology were identified and chose for this study includes solar energy, wastewater treatment, rainwater harvesting, window shading, housing landscape, natural ventilation, smart PDLC film glasses, roof thermal insulation and smart home control panel. All of them were found significantly contributed to the sustainability of a building in many ways as shown in Table 1.

Table 1. Contribution of green engineering-based technology towards sustainability.

Green engineering-based technology	Contribution towards sustainability
Solar panel	The most environmental-friendly to generate power [13] Enhance ecofriendly value of building [14] Produce more renewable energy [15] Reduce consumption of raw materials by replacing roof with building integrated photovoltaics (BIPV) [16] Promote production of solar energy which considered as finest renewable energy [17]
Wastewater treatment	Conserve water consumption [18] Reduce pollutants to be released into catchment areas [19] Curb scarcity of water supply [20] Reduce power consumption and substitute for fertilizers [21] Microfiltration to treat laundry wastewater require less expenses and polluted water not generated [22]
Rainwater harvesting	Decrease consumption of water [23] Prevent water wastage [24] Decrease non-renewable energy usage [25]

Green engineering-based technology	Contribution towards sustainability
	Lessen utilization of clean water [26] Conservation of water [27]
Window shading	Reduce energy consumption of cooling load [28] Enables conservation of energy from consumption of artificial lighting [29] Decrease thermal loading decrease, therefore consequently requires less cooling load [30] Minimise cooling loads consumption [31] Provide optimum natural lighting, thus save energy usage on electricity [32]
Housing landscape	Contribute to less gain of heat energy that requires high power consumption [33] Reduce consumption of cooling energy [34] Provide cooling effect to surrounding area consequently provide natural cooling [35] Provide catchment areas for runoff and contribute to indoor comfort temperature with minimal power required [36] Minimize heat gain, therefore requiring less air conditioning usage [37]
Natural ventilation	Decrease usage of energy in building [38] Provide comfort temperature, consequently reduce utilization of air conditioning and fans [39] Improve air flow without using so much electricity for cooling [40] Better efficiency of air flow and reduce power usage [41] Reduce cooling loads [42]
Smart PDLC	Decrease heat gain, then reduce usage of air conditioning [43] Lower the usage of lighting [44] Regulate thermal gain of building consequently lessen cooling loads [45] Promote energy efficiency [46] Enhance the performance of cooling loads with minimal power [47]
Roof thermal insulation	Decrease thermal radiation [48] Enhance thermal reflectance and requires less energy to cool the building [49] Enhance efficiency of building energy [50] Declining temperature in building for optimum usage of mechanical cooling [51] Minimise heat gain and deduction of energy required in building [52]
Smart home control panel	Saves energy consumption at home [53] Reduce unnecessary wastage of electrical consumption [54] Reduce the usage of artificial lighting [55] Decrease the usage of energy [56] Managing energy consumption with integrated automation system monitors power consumption of building [57]

Solar energy is recognized for its practicality as green energy. It is stated as the most environmental-friendly in generating electricity [58]. Solar energy can be produced by integrating solar panels at home which able to harvest sunrays. With continuous evolving for the betterment of solar panels, the generation of electricity will be more significant with solar energy causing less dependent on fossil fuels. Hence, a significant reduction of carbon and greenhouse gas emissions can be achieved.

For wastewater treatment, reusing and treating wastewater enable to conserve the usage of freshwater [20]. Wastewater can also be reused for watering crops, even can be converted to fertilizers[18,21]. On the other hand, treating wastewater could reduce pollutants to be released into catchment areas and this prevents water contamination [19].

Besides that, consumption of water can be reduced by implementing rainwater harvesting [23][27]. The usage of daily water demand supplied from harvesting rainwater causing less burden on freshwater supply. With the demanding usage of water due to the increasing of the human population, practicing rainwater harvesting might conserve water for important usage.

Contributing to indoor comfort, window shading is significant to control sunrays entering the building. This is proven when upgrading window with double glazing and solar film reduces energy consumption of cooling loads [28]. Besides, optimization of blinds and louvers provides optimum natural lighting during daytime [32]. Thus, window shading is required to reduce the usage of cooling loads and artificial lighting in the building [29].

For the housing landscape, thermal gain inside the building can be controlled with a better shading mechanism [37]. Suitable tress can provide shading for housing from direct rays of sunlight. Another point, evapotranspiration from plants provides a cooling effect to the surrounding area of the building [35]. Hence, housing landscape is beneficial to ensure better efficiency of mechanical cooling performance in the building, thus contributes to lesser energy consumption.

Furthermore, natural ventilation is identified as a way to naturally cool the house with minimal usage of mechanical ventilation [38]. Night-time ventilation is identified as the best time to allow cool air from outside to cool the building [39]. Thus, incorporating natural ventilation and mechanical ventilation separately or together will reduce the power consumption of cooling loads required [42]. This lead building to be more energy efficient in operation.

Next, smart PDLC film glasses are advancements of green technology which able to regulate the thermal gain inside a building with minimal power required to operate. It is shown that utilizing PDLC contribute to lesser heat gain in a building [43,46]. The adoption of smart PDLC film glasses is beneficial to achieve thermal comfort in buildings with reducing consumption of cooling loads.

For roof thermal insulation, heat gain in a building can be reduced [49][51]. It was found that the adoption of aluminium insulation is better in efficiency to reduce the absorption of heat compared to low emissivity paint [48]. By reducing the thermal gain, excessive cooling energy needed can be reduced to achieve indoor thermal comfort.

Another technology, a smart home control panel is an integrated technology to control the usage of electrical appliances at home. More importantly, energy monitoring is possible with this technology. This enables the building occupants to optimize the energy consumption and manage the operating electrical appliances [54]. Thus, the integration of building automation system into homes or buildings is believed could significantly minimize the usage of electricity [54][56].

3.2. Users' preference of the green engineering-based technology

Data of the preferred green engineering-based technology of the users to be incorporated into their homes is shown in Figure 1. Based on the mean value shown, the highest mean reflects the most preferred green engineering-based technology while the lowest mean reflects the least preferred green engineering-based technology in terms of upgrading existing houses to be more sustainable. According to the analysis shown, the highest mean value is 4.40 which is housing landscape, ranked as first. The second rank with a mean of 4.33 is a daylighting system. Both of these are more preferred among other technologies perhaps because they are easy to execute. Besides, they are also a low-cost solution. On the contrary, even the solar panel known as one of the expensive technology to adopt, it was ranked third with mean value 4.27. Probably this might due to the potential of electrical savings that could be earned by the users.

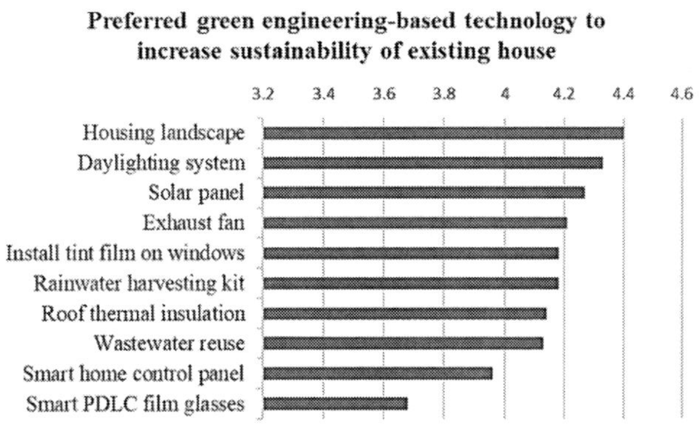

Figure 1. Mean for the preferred green engineering-based technology.

The least mean ranked is smart PDLC film glasses with a value of 3.68. This shows the retrofitting with this technology is the lowest preferred among respondents. The high cost and unfamiliar technology of smart PDLC film glasses might be reasons for respondents to not install it at their houses. Besides, most respondents are living in terrace houses and they perceived this technology is not necessary to be installed as their houses are not categorized as luxury type.

Table 2 shows the willingness to spend on green engineering-based technology among respondents. The price ranges for each item are shown in the table. For the solar panel, the highest chosen price range is 'RM5000-RM20,000' with 65.10%. This is the most chosen among respondents due to the affordable range to spend on solar panels. Only 0.8% willing to spend at the highest price range. For other items such as rainwater harvesting kit, tint film, housing landscape, exhaust fan, roof insulation and smart home system, high percentage of respondent are also willing to spend but with minimal range of spending. Differently for smart PDLC film glasses, it was found the percentage of respondents that unwilling to spend and willing to spend with minimal range is quite closed which are 43% and 40.1% respectively. This might because the cost of the technology is high and expensive. Most respondents probably are not familiar with the advanced technology of smart glasses. Consider among the cheapest solution, for wastewater treatment and daylighting system, the majority of respondents willing to spend as required. Perhaps both fall under affordable expenses.

Table 2. User's willingness to spend on green engineering-based technology.

Cost to spend on Green Engineering-Based Technology	Frequency Percentage (%)
Solar panel	
Not willing to spend	29.40
RM5000-RM20,000	65.10
RM20,000-RM40,000	4.70
>RM40,000	0.80
Wastewater treatment	
Not willing to spend	17.70
Spend as required	82.30
Rainwater harvesting kit	
Not willing to spend	24.00
RM1000-RM5000	70.80
RM5000-RM10,000	4.70
>RM10,000	0.50
Install tint film on windows	
Not willing to spend	15.60
RM100-RM300	47.90
RM300-RM500	24.20
>RM500	12.20
Daylighting system	
Not willing to spend	16.10
Spend as required	83.90
Housing landscape	
Not willing to spend	11.50
RM100-RM300	44.80
RM300-RM500	27.30
>RM500	16.40
Exhaust fan	
Not willing to spend	11.50
RM100-RM200	55.20
RM200-RM300	22.40

Cost to spend on Green Engineering-Based Technology	Frequency Percentage (%)
>RM300	10.90
Smart PDLC film glasses	
Not willing to spend	43.00
RM1000-RM3000	40.10
RM3000-RM5000	11.20
>RM5000	5.70
Roof thermal insulation	
Not willing to spend	26.80
RM1000-RM2000	53.40
RM2000-RM3000	19.80
Smart home control panel	
Not willing to spend	32.60
RM1000-RM2000	46.10
>RM2000	21.40

4. Conclusion

This study emphasizes on retrofitting existing residential buildings with practical green engineering-based technology for building sustainability. The green engineering-based technology that has been recognized are solar energy, wastewater treatment, rainwater harvesting, window shading, housing landscape, natural ventilation, smart PDLC film glasses, roof thermal insulation, and smart home control panel as comprehensively discussed. From a total of 384 feedbacks, it was found that the most three preferred green engineering-based technology are housing landscape, daylighting system, and solar panel with mean values 4.40, 4.33, and 4.27 respectively. From the willingness-to-spend survey, respondents are likely to spend as required for the daylighting system. Meanwhile for both housing landscape and solar panels, most of them only willing to spend in a minimum range. Thus, it can be concluded that the low cost and familiarity among users were the main criteria that contributed to green engineering-based technology preferred in this study. The findings of the study may be useful for construction practitioners to embed the green engineering-based technology to retrofit the existing residential buildings to enhance energy efficiency and sustainability.

5. References

[1] McDonough W, Braungart M, Anastas P T and Zimmerman J B 2003 Applying the Principles of Green Engineering to Cradle-to-Cradle Design, *Environmental Science and Technology* **37** 434A-441A

[2] Chi B, Lu W, Ye M, Bao Z and Zhang X 2020 Construction waste minimization in green building: A comparative analysis of LEED-NC 2009 certified projects in the US and China, *J. Clean. Prod.* **256** 120749

[3] Shen L, Zhang Z and Long Z 2017 Significant barriers to green procurement in real estate development, *Resour. Conserv. Recycl.* **116** 160-168

[4] ISO 14006:2020 2011 *Environmental management systems - Guidelines for incorporating ecodesign* (London: International Organization for Standardization)

[5] Triebswetter U and Wackerbauer K 2008 Integrated environmental product innovation in the region of Munich and its impact on company competitiveness, *J. Clean. Prod.* **16** 1484-1493

[6] Quist J and Tukker A 2013 Knowledge collaboration and learning for sustainable innovation and consumption: Introduction to the ERSCP portion of this special volume, *Journal of Cleaner Production* **48** 167-175

[7] Liu Y, Chen H and Wang X J 2021 Research on green renovations of existing public buildings based on a cloud model –TOPSIS method, *J. Build. Eng.* **34** 101930

[8] Bohari A A M, Skitmore M, Xia B, Teo M and Khalil N 2020 Key stakeholder values in encouraging green orientation of construction procurement, *J. Clean. Prod.* **270** 122246
[9] Nasier S 2020 Sustainable green materials for new construction in *Materials Today: Proceedings* **37**
[10] Brambila-Macias S A and Sakao T 2021 Effective ecodesign implementation with the support of a lifecycle engineer, *J. Clean. Prod.* **279** 123520
[11] Dekoninck E A, DomingoL, O'Hare J A, Pigosso D C A, Reyes T and Troussier N 2016 Defining the challenges for ecodesign implementation in companies: Development and consolidation of a framework, *J. Clean. Prod.* **135** 410-425
[12] Krejcie R V and Morgan D W 1970 Determining Sample Size for Research Activities, *Educational and Psychological Measurement* **30** 607-610
[13] Dixit S 2020 Solar technologies and their implementations: A review, *Mater. Today Proc.* **28** 2137–2148
[14] Magadley E, Kabha R and Yehia I 2021 Outdoor comparison of two organic photovoltaic panels: the effect of solar incidence angles and incident irradiance, *Renew. Energy* **173** 721–732
[15] Sun W, Yuan P, Wang S, Du Y, Zou J, Cao Fengxian, Lan Z and Wu J 2021 Kalium persulfate as a low-cost and effective dopant for spiro-OMeTAD in high performance and stable planar perovskite solar cells, *Electrochim. Acta* **380** 138233
[16] Strong S, *Building Integrated Photovoltaics (BIPV)* Retrieved on Dec 10, 2020 from https://www.wbdg.org/resources/building-integrated-photovoltaics-bipv
[17] Hafez A Z, Soliman A, El-Metwally K A and Ismail I M 2017 Tilt and azimuth angles in solar energy applications – A review, *Renew. Sustain. Energy Rev.* **77** 147–168
[18] Radingoana M P, Dube T and Mazvimavi D 2020 Progress in greywater reuse for home gardening: Opportunities, perceptions and challenges, *Phys. Chem. Earth* **116** 102853
[19] Salgot M and Folch M 2018 Wastewater treatment and water reuse, *Curr. Opin. Environ. Sci. Heal.* **2** 64–74
[20] Bilad M R, Mat Nawi N I, Subramaniam D D, Shamsuddin N, Khan A L, Jaafar J and Nandiyanto A B D 2020 Low-pressure submerged membrane filtration for potential reuse of detergent and water from laundry wastewater, *J. Water Process Eng.* **36** 101264
[21] Akpan V E, Omole D O and Bassey D E 2020 Assessing the public perceptions of treated wastewater reuse : opportunities and implications for urban communities in developing countries, *Heliyon* **6** e05246
[22] Manouchehri M and Kargari A 2017 Water recovery from laundry wastewater by the cross flow microfiltration process : A strategy for water recycling in residential buildings, *J. Clean. Prod.*, **168** 227–238
[23] Pala G K, Pathivada A P, Velugoti S J H, Yerramsetti C and Veeranki S 2021 Rainwater harvesting - A review on conservation, creation & cost-effectiveness, *Mater. Today Proc.* **45** 6567-6571
[24] Hanley S 2017 *Inverted Umbrella Brings Clean Water & Clean Power To India* Retrieved on Feb 14, 2021 from https://cleantechnica.com/2017/12/04/inverted-umbrella-brings-clean-water-clean-power-india/
[25] Pham D 2017 *New rooftop solar hydropanels harvest drinking water and energy at the same time* Retrieved on Jan 7, 2020 from https://inhabitat.com/new-rooftop-solar-hydropanels-harvest-drinking-water-and-energy-at-the-same-time/
[26] López Zavala M A, Prieto M J C and Rojas C A R 2018 Rainwater harvesting as an alternative for water supply in regions with high water stress, *Water Sci. Technol. Water Supply* **18** 1946–1955
[27] Kashiwar S R and Dongarwar U R 2016 An Overview on the Ground Water Recharge by Rain Water Harvesting, *Journal of Energy Research and Environmental Technology* **3** 146–148
[28] Somasundaram S, Thangavelu S R and Chong A 2020 Improving building efficiency using low-e coating based retrofit double glazing with solar films, *Appl. Therm. Eng.* **171** 115064
[29] Hee W J, Alghoul M A, Bakhtyar B, Elayeb O, Shameri M A, Alrubaih M S and Sopian K 2015 The role of window glazing on daylighting and energy saving in buildings, *Renew. Sustain. Energy Rev.* **42** 323–343

[30] Li C, Tan J, ChowT T and Qiu Z 2015 Experimental and theoretical study on the effect of window films on building energy consumption, *Energy Build.* **102** 129–138

[31] Edeisy M and Cecere C 2017 Envelope Retrofit in Hot Arid Climates, *Procedia Environ. Sci.* **38** 264–273

[32] Reffat R M and Ahmad R M 2020 Determination of optimal energy-efficient integrated daylighting systems into building windows, *Sol. Energy* **209** 258–277

[33] Misni A 2016 Strategically Designed of Landscaping around Houses Produce an Extensive Cooling Effect, *Procedia - Soc. Behav. Sci.* **222** 693–701

[34] Zheng X,Dai T and Tang M 2020 An experimental study of vertical greenery systems for window shading for energy saving in summer, *J. Clean. Prod.* **259** 120708

[35] Georgi J N and Dimitriou D 2009 The contribution of urban green spaces to the improvement of environment in cities : Case study of Chania, Greece, *Build. Environ.* **45** 1401–1414

[36] Larson K L, Andrade R, Nelson K C, Wheeler M M, Engebreston J M, Hall S J, Avolio M L, Groffman P M, Grove M, Heffernan J B, Hobbie S E, Lerman S B, Locke D H, Neill C, Roy R and Trammell T L E 2020 Municipal regulation of residential landscapes across US cities : Patterns and implications for landscape sustainability, *J. Environ. Manage.* **275** 111132

[37] Misni A 2013 Modifying the Outdoor Temperature around Single-Family Residences: The Influence of Landscaping, *Procedia - Soc. Behav. Sci.* **105** 664–673

[38] Bekkouche S M A, Benouaz T, Belarbi R, Cherier M K, Ferrão P, Corre O Le and I U De 2017 The Study Natural Ventilation by Using Buildings Windows : Case Assessing feasibility of using the heat demand-outdoor Study the in a Hot Dry Climate, Chardaïa, Algeria, *Energy Procedia* **139** 475–480

[39] Kyritsi E and Michael A 2019 An assessment of the impact of natural ventilation strategies and window opening patterns in office buildings in the mediterranean basin, *Build. Environ.* **175** 106384

[40] Zhang H, Yang D, TamV Y W, Tao Y, Zhang G and Setunge S 2021 A critical review of combined natural ventilation techniques in sustainable buildings, *Renew. Sustain. Energy Rev.* **141** 110795

[41] Meng X, Wang Y, Xing X and Xu Y 2020 Experimental study on the performance of hybrid buoyancy-driven natural ventilation with a mechanical exhaust system in an industrial building, *Energy Build.* **208** 109674

[42] Ezzeldin S and Rees S J 2013 The potential for office buildings with mixed-mode ventilation and low energy cooling systems in arid climates, *Energy Build.* **65** 368–381

[43] Shaik S, Gorantla K, Venkata Ramana M, Mishra S and Kulkarni K S 2020 Thermal and cost assessment of various polymer-dispersed liquid crystal film smart windows for energy efficient buildings, *Constr. Build. Mater.* **263** 120155

[44] Baetens R, Jelle B P and Gustavsen A 2020 Properties, requirements and possibilities of smart windows for dynamic daylight and solar energy control in buildings: A state-of-the-art review, *Sol. Energy Mater. Sol. Cells* **94** 87–105

[45] Ghosh A and Mallick T K 2018 Evaluation of optical properties and protection factors of a PDLC switchable glazing for low energy building integration, *Sol. Energy Mater. Sol. Cells* **176** 391–396

[46] Hemaida A, Ghosh A, Sundaram S and Mallick T K 2020 Evaluation of thermal performance for a smart switchable adaptive polymer dispersed liquid crystal (PDLC) glazing, *Sol. Energy* **195** 185–193

[47] Ghosh A and Norton B 2018 Advances in switchable and highly insulating autonomous (self-powered) glazing systems for adaptive low energy buildings, *Renew. Energy* **126** 1003–1031

[48] Fantucci S and Serra V 2019 Investigating the performance of reflective insulation and low emissivity paints for the energy retrofit of roof attics, *Energy Build.* **182** 300–310

[49] Pisello A L, Castaldo V L, PignattaG, Cotana F and Santamouris M 2016 Experimental in-lab and in-field analysis of waterproof membranes for cool roof application and urban heat island mitigation, *Energy Build.* **114** 180–190

[50] Filho J P B and Santos T V O 2014 Thermal analysis of roofs with thermal insulation layer and reflective coatings in subtropical and equatorial climate regions in Brazil, *Energy Build.* **84** 466–474
[51] YewM C, Ramli Sulong N H, Chong W T, Poh S C, Ang B C and Tan K H 2013 Integration of thermal insulation coating and moving-air-cavity in a cool roof system for attic temperature reduction, *Energy Convers. Manag.* **75** 241–248
[52] Lee S W, Lim C H and Salleh E @ I 2016 Reflective thermal insulation systems in building: A review on radiant barrier and reflective insulation, *Renew. Sustain. Energy Rev.* **65** 643–661
[53] Grady T O, Chong H and Morrison G M 2021 A systematic review and meta-analysis of building automation systems, *Build. Environ.* **195** 107770
[54] Gomes L, Ramos C, Jozi C, Serra B, Paiva L and Vale Z 2019 IoH: A Platform for the Intelligence of Home with a Context Awareness and Ambient Intelligence Approach, *Future Internet* **3** 11030058
[55] Babu S, Zhou J, Wan M P, Lamano A S, Sarvaiya J N, Zhang Z, Kumar D E V S K, Gao C ping, Valliappan S, Goh A and Seoh A 2019 Investigation of an integrated automated blinds and dimmable lighting system for tropical climate in a rotatable testbed facility, *Energy Build.* **183** 356–376
[56] AghemoC, Blaso L and Pellegrino A 2014 Building automation and control systems: A case study to evaluate the energy and environmental performances of a lighting control system in offices, *Autom. Constr.* **43** 10–22
[57] Mataloto B, Ferreira J C and Cruz N 2019 LoBEMS - IoT for Building and Energy Management Systems, *Energy Management Based on Internet of Things* **8** 763
[58] Dixit M K, Fernández-Solís J L, Lavy S and Culp C H 2012 Need for an embodied energy measurement protocol for buildings: A review paper, *Renewable and Sustainable Energy Reviews* **16** 3730-3743

Acknowledgments
The authors would like to thank the Faculty of Civil Engineering and Built Environment, Universiti Tun Hussein Onn Malaysia for its support.

IOP Conf. Series: Materials Science and Engineering 1200 (2021) 012030 doi:10.1088/1757-899X/1200/1/012030

Triaxial Testing on Geogrid-reinforced Granular Soils

Tigo Mindiastiwi[1], Po-Kai Wu[2], Agus Bambang Siswanto[3] and Mukhamad Afif Salim[3]

[1]Department of Civil Engineering, 17 Agustus 1945 Semarang University, Semarang, Indonesia
[2]Department of Civil and Construction Engineering, National Yunlin University of Science & Technology, Yunlin, Taiwan
[3]Department of Civil Engineering, 17 Agustus 1945 Semarang University, Semarang, Indonesia
Corresponding author: tigomindiastiwi@untagsmg.ac.id

Abstract. Laboratory triaxial compression tests were carried out to investigate the mechanical behavior of dense sand and geogrid-reinforced granular soils. The tested sand having its mean particle size (D_{50}) equal to 0.6 mm was adopted. Three geogrids with different longitudinal and transverse nominal strengths were used. The dimensions of the cylindrical soil specimen were 70 mm (diameter) × 160 mm (height). The relative density was equal to 70% for all tests. The reinforced sand specimens with one or two geogrid layers were sheared under effective confining pressures (σ'_3) equal to 50 kPa. The test results of unreinforced sand indicate the general stress-strain behavior of dense sand when sheared, whereas the deviatoric stress reaches its peak value, after which it gradually decreases to ultimate value $(\sigma_1 - \sigma_3)_{ult}$. The difference of effective confining pressure indicates that the peak of deviatoric stress $\Delta\sigma_d = (\sigma_1 - \sigma_3)$ increases with the increase in effective confining pressure (σ'_3), while the peak principal stress ratio (σ'_1/σ'_3) decreases with the increase (σ'_3). The friction angle (ϕ')and cohesion (c'), defined by analytical and graphical methods for unreinforced sand. Geogrid as reinforcement increasing peak shear strength. The increasing peak shear strength is more pronounced with a higher number of geogrid and the geogrid with higher stiffness. Increased in confining stress inside reinforced soil mass ($\Delta\sigma_{3R}$) can be interpreted by cohesive reinforced soil (C_R).

1. Introduction

Geogrids have been spread in civil engineering projects such as slope stabilization, subgrade stabilization, bridge abutments, and drainage. This is due to geogrid high-strength material, simple construction techniques, and low cost. To optimized the designs of geogrid reinforced soils in civil engineering projects, laboratory testing such as triaxial test, direct shear test, and plane strain test were conducted. Many researchers have investigated the mechanism of reinforcing granular soils through a laboratory test using triaxial compression [1][2][3][4], direct shear test [5], plane strain test [6] [7]. The previous studies showed that geogrid as reinforcing materials improved the shear strength compare with unreinforced sand. The behavior of reinforced soils is affected by the type of granular material, effective confining pressure, the stiffness of reinforcement, and the arrangement layers. Until now, little attention has been observations to the shear strength parameter unreinforced sand and Increased confining stress inside reinforced soil mas. This research study focused to observe the shear strength parameter sand and behavior of unreinforced and reinforced granular soils with varying types of geogrid stiffness and the configuration of geogrid layers under the triaxial test.

2. Experimental program

Standard triaxial compression tests were adopted to investigate the behavior of unreinforced sand and reinforced sand. The tested sand having its mean particle size (D_{50}) equal to 0.6 mm was adopted. Three geogrids with different longitudinal and transverse nominal strengths were used. The relative density was equal to 70% for all tests. The reinforced sand specimens with one or two geogrid layers were sheared under effective confining pressures (σ'_3) equal to 50 kPa. The dimensions of the cylindrical soil specimen were 70 mm (diameter) × 160 mm (height). Table 1 shows the detailed experimental program.

Table 1. The experimental program of consolidated drained (CD) triaxial test

Group	Test Number	Reinforcement Layer	γ_d (kN/m^3)	Dr (%)	Confining pressure, σ_3' (kPa)
UR	UR01	None	18.45	70	50
	UR02	None	18.45	70	100
R1	R1L1	1	18.45	70	50
	R1L2	2	18.45	70	50
R2	R2L1	1	18.45	70	50
	R2L1	2	18.45	70	50
R3	R3L1	1	18.45	70	50
	R3L2	2	18.45	70	50

Note: 1 mm/min strain rate for all tests

2.1. Tested materials

A poorly graded sand (SP) with particle size between No.10 and No. 200 sieves was used in this study. The tested sand has a specific gravity (GS) 2.53 coefficient of uniformity (CU) 4.866 and gradation (CC) 0.94. The dry unit weights of sand were $\gamma_{d,max}$ = 19.39 kN/m^3 and $\gamma_{d,min}$ = 16.82 kN/m^3. Three geogrids with different longitudinal and transverse nominal strengths were used. The specification of three geogrids (R1, R2, R3) were shown in figure 1.

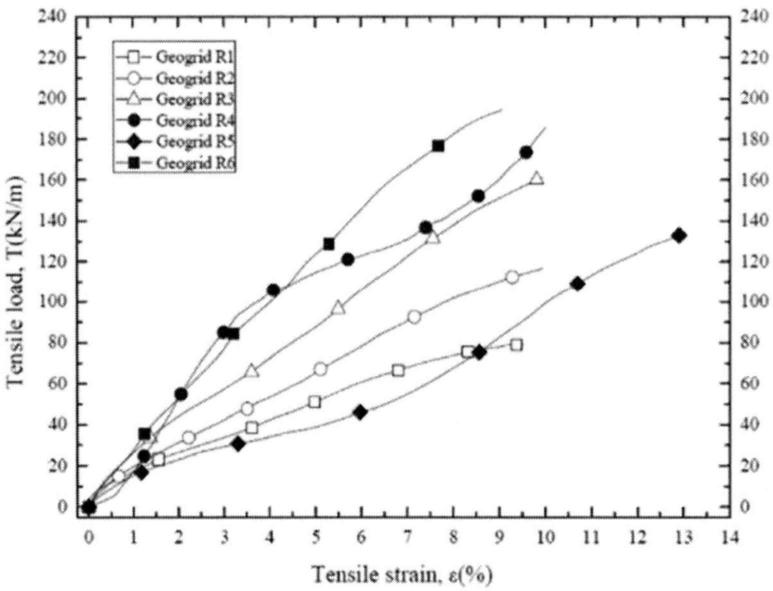

Figure 1. Tensile strength test of geogrid (standard specifications: ASTM D6637-A)

2.2. Test procedure

The dimensions of the cylindrical soil specimen were 70 mm (diameter) × 160 mm (height). A small tamper was prepared to compact dense sand into four layers. The relative density was equal to 70% for all tests, unit weight was 18.45 45 kN/m^3. The geogrid was arranged in the specimen after compacted. Figure 2 shows the configuration of the reinforcement layer. The specimen sheared under effective confining pressures (σ'_3) equal to 50 kPa. The tests stop when the axial strain reached 15%.

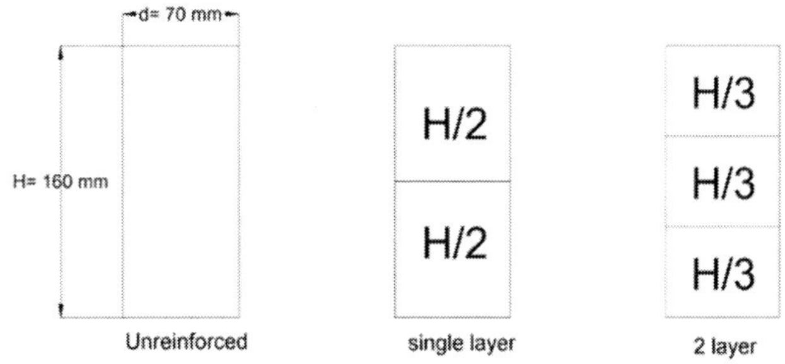

Figure 2. Number of reinforcement layers for triaxial compression tests unreinforced sand; (b) 1 layer of geogrid; (c) 2 layer of geogrid.

3. The behavior of unreinforced sand

3.1. The stress and volumetric behavior of unreinforced sand

To investigate the behavior of unreinforced sand, a total of two samples were conducted. By testing the same specimens at the same relative densities (Dr = 70%) and with different confining pressures, the relationship between deviatoric stress $\Delta\sigma_d = (\sigma_1 - \sigma_3)$, effective principal stress ratio (σ_1'/σ_3') and volumetric strain against axial strain (ε_1) could be determined. The test results indicate the general stress-strain behavior of dense sand when sheared, whereas the deviatoric stress reaches its peak value, after which it gradually decreases to ultimate value $(\sigma_1 - \sigma_3)_{ult}$. The difference of effective confining pressure shows that the peak of deviatoric stress $\Delta\sigma_d = (\sigma_1 - \sigma_3)$ increases with the increase in effective confining pressure (σ'_3) as shown in figure 3(a), while the peak principal stress ratio (σ'_1/σ'_3) decreases with the increase (σ'_3) as shown in figure 3(b). The volumetric behavior for two samples of unreinforced sand in different (σ'_3) given in figure 4. The behavior of dense sand normally decreases (compression) in volumetric strain slightly at first, then the volumetric strain increases or dilates up when shearing progressed.

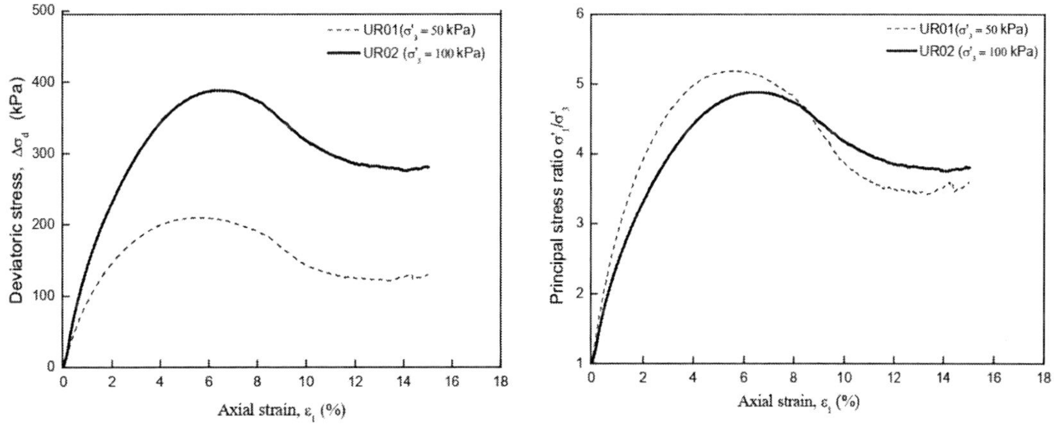

Figure 3. (a) Deviatoric stress; (b) Principal stress ratio.

IOP Conf. Series: Materials Science and Engineering 1200 (2021) 012030 doi:10.1088/1757-899X/1200/1/012030

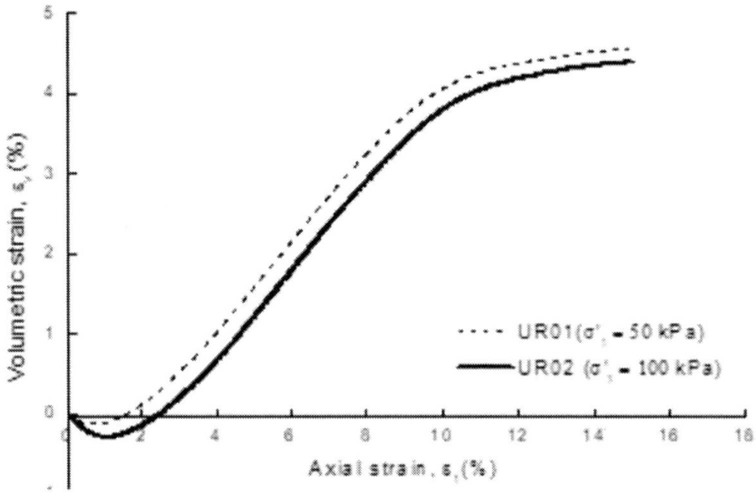

Figure 4. Volumetric strain-strain of unreinforced sand tests

3.2. Shear strength parameters of unreinforced sand

To analysis shear strength parameters such as effective friction angle (ϕ') and cohesion (c'), the summary of unreinforced sand test results was shown in table 2.

Table 2. The summary of unreinforced sand tests results

Group	Test	σ_c (kPa)	u_B (kPa)	$(\Delta\sigma_d)_f$ (kPa)	σ_{3f} (kPa)	σ_{1f} (kPa)	σ'_{3f} (kPa)	σ'_{1f} (kPa)	p'_f (kPa)	q'_f (kPa)
UR	UR01	250	200	210	250	460	50	260	155	294
	UR02	320	220	388	320	709	100	488	105	194

Friction angle ($\emptyset'$) is calculated by using the analytical and graphical method (Kf line regression method). The p'_f and q'_f of unreinforced sand, tests are parameters to draw the K_f line regression, as shown in figure 5.

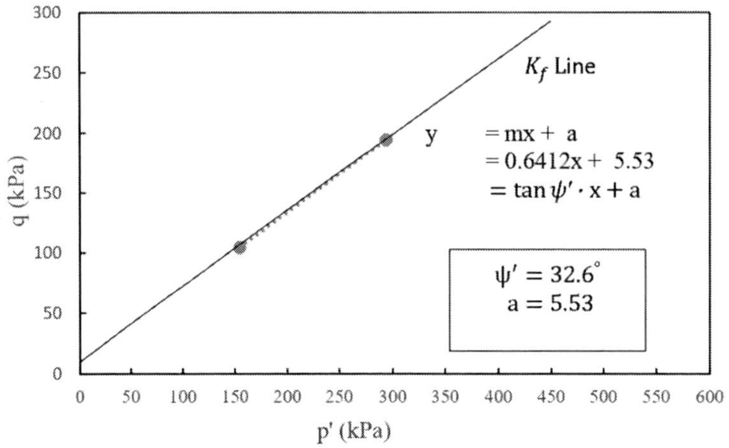

Figure 5. K_f line regression of unreinforced sand tests.

The stress paths of the consolidated drained (CD) test were presented in figure 6. No excess pore water pressure for the drained triaxial test, the drained total stresses path (T-u) SP is equal to the effective stresses path (ESP). The circle shown in Figure 6 represents a failure in terms of the p-q diagram. The identical failure, as shown in figure 7 is the same failure on the Mohr τ- σ diagram. The equation of the K_f line is

$$q_f = p_f \tan \psi' + a \quad (1)$$

Where a = the intercept on the q-axis, in the stress units, and
ψ'= the angle of the K_f line concerning the horizontally, in degrees.

The equation Mohr-Coulomb failure envelope is

$$\tau_{ff} = \sigma_{ff} \tan \phi' + c \quad (2)$$

From the Mohr-Coulomb, it can be shown that

$$\sin \phi' = \tan \psi' \quad (3)$$

$$\phi' = \sin^{-1} \tan \psi' \quad (4)$$

$$\phi' = \sin^{-1} \tan \cdot 32.6^0 = 40^0 \quad (5)$$

And then,

$$C'_{UR} = \frac{a}{\cos \phi'} \quad (6)$$

$$C'_{UR} = \frac{5.53}{\cos \phi'} = 7.2 \quad (7)$$

From the p-q diagram ϕ' and c' can be calculated based on the obtained a and ψ'. The theory for calculating shear strength parameters of soil in this research study was based on that developed by Head and Epps [8].

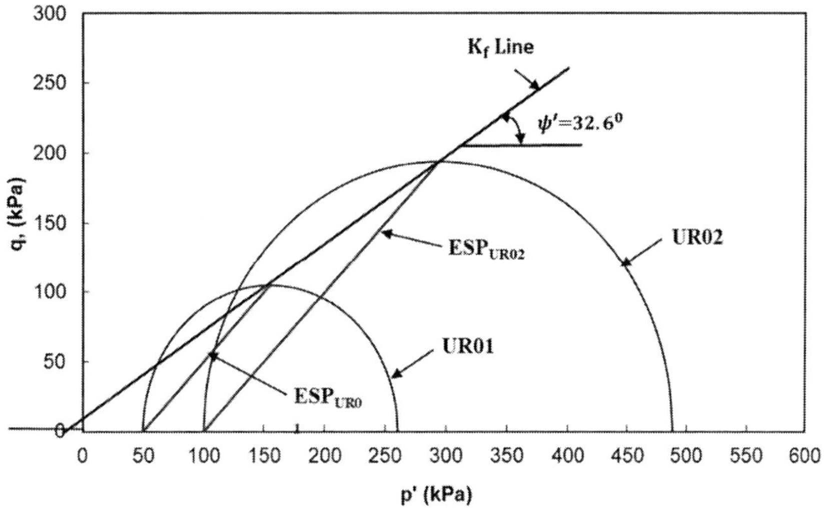

Figure 6. K_f line of CD tests of unreinforced sand.

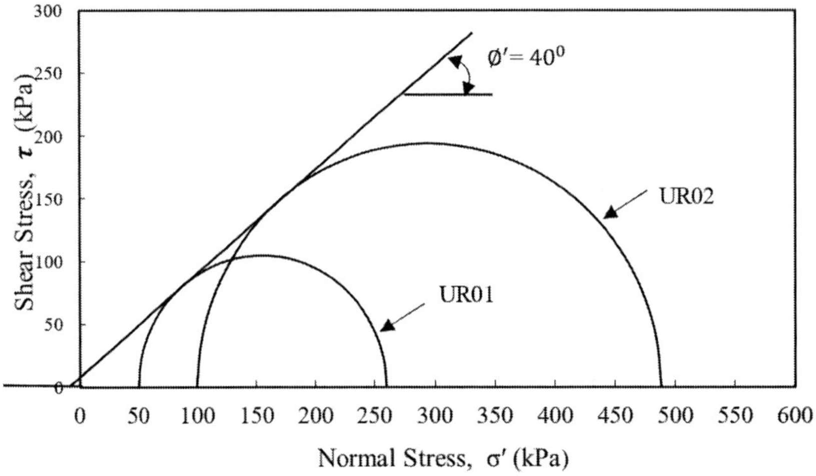

Figure 7. Mohr-Coulomb failure envelope of CD tests of unreinforced sand.

Figure 8 shown illustration between the K_f line and the Mohr-Coulomb failure envelope.

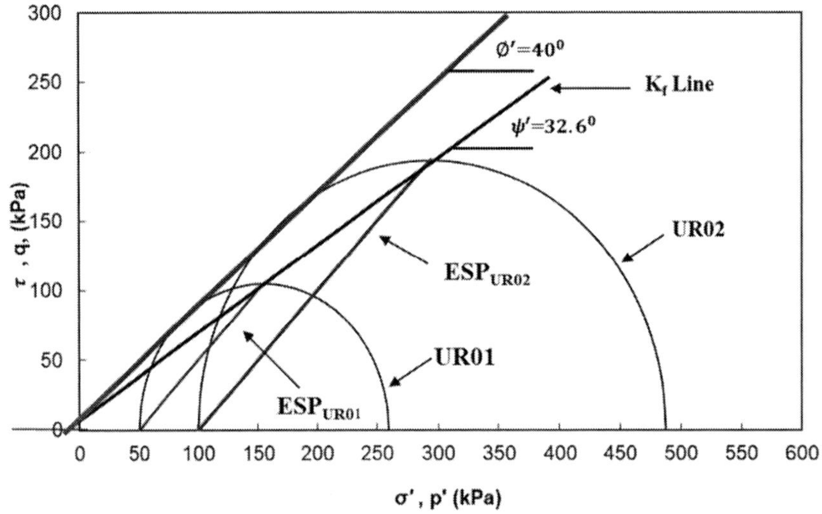

Figure 8. Illustration K_f line and Mohr-Coulomb failure envelope.

4. The behaviour of reinforced sand

4.1. Stress-strain and volumetric behaviour of reinforced sand

The behavior of geogrid-reinforced sand with a different type of stiffness and number of geogrid layers are shown in figure 9, compared with unreinforced specimens, all reinforced specimens increase in the peak of stress-axial strain behavior in terms of increases in peak deviatoric $\Delta\sigma_d = (\sigma_1 - \sigma_3)$ and principal stress ratio (σ'_1/σ'_3). In addition, geogrid inclusion increases peak strength significantly. An increase in peak deviatoric stress and principal stress ratio is more pronounced as a higher number of geogrid and the nominal geogrid strength increases (higher stiffness). In the initial stage of shearing volumetric strain both unreinforced and reinforced dense specimen decrease (compression) slightly, then the volumetric strain increased (dilation) as shearing progressed.

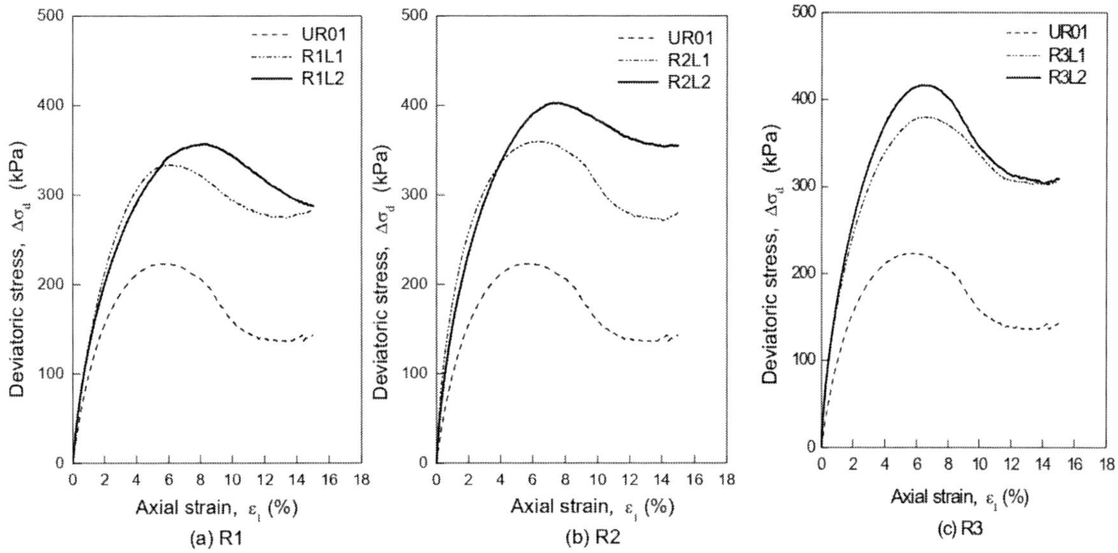

Figure 9. Effect the number of reinforcement layers with types of geogrid (a) R1, (b) R2 (c) R3.

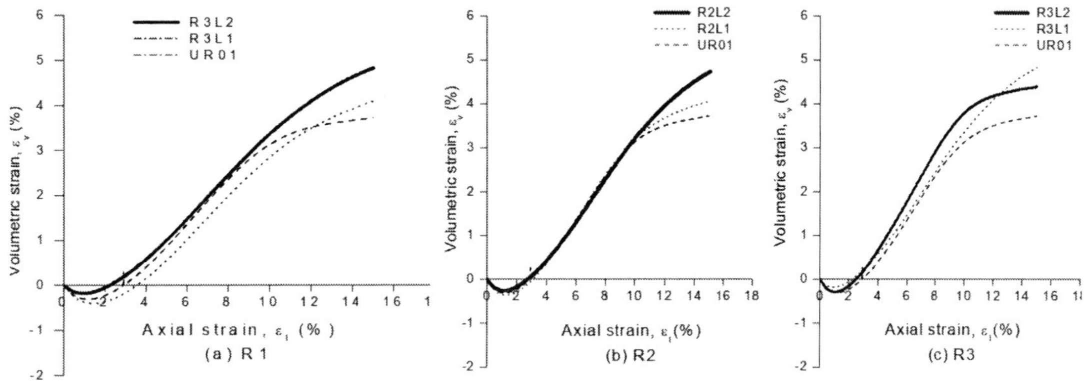

Figure 10. Volumetric strain behavior of reinforced sand with geogrid types (a) R1, (b) R2, and (C) R3

4.2. The effect of stiffness and number of geogrid layer

The effect of stiffness and the number of geogrid layers are shown in figure 11. The curve shows that specimens reinforced with geogrid type 3 (R3) have a higher peak of deviatoric stress than specimens reinforced with geogrid type 2 (R2) and finally specimens reinforced with geogrid type 2 (R2) have a higher peak deviatoric stress than geogrid type 1 (R1). The results showed that the increasing peak shear strength is more pronounced with specimens that have a higher number of geogrid layers and the geogrid with higher stiffness.

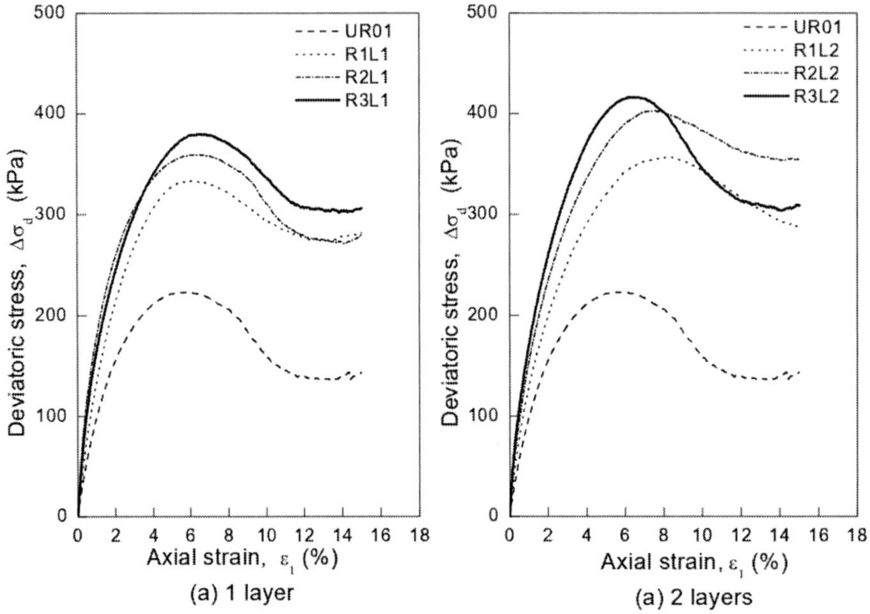

Figure 11. Effect of geogrid stiffness on deviatoric stress ratio-strain curves for reinforced sand configuration layers (a) single layer and (b) two layers.

4.3. Analysis of Pseudo-cohesion concept theory

Vidal [9] proposed the principle of reinforcing earth and the pseudo-cohesion concept for reinforced soil. The pseudo-cohesion concept theory was implemented to interpret the increased shear strength of geogrid-reinforced soil. The illustration apparent cohesion (C_R) using test R3L2 shown in Figure 12. The increased in confining stress inside reinforced soil mass ($\Delta\sigma_{3R}$) can be interpreted by cohesive reinforced soil C_R). The mechanism of the development of cohesion is due to the increased confining stress generated in the soil due to the membrane stress in the wall of geogrids.

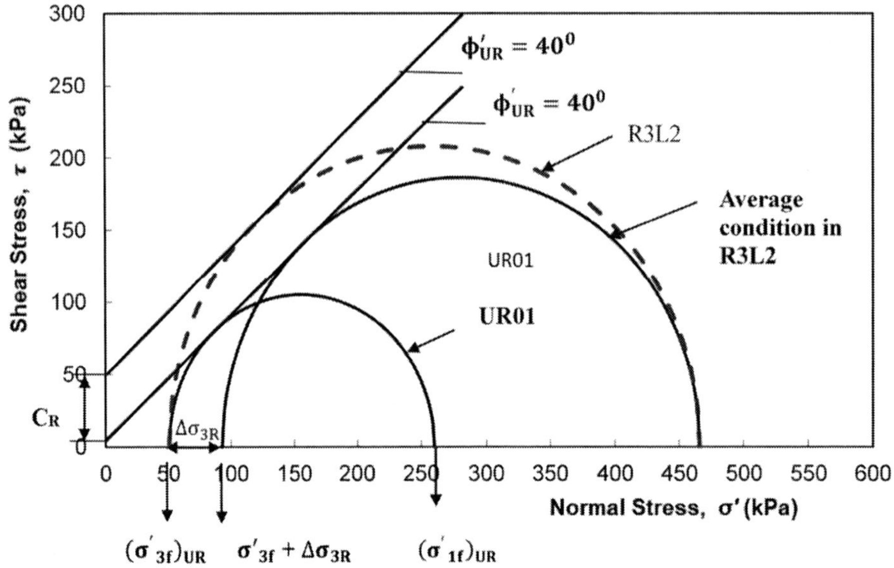

Figure 12. Illustration of the apparent cohesion CD triaxial tests.

To analyze the cohesive reinforced were used the following equation

$$\sigma'_{1f} = \sigma'_{3f} \cdot \tan^2\left(45 + \frac{\phi'_{UR}}{2}\right) + 2C'_{New} \cdot K_P \quad (8)$$

From equation 15 we get the C'_{New} and we input the value the equation 16

$$C_R = C'_{New} - C'_{UR} \quad (9)$$

The summary result of cohesive reinforced is given in Table 3.

Table 3. The Summary of apparent cohesion (C_R).

Group	Test	The number of geogrid layers	σ'_{3f} (kPa)	σ'_{1f} (kPa)	Kp	C'_{New}	C'_{UR}	C_R
R1	R1L1	1	50	384	4.6	36	7.2	29
	R1L2	2	50	407	4.6	41	7.2	34
R2	R2L1	1	50	410	4.6	42	7.2	35
	R2L2	2	50	453	4.6	52	7.2	45
R3	R3L1	1	50	430	4.6	47	7.2	40
	R3L2	2	50	466	4.6	55	7.2	48

4.4. Failure Pattern

The typical image before failure is shown in figure 13 a and the typical image after failure are shown in figure 13 b to 13 d. The investigation of the failed specimen unreinforced sand shows in figure 13 b. The results indicate that unreinforced specimens fail along at the angle $(45 + \emptyset'/2)$ while the geogrid-reinforced specimen bulging occurred between the surface of the geogrid layer (figure 13 c to 13 d).

Figure 13. The failed samples (a) unreinforced specimen before loaded; (b) unreinforced specimen after loaded; (c) reinforced 1 layer; (d) Reinforced 2 layer.

5. Conclusions

The conclusion is drawn as follows:

1. The test results of unreinforced sand indicate the general stress-strain behavior of dense sand when sheared. The peak of deviatoric stress $\Delta\sigma_d = (\sigma_1 - \sigma_3)$ increases with the increase in (σ'_3), while the peak of principal stress ratio (σ'_1/σ'_3) decreases with the increase in (σ'_3).
2. The shear strength parameters sand in terms ϕ' and c' can be defined by analytical and graphical methods.
3. Geogrid as reinforcement increasing peak shear. The increasing peak shear strength is more pronounced with a higher number of geogrid and the geogrid with higher stiffness.
4. In the initial stage of shearing volumetric strain both unreinforced and reinforced dense specimen decrease (compression) slightly, then the volumetric strain increased (dilation) as shearing progressed.

5. The increased in confining stress inside reinforced soil mass ($\Delta\sigma_{3R}$) can be interpreted by cohesive reinforced soil (C_R).
6. The investigation of the failed specimen indicates that the unreinforced specimen failed along with a planner at the angle $\left(45 + \frac{\emptyset'}{2}\right)$, while the geogrid-reinforced specimen bulging between the surface of the geogrid layer.

6. References

[1] Bathurst R J and Karpurapu R 1993 Large Scale Triaxial Compression Testing of Geocell Reinforced Granular Soils, *Geotechnical Testing Journal* **16** 293-303

[2] Haeri S M, Noorzad R and Oskoorouchi A M 2000 Effect of Geotextile Reinforcement on The Mechanical Behavior of Sand, *Geotextile and Geomembranes* **18** 385-402

[3] Zhang M X, Javadi A A and Min X 2016 Triaxial tests of sand reinforced with 3D inclusions, *Geotextile, and Geomembranes* **24** 201-209

[4] Nguyen M D, Yang K H, Lee SH, Wu C S and Tsai M H 2013 Behaviour of nonwoven-geotextile-reinforced sand and mobilization of reinforcement strain under triaxial compression, *Geosynthetics International* **20** 207-225

[5] Jewell R A and Wroth C P 1987 Direct shear tests on reinforced sand, *Geotechnique* **37** 53-68

[6] Kongkitkul W, Tatsuoka F and Hirakawa D 2007 Effects of reinforcement type and loading history on the deformation of reinforced sand in-plane strain compression, *Soils, and Foundations* **47** 395-414

[7] Liu C N, Yang K H and Nguyen M H.2014 Behaviour of geogrid-reinforced sand and effect of reinforcement anchorage in large-scale plane strain compression, *Geotextile and Geomembranes* **42** 479-493

[8] Head K H and Epps R J 2014 *Manual of soil laboratory testing* (Scotland, United Kingdom: Whittles Publishing)

[9] Vidal H 1969 The principle of reinforced earth, *Highways Res. Record* **282** 1-16

Soil Stabilization using Polypropylene Clamshell Food Containers

N A C M Salwi[1] and N Hamzah[1]

[1] School of Civil Engineering, College of Engineering, Universiti Teknologi MARA, 40450, Shah Alam, Selangor, Malaysia
Corresponding author: noorfaizah1209@uitm.edu.my

Abstract. Soil stabilization is the method of improving the physical properties of soil, such as shear strength and bearing capacity of the soil, by using controlled compaction or the addition of admixtures to produce an improved soil material that has all the desired engineering properties. The new technique of soil stabilization uses plastic waste as an alternative material is of outmost crucial since plastic wastes are non-biodegradable and remain intact after being buried in soil for many years. The present study is focused on investigating the effectiveness of utilizing polypropylene clamshell food containers as soil stabilizers. The physical properties of the untreated clayey soil are determined by conducting moisture content, specific gravity, particle size distribution, and the Atterberg limit test. Also, the Standard Proctor compaction test, as well as the unconfined compressive strength test, are carried out to determine the compaction and strength parameters of the soil sample before and after reinforcing with different percentages of polypropylene clamshell food container strips such as 0.4%, 0.8%, and 1.2%. Findings from this study indicate that the addition of polypropylene clamshell food container strips in the clayey soil is capable of becoming a soil stabilizer agent as the optimum compressive strength of the soil was achieved with replacement of 0.8% of plastic strips, along with increasing the Optimum Moisture Content (OMC) while decreasing the Maximum Dry Density (MDD). Successful implementation of polypropylene plastic in soil stabilization can help minimize the volume of plastic waste in the environment, which then leads to developing a sustainable future by utilizing recyclable material as alternative sources in the geotechnical field.

1. Introduction

Soils substantially have low shear and tensile strength, where their characteristics highly depend on the surrounding area. Consequently, soil requires adequate reinforcement to enhance the basic properties in order for construction to be built over the soil structure. Soft clay soil, also known as saturated clay soils, has low shear strength values but high values for compression and secondary compression and creep behaviour. Clay soils are considered problematic soil due to the high compressibility properties of clay soils, which eventually resulted in settlement under loading, with swelling or compression when in contact with water [1].

Stabilization can improve the shear strength and overall bearing capacity of soil with the addition of stabilizing agents. Once stabilized, the permeability of soil decreases, which in turn reduces the shrink/swell potential of soil where the soil volume change depending on temperature and moisture content. Results of the California Bearing Ratio (CBR) and Unconfined Compressive Strength (UCS) test deduced that the bearing capacity and compression strength of soil is improved with the addition of 1.0% and 0.50% combination of plastic waste and plastic granules, respectively [2].

The soil stabilization process involves determining proper additives and admixtures rates to be applied to attain the desired engineering properties as a result of improving the parameters such as shear strength, compressibility, and density. All these years, many road engineers have applied traditional admixtures such as lime, cement, and fly ash to increase the properties of the soil. Thereby laboratory and field performance testing had proved that the inclusion of such stabilizer agents could improve the strength and stability of the soils. Nevertheless, the application of these additives may result in a costly method, as stated by [3], where the cost of soil stabilization is minimized if the plastic strips prepared from waste plastic bottles are used for soil stabilization instead of those expensive admixtures such as cement and lime, etc. As construction on weak soil is a great challenge in the field of geotechnical engineering, the need to improve the quality of soil using soil stabilization is becoming more crucial. As a result, new methods are continuously being explored to improve the strength properties and decrease the swell behaviour of soil. Reusing plastic waste as an admixture for soil stabilization is not only an eco-friendly method for improving the subgrade soil of pavement besides reducing the amount of plastic waste efficiently but also a cost-effective and energy-saving method. Several studies have proved that the application of plastic waste in soil will contribute to the development of soil's strength characteristics.

Nowadays, people would prefer to use clamshell containers as an ideal packaging material for food since such packaging is not only lightweight and durable enough to keep the food inside safe but also comes with secure protection where the containers are more easily opened and can be resealed [4]. The rise in food delivery services in the present moment during Covid-19 lockdowns worldwide will further drive the usage of plastic food containers [5]. Due to that, people usually use them only once before being thrown away into bins or littering along the streets, which contributes to the exponential increase in the use of single-use plastic [6]. Therefore, an attempt is made in this paper to investigate the effectiveness of utilizing polypropylene clamshell food containers as soil stabilizer.

2. Methodology

2.1. Sample collection

2.1.1. Clayey soil. About 30 kg of clayey soil samples were collected by using a hoe and shovel from Kampung Parit Satu Barat, Sungai Besar, Selangor and brought to the soil mechanics laboratory for testing. The soil sample was placed in polyethylene bags and was tied tightly to prevent the loss of moisture content. A soil index test was conducted to determine the physical properties of soil such as moisture content, particle size distribution, specific gravity, plasticity, compaction, and strength characteristics.

2.1.2. Polypropylene clamshell food containers. The clamshell plastic food containers used in this study were made of polypropylene which is categorised among the types of plastic with resin number 5. The strips were prepared from polypropylene clamshell food containers bought from a convenience store, 99 Speedmart in Taman Seroja, Bandar Baru Salak Tinggi, Sepang, Selangor. The food containers were cut manually into strips with a dimension of 2 mm x 20 mm using a scissor. The strips that pass through the 10mm sieve test are used as a stabilizer to reinforce the clayey soil with different proportions of 0.4%, 0.8%, and 1.2% by weight of the soil, respectively.

2.2. Sample preparation

The soil sample was prepared for compaction and strength test by mixing with various percentages of polypropylene clamshell food container strips, i.e., 0.4%, 0.8%, and 1.2%. Table 1 shows the proportions of polypropylene clamshell food container strips that were added into clayey soil as an additive for sample preparation. The percentages of plastic strips used in this study are based on a previous study carried out on discarded plastic waste [7].

Table 1. Proportions of polypropylene clamshell food container strips reinforced with clayey soil.

Batch	Sample	Laboratory tests
1	Clayey soil (100%)	- Moisture content - Specific gravity - Particle size distribution - Atterberg limits - Standard Proctor compaction test - Unconfined compressive strength test
2	Clayey soil + 0.4% Polypropylene clamshell food container strips	- Standard Proctor compaction test - Unconfined compressive strength test
3	Clayey soil + 0.8% Polypropylene clamshell food container strips	
4	Clayey soil + 1.2% Polypropylene clamshell food container strips	

2.3. Soil index test

2.3.1. Moisture content. Moisture content is determined as the ratio of the water mass contained in the soil to the mass of the dry soil. The soil sample was oven-dried for 24 hours at the temperature of 110°C±5°C to obtain the percentage of water contained in the soil sample in accordance with BS 1377: Part 2:1990:3.2. The moisture content of the soil sample is determined by obtaining the average reading of three samples.

2.3.2. Particle size distribution & hydrometer test. Sieve analysis test is carried out to measure the distribution of particle size in the soil sample. 2kg of subsample is used in the wet sieve method based on BS 1377: Part 2:1990:9.2 to separate coarse-grained particles from fine-grained particles that passed through 63 μm mesh sieve. In the dry sieve method, the retained fractions were oven-dried then passed again into a nest of sieves; 5mm, 4.75mm, 3.35mm, 2mm, 1.18mm, 0.6mm, 0.425mm, 0.3mm, 0.212mm, 0.15mm, and 0.063mm in accordance with BS 1377: Part 2:1990:9.3. Hydrometer test is conducted in accordance with BS 1377: Part 2:1990:9.5 for the particle size smaller than 0.063 mm where 50g of soil sample were used.

2.3.3. Specific gravity. Specific gravity is determined as the ratio of the weight of a given volume of soil to the weight of water displaced by soil. The pycnometer method is used to determine the particle density of soil sample in accordance with BS 1377: Part 2:1990:8.3. This test was performed using three density bottles, each filled with 20 g of oven-dried soil sample and de-aired distilled water. The density bottles were placed in a desiccator for one hour to remove entrapped air. The specific gravity of the soil sample is obtained by calculating the average specific gravity of the three samples tested.

2.3.4. Atterberg limit. Cone penetration test based on BS 1377: Part 2:1990:4.3 is carried out to determine the liquid limit of the fine-grained soil sample meanwhile, the plastic limit is measured as the gravimetric moisture content in accordance with BS 1377: Part 2:1990:5.3, at which soil sample was rolled until the diameter of the thread reaches 3.2 mm without splitting. The difference in values of liquid limit and plastic limit is determined as plasticity index, as a means of classifying the soil plasticity.

2.4. Standard Proctor compaction test

In accordance with BS 1377: Part 4:1990:3.3, A Standard Proctor compaction test is conducted to determine the compaction parameters, i.e., Optimum Moisture Content (OMC) and Maximum Dry Density (MDD). A 2.5kg rammer falling through a height of 300mm is used to compact the soil sample mixed with 100mL of water in three layers with 25 evenly distributed blows on each layer into a compaction mould, as shown in figure 1. This test was conducted repeatedly for untreated clayey soil sample and treated clayey soil sample with varying proportions of 0.4%, 0.8%, and 1.2% of polypropylene clamshell food container strips, as shown in figure 2.

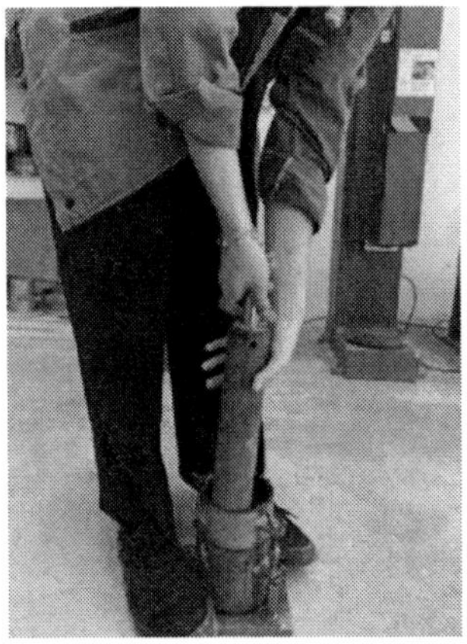

Figure 1. Soil sample is compacted by using a rammer.

Figure 2. Polypropylene clamshell food container strips are mixed with soil sample.

2.5. Unconfined compressive strength test (UCT)

The unconfined compressive strength of soil is measured as the maximum value of compressive force per unit area that the soil sample can sustain. The unconfined compressive strength test in accordance with BS 1337: Part 7:1990: 7 was carried out on four cylindrical specimens with each dimension of 100 mm in length and 50 mm in diameter that was extruded directly from compaction specimens as shown in figure 3 then they were tested immediately. This test was conducted using a load frame apparatus, as shown in figure 4, to determine the maximum compressive strength of untreated clayey soil sample and treated clayey soil sample with varying proportions of 0.4%, 0.8%, and 1.2% of polypropylene clamshell food container strips. The readings of the force measuring device and the axial deformation gauges were recorded simultaneously at regular intervals of compression. Several sets of readings were obtained in order to define the stress-strain curve.

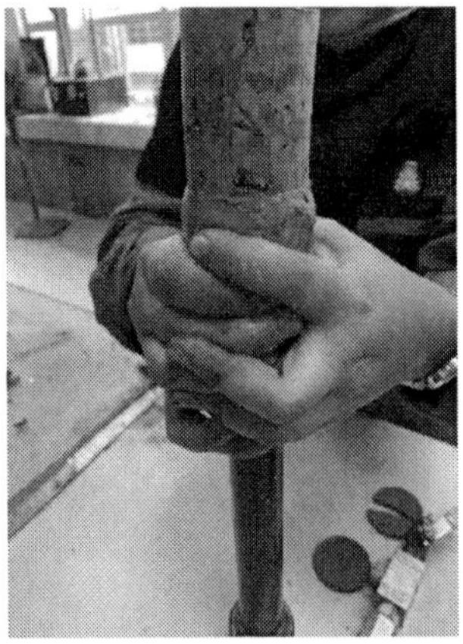

Figure 3. A soil specimen is extruded from a steel cylinder.

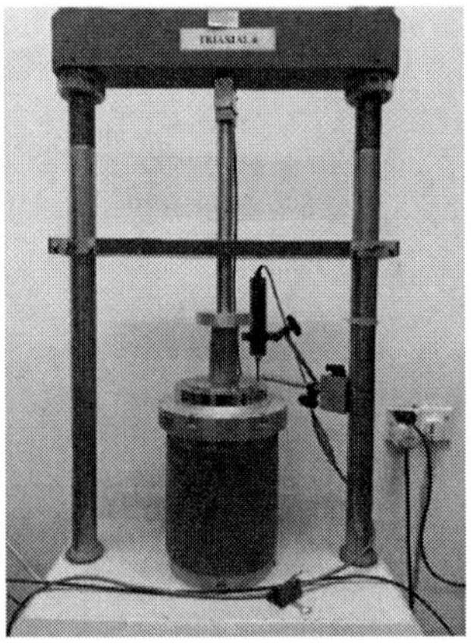

Figure 4. A soil specimen is tested using a load frame apparatus.

3. Results and discussion

3.1. Physical properties of soil sample

The physical properties of untreated clayey soil are determined by conducting moisture content, specific gravity, particle size distribution, and the Atterberg limit test. Based on the physical properties test conducted, the soil sample is classified as clayey SILT of Intermediate Plasticity (CI) according to British Soil Classification System (BSCS). Details of the soil index properties are shown in table 2.

Table 2. Physical properties of soil sample.

Soil Properties		Test Values
Colour		Dark brown
Moisture content	%	39.48
Specific gravity		2.26
Particle size distribution		
(a) Gravel	%	0.40
(b) Sand	%	0.52
(c) Silt	%	77.58
(d) Clay	%	21.50
Liquid Limit, LL	%	35.76
Plastic Limit, PL	%	13.99
Plasticity Index, IP	%	21.77
Soil Classification		CI

3.2. Standard Proctor compaction test

The plot of compaction curves where dry density against moisture content of soil sample reinforced with varying proportions of 0.4%, 0.8%, and 1.2% of polypropylene clamshell food container strips as shown in figure 5.

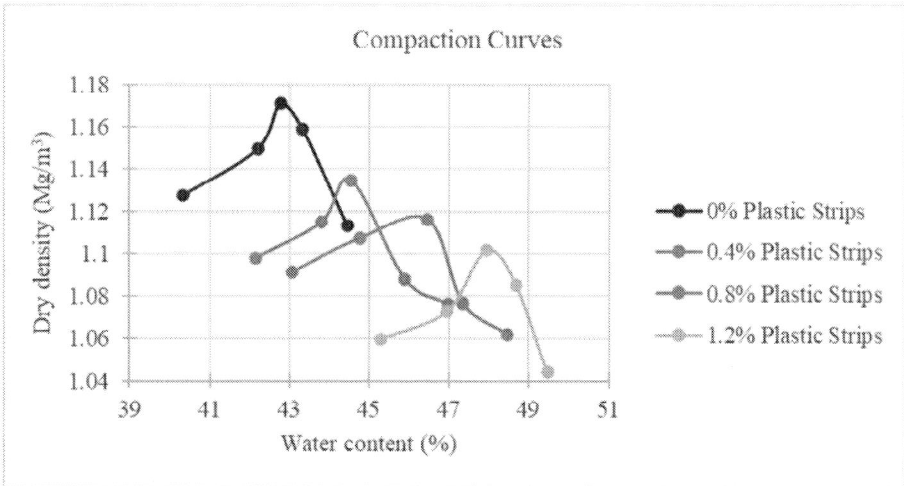

Figure 5. Compaction curves of soil samples.

It is proven based on a series of conducted compaction test that by mixing plastic strips with clay soil at different replacement of 0.4%, 0.8%, and 1.2% by weight of soil respectively, the Maximum

Dry Density (MDD) decrease concurrently with the increase of Optimum Moisture Content (OMC) in comparison to untreated soil due to the presence of plastic strips that filled up the voids in between soil particles thereby reducing the soil permeability. Table 3 shows the Maximum Dry Density (MDD) significantly fell from 1.171Mg/m^3 to 1.102Mg/m^3 with an increase in the proportion of plastic strips from 0% to 1.2% by weight of the soil. Conversely, the Optimum Moisture Content (OMC) gradually increased from 42.83% for untreated soil to 48.04% for soil stabilised with 1.2% plastic strips replacement. It may be disclosed based on the findings that the compacted soil samples become denser as a result of the volume of air in the soil sample slowly depleted, which also contributed by the reduction in soil content with respect to the increment of the plastic percentage by 0.4%, thereby caused reduction in dry density of soil.

Table 3. Optimum moisture content and maximum dry density of soil samples.

Sample	Optimum Moisture Content OMC (%)	Maximum Dry Density MDD (Mg/m^3)
Clayey soil (100%)	42.83	1.171
Clayey soil + 0.4% Polypropylene clamshell food container strips	44.45	1.135
Clayey soil + 0.8% Polypropylene clamshell food container strips	46.25	1.117
Clayey soil + 1.2% Polypropylene clamshell food container strips	48.04	1.102

Figure 6 and figure 7 reveal the results of compaction tests on untreated and treated soils with plastic waste. Both figures also indicate the comparison results of this study with previous research results of soil mixed with different plastic waste. Previous studies conducted by [2], [8] and [9] revealed that the Maximum Dry Density (MDD) increased at 0.5% of plastic waste replacement by weight of the soil meanwhile, [7] proved that with the addition of 0.4% of discarded plastic waste but further increasing the percentage of plastic will result in the reduction of Maximum Dry Density (MDD). However, the current study shows a declining trend in Maximum Dry Density (MDD) values from 0% to 1.2% of polypropylene clamshell food container strips by weight of the soil.

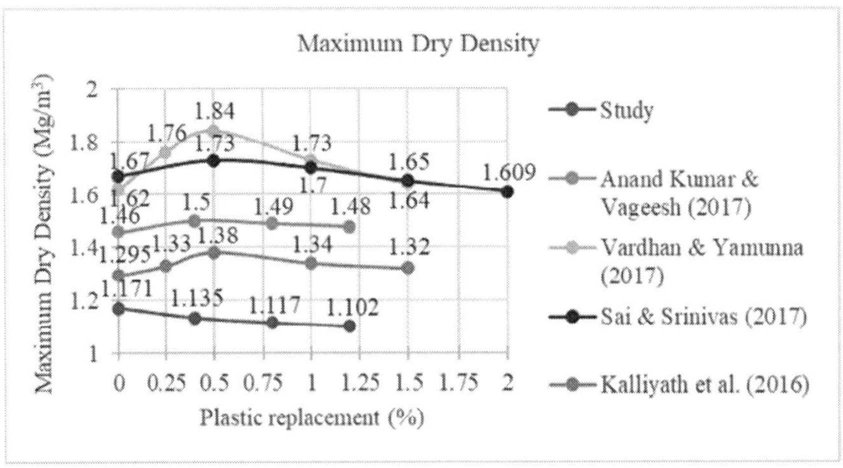

Figure 6. Maximum dry density comparison between study and previous research.

The studies done by [2] and [7] revealed that the Optimum Moisture Content (OMC) values initially decrease with 0.5% and 0.4% plastic waste replacement, respectively. However, beyond those percentages of plastic replacement, the Optimum Moisture Content (OMC) was observed to increase gradually. Meanwhile, [8] and [9] gained the highest Optimum Moisture Content (OMC) when reinforced with 2% of plastic waste. However, this study found that the Optimum Moisture Content (OMC) values gradually increased from low to high percentage of polypropylene clamshell food container strips.

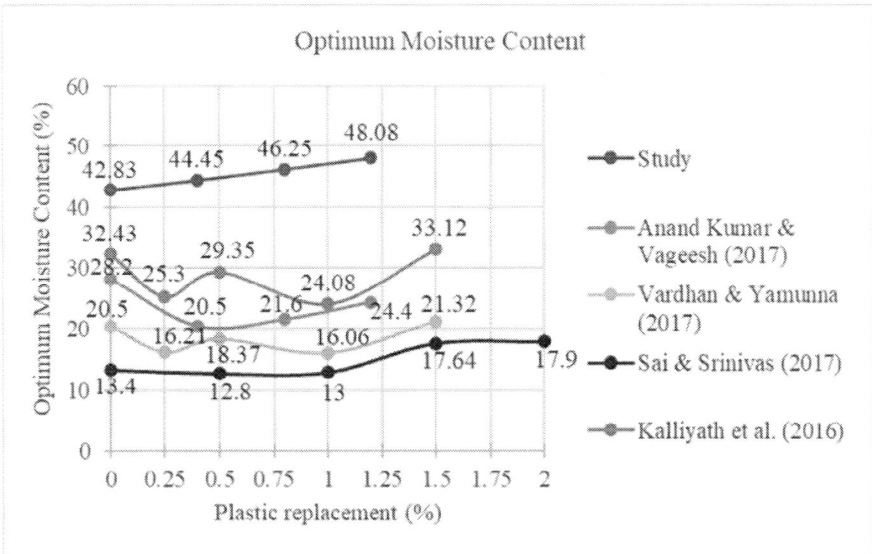

Figure 7. Optimum moisture content comparison between study and previous research.

3.3. Unconfined compressive strength test

The stress-strain curves are plotted as shown in figure 8 for soil sample reinforced with varying proportions of 0.4%, 0.8%, and 1.2% of polypropylene clamshell food container strips. The stress at first increases quickly while the strain increases slowly, but as the soil yields, the strain increases dramatically while the stress levels are off.

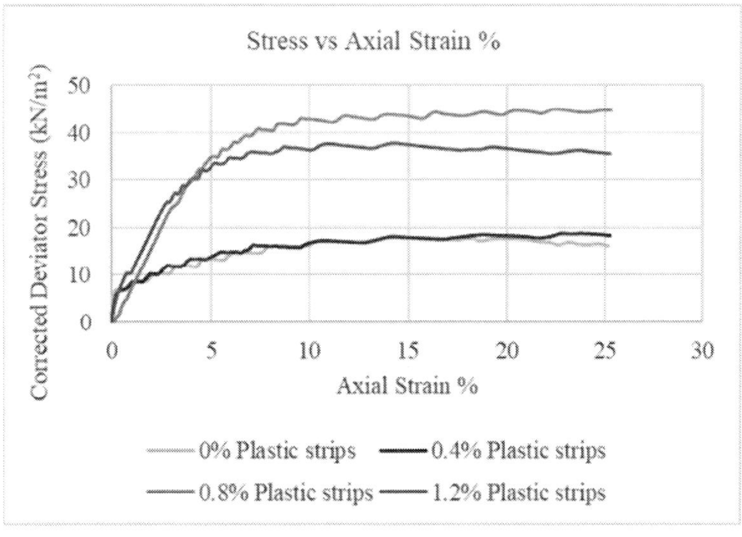

Figure 8. Stress against strain curves of soil samples.

It is proven based on a series of unconfined compressive strength tests that by mixing polypropylene clamshell food container strips with clayey soil at different replacements of 0.4%, 0.8%, and 1.2% by weight of soil, respectively, the unconfined compressive strength values keep increasing then decreasing after 1.2% of plastic strips was added to the soil. The unconfined compressive strength value of untreated clayey soil was found to be 18.2kN/m^2. With the addition of 0.4% of polypropylene clamshell food container strips, the unconfined compressive strength value slightly increased to 18.8kN/m^2. After that, the unconfined compressive strength value significantly rose to 44.9kN/m^2 with the addition of 0.8% of plastic strips. Further increase in the plastic replacement at 1.2% of plastic strips by weight of the soil showed a decrease in the unconfined compressive strength value, which was determined to be 37.8kN/m^2. This is because clay content replacement by polypropylene clamshell food container strips caused a reduction of cohesive force in the soil sample. Compaction reduces the gaps between the soil particles when the water trapped inside the soil matrixes expels from soil pore holes caused a reduction in the dry density of soil and enhanced the bond between the particles, thereby increasing the compressive strength of the soil.

Figure 9 indicates the comparison results of unconfined compression tests of this study with previous research results on untreated and treated soils with different plastic waste. Previous studies conducted by [2] and [8] revealed that the unconfined compressive strength values increased at 0.5% of plastic waste replacement by weight of the soil. Meanwhile, [9] proved that the strength values increased with an addition of 1% discarded waste plastic covers, such as milk and curd packets but afterward, it decreases in unconfined compressive strength values with further addition of plastic waste. However, this study found that the unconfined compressive strength values significantly increased at the insertion of 0.8% of polypropylene clamshell food container strips by weight of the soil.

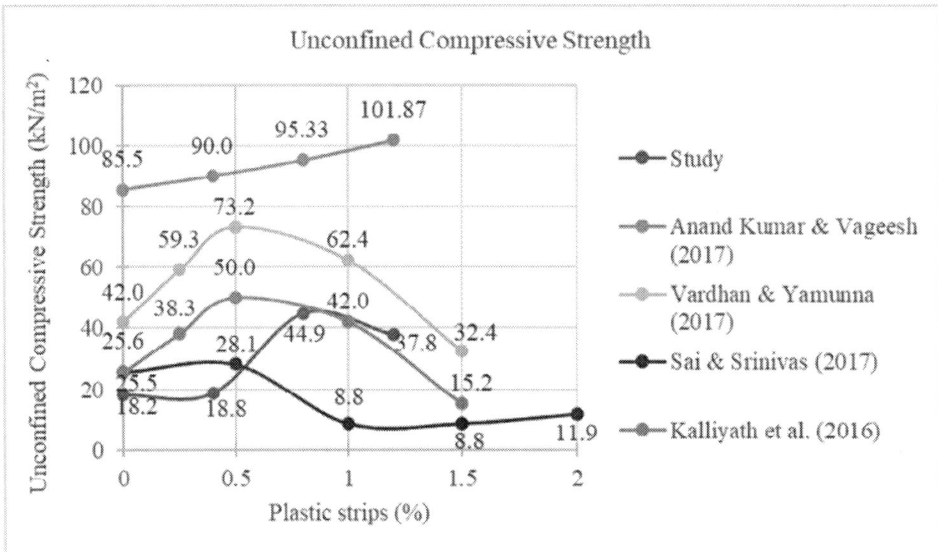

Figure 9. Unconfined compressive strength comparison between study and previous research.

Table 4 reveals the findings on strength parameters of soil samples for unconfined compressive strength and shear strength.

Table 4. Unconfined compressive strength and shear strength of soil samples.

Sample	Unconfined Compressive Strength q_u (kN/m^2)	Shear Strength S_u (kN/m^2)
Clayey soil (100%)	18.2	9.10
Clayey soil + 0.4% Polypropylene clamshell food container strips	18.8	9.40
Clayey soil + 0.8% Polypropylene clamshell food container strips	44.9	22.45
Clayey soil + 1.2% Polypropylene clamshell food container strips	37.8	18.90

This study proved that the unconfined compressive strength was clearly affected by a further increment of polypropylene clamshell food container strips. There is an increase in soil cohesion with the addition of plastic strips in the soil sample, which leads to an increase in unconfined compressive strength. However, further increase in plastic content led to a decrease in cohesion and thereby decrease in soil strength. In addition, the shear strength of soil, which is directly related to cohesion, shows a similar pattern such that a maximum value of 22.45kN/m^2 is obtained at 0.8% plastic replacement, but thereafter, the value decreases. It may be inferred from the results of the study that the strength of the treated soil up to the addition of 0.8% of plastic strips had increased, however thereafter, it decreased eventually at 1.2% of plastic replacement. Hence, the optimum percentage of polypropylene clamshell food container strips to be added as a stabilizer is found at 0.8% by weight of dry soil.

4. Conclusion

The engineering properties of the soil sample collected from the study area are classified as clayey SILT of Intermediate Plasticity (CI). The effect of adding the polypropylene clamshell food container strips with soil samples at different replacement of 0.4%, 0.8%, and 1.2% by weight of soil, respectively, has caused the Maximum Dry Density (MDD) to decreased. At the same time, the Optimum Moisture Content (OMC) increased. Meanwhile, the unconfined compressive strength of soil was found to have increased for soil treated up until the addition of 0.8% of plastic strips. Still, afterward, it eventually decreased at 1.2% of plastic replacement. Based on the observations on strength characteristics, it is noted that the optimum proportions of polypropylene clamshell food container strips to stabilize the clayey soil is found at 0.8% by weight of dry soil. Overall, it can be deduced that the polypropylene clamshell food container strips are suitable as a soil stabilizer in improving the strength characteristics of the soil. Successful implementation of polypropylene plastic in soil stabilization can help to minimize the volume of plastic waste in the environment, leading to a sustainable future by utilizing recyclable material as alternative sources in the geotechnical field.

5. References

[1] Mohamad N O, Razali C E, Hadi A A A, Som P P, Eng B C, Rusli M B and Mohamad F R 2016 Challenges in Construction Over Soft Soil – Case Studies in Malaysia *IOP Conf. Series: Materials Science and Engineering* **136** 1-8

[2] Sai M and Srinivas V 2019 Soil Stabilization by Using Plastic Waste Granules Materials *IOSR Journal of Computer Engineering (IOSR-JCE)* **21** 42-51
[3] Ashraf A, Sunil A, Dhanya J, Joseph M, Varghese M and Veena M 2011 SOIL STABILIISATION USING RAW PLASTIC BOTTLES *Proceedings of Indian Geotechnical Conference, Kochi (Paper No. H-304)* 489-492
[4] Helmke R 2015 The Advantages of Clamshell Packaging *Plastic Ingenuity* Retrieved on July 13, 2021 from https://www.plasticingenuity.com/blog/advantages-clamshell-packaging
[5] Zaki A 2020 Urgent need to deal with food packaging waste *TheStar* Retrieved on July 13, 2021 from https://www.thestar.com.my/metro/metro-news/2020/09/04/urgent-need-to-deal-with-food-packaging-waste
[6] The FoodPrint of Food Packaging 2019 *FoodPrint* Retrieved on July 13, 2021 from https://foodprint.org/reports/the-foodprint-of-food-packaging/#main-content
[7] Anand Kumar B G and Vageesh H P 2017 Effect of Discarded Plastic Waste as Stabilizer on Engineering Properties of Cohesive Soil, *International Journal of Engineering Technology Science and Research (IJETSR)* **4** 779-786
[8] Vardhan P H and Yamunna M 2017 Stabilization of Black Cotton Soil Using Plastic Waste *International Journal of Innovative Research in Science, Engineering and Technology (IJIRSET)* **6** 66-70
[9] Kalliyath J V, Joy J T, Paul J M and Vadakkel A M 2016 Soil Stabilization using Plastic Fibers Materials, *IJSTE-International Journal of Science Technology & Engineering* **2** 484-487

Acknowledgments
Sincere gratitude to the technician in the soil mechanics laboratory for their help in undertaking the experimental works. The author also wishes to extend appreciation to School of Civil Engineering, College of Engineering, Universiti Teknologi MARA Shah Alam for providing the required facilities for this research.

Numerical Modelling Observations of Settlement for Pad Footings Supported on Soft Clay Soil

O A Arab[1], A J M S Lim[1], S Y Sim[2] and N A A Guntor[1]

[1] Department of Civil Engineering, Research Centre for Soft Soil, Faculty of Civil Engineering and Built Environment, Universiti Tun Hussein Onn Malaysia
[2] Department of Electrical Engineering Technology, Faculty of Engineering Technology, Universiti Tun Hussein Onn Malaysia
Corresponding author: alvin@uthm.edu.my

Abstract. Settlement calculation is an important part in the design of shallow foundations resting on soft soils. The size of the foundation, the depth of the footings, and the rise in ground water level are thought to influence settlement and have been the subject of much research for many years. Thus, this study compared several pad footing sizes using numerical techniques as the basis. The first objective of this study was to analyse soil and pad footing settlement, and to determine the optimal size of footing that withstands excessive settlement due to variation in the water table and the depth of the foundation. Three footing embedment depths of 1.5, 2, and 3m with three water table positions, at the GL (0m), 1.5, and 3m with an applied foundation concentrated load of 440 kN using five footing models of 1.5mx1.5m, 2mx1.5m, 2 m x 2 m, 2.5x1.5m, and 2.5x1.5m pad footing with a uniform thickness of 0.5m were considered. In this study, a 3D Plaxis simulation is used for predicting the settlement of shallow foundations on soft clay soils. Settlement results were discovered at various water table positions and foundation depths. The study found that the 2.5x2m footing was deemed the best among the simulated foundations, and the 3m foundation embedment was considered the best at shallow depths due to less excessive settlement than the other tested foundations. The settlement had a significant impact on the size of the foundation and the depth of the footing. The depth of the water table has a small impact on the settlement. Parametric analysis is also being used to gain a better understanding of the behaviour of the elastic settlement of various shallow foundations. It is found that the footing area increases, settlement decreases and vice versa.

1. Introduction

A foundation is one of the main building components that transfers the structure's load to the ground. When designing a foundation, it is necessary to examine the type of soil, its behaviour, as well as its settlement and bearing capacity. The foundation is a key element in design and construction. To avoid soil failure, it is essential to ensure that the load distribution on the soil is adequate. A foundation is important because it serves as the contact between a structure's load-bearing components and the ground. Two types of foundations exist. There are two types of foundations: shallow foundations and deep foundations. To provide for adequate load distribution, the size and kind of the foundation should be precisely calculated. The increased size of the footing relative to the column support results in an increased area of contact between the footing and the soil [1]. Settlement is an essential parameter in the design of the pad footing, particularly when the footing width exceeds one meter. Excessive settlements can result in serviceability issues, making settlement a more crucial parameter to consider when designing foundations than the bearing capacity of the soil [2]. Locally and internationally, the number of project failures due to soil settlement and structural deformation has grown. Several of the failures were attributed to the soft soil's poor shear strength and excessive compressibility. Settlement may be considered as the downward movement of the ground in the soil as a result of applied pressures. As a result of soil settlement, the load carrying system will be change. The masonry buildings are relatively intolerant of settlement and may undergo structural instability (cracking) when angular distortions exceed 1/300. In practise, bearing stresses are generally limited to a maximum of 25 mm (1 in) of total settlement and roughly half of the total settlement as differential settlement [3]. The research focuses on numerical analysis through the PLAXIS 3D software. The Finite Element Method (FEM) is a numerical analysis technique for theoretical systems that transforms the continuous into a finite structure comprised of a finite number of units and then analyses the stress and deformation of each unit [4]. A 3D finite element mesh is produced from this geometry. Boreholes are used to define soil strata. Horizontal work planes were used to define structures. [5]. Plaxis 3D performs calculations and outputs the results in animation and/or numerical form using its calculating features [6]. Construction on soft soils is a significant problem in the field of geotechnical engineering. The settlement of five different footing sizes will be evaluated in this study, and the data will be examined to identify the settlement behaviour of pad footings resting on soft clay, as well as the optimum dimension of Pad footing on soft Clay. This study helps to perform analysis of settlement of different sizes of pad footings on soft clay soil beds using PLAXIS 3D software. The importance of this study is to provide the optimum size of the pad footing with the best depth of the foundation in a soft clay bed. In soils with a high compressibility, significant settlement and compression will occur, and these values will likely exceed acceptable limits, resulting in serious damage to the building. Accurate settlement prediction is important because settlement, rather than bearing capacity, typically determines foundation design. The outcome of this study will contribute knowledge about the settlement behaviour of pad footings of various sizes, depths, and water table positions; the importance of knowing the proper size of footing and depth of foundation results in the use of an appropriate amount of concrete in the foundation and the reduction of foundation trench excavation, which contributes to the economy. The importance of this research is to apply or use the effectiveness of geotechnical design throughout the design process in order to ensure the structure's stability. Other researchers may be able to apply the findings from this study to conduct future research that will have a significant influence, particularly in the geotechnical sector. As a result, it predicted settlements with a variety of footing sizes, water table positions and foundation depths. This study will help designers in predicting settlement for future projects. This research will provide the importance of pad foundation size selection.

2. Methodology

This study used soft clay soil, it emphasizes the application of the Mohr Coulomb models of soil behaviour. The soil characteristics used in this study are taken from the properties of soft clay soil based on previous research done by [7], the study investigated a bearing capacity of the gravelly sand column, installed in soft clay bed. The soil properties was given in table 1.

2.1. Numerical analysis

Plaxis 3D was used to conduct the numerical study of foundation settlement. Plaxis 3D uses index, elastic, and strength characteristics as input data. It produces unstructured three-dimensional finite element meshes with the ability to adjust the mesh globally or locally. Plaxis 3D connect performed calculations and outputs the results.

Table 1. Properties of soft soil [7].

Properties	Unit	Soft Clay
Specific gravity	....	2.65
Moisture content	%	42
Un-drained shear strength	kPa/ kN/m2	15
Cohesion	KN/m2	8.87

Angle of internal friction	o	31.2
Dilatancy angle	o	10.4
Permeability	m/day	0.0006048
Young modulus	KN/m2	1800
Poisson ratio	....	0.35
Unit weight saturation	KN/m3	17.38
Unit weight unsaturation	KN/m3	12.24

The three-dimensional finite element program PLAXIS is used to simulate pad foundations to determine the settlement. During the modelling in PLAXIS, various trial analyses were performed to assess the model behaviour. The variation in strength parameters ' c and ϕ' done since the settlement of a soil depends on strength parameters. To carry out a finite analysis using PLAXIS, a finite element mesh was created and the material properties and boundary conditions were specified. To set up a FEM, a geometry model composed of points, lines, and other components was created in the XYZ-space for PLAXIS 3D. The finite element mesh was generated in PLAXIS 3D after the material properties and boundary conditions are assigned. The Mohr-Coulomb failure criteria was used to model the foundation clay soil as an isotropic elastic-perfectly plastic material. The Plaxis programme inputs the constitutive parameters for clay soils, which include the un-drained shear strength (C), un-drained elastic modulus (E), poison ration (v), and bulk unit weight (γ). The FEM analysis is done with Plaxis 3D numerical application. Soil limits are Xmin = 0 m, Xmax = 10 m, Ymin = 0 m and Ymax = 10 m, Zmin = 0 m and Zmax = 10 m respectively. The footings were placed at the centre of soft clay boundaries in X-Y plane with varying depths and water table positions.

2.2. Structural geometry parameters

The dimensions of the footings were taken as assumptions. They are rectangle, square shapes. The Pads used in this project are concrete pads, therefore, table 2 below indicated some characteristics of concrete that needed for the Plaxis software input during the model's simulation including concrete density, modulus of elasticity, poisons ratio, and shear modulus.

Table 2. Parameters of Pad footing foundation.

Geometry parameters (Assumption)			Material Property [8].			
Footing Shape	Footing Dimension (LXB) in (m)	Thickness (m)	Density of material γ(KN/m^3)	Elastic Modulus E, (KN/m^2)	Poisson's Ratio.	Shear Modulus, G (KN/m^2)
Square	1.5x1.5	0.5	24	3x10^7	0.2	1.25x10^7
Rectangle	2.0x1.5	0.5	24	3x10^7	0.2	1.25x10^7
Square	2.0x2.0	0.5	24	3x10^7	0.2	1.25x10^7
Rectangle	2.5x1.5	0.5	24	3x10^7	0.2	1.25x10^7
Rectangle	2.5x2.0	0.5	24	3x10^7	0.2	1.25x10^7

2.3. Load

After completely defining the geometry, the load used in this study was calculated for a light-weight mono-family house consists of one floors and has a spread type of foundation resting on a soft clay soil shown in Figure 1. The calculated load was 440KN. The thickness of the spread foundations is 0.5 m each, and the foundation is applied vertically to the center of the footing.

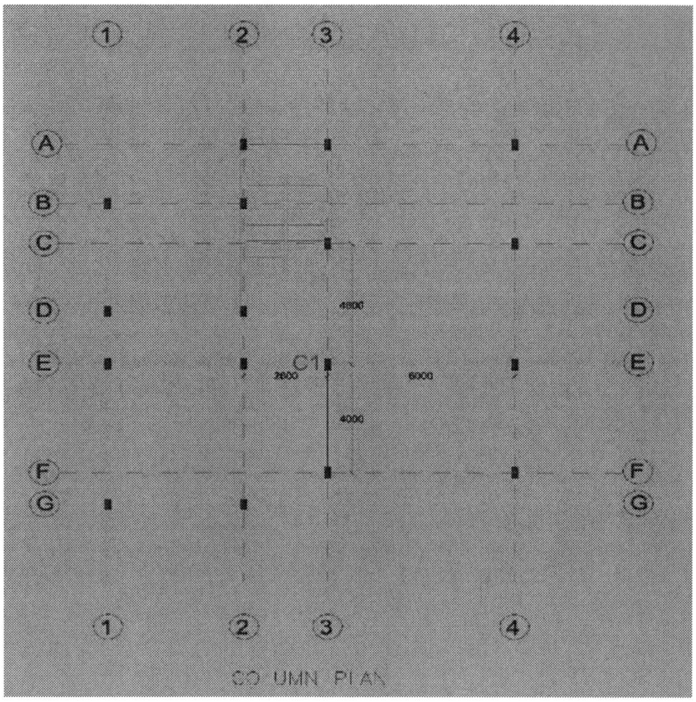

Figure 1. Column plan.

3. Result and discussion

The finite element program PLAXIS 3D connected version, it is used to perform an analysis of pad footing on a clay bed. The results of the numerical analysis of five footings applied to a vertical load of 440kN at three different foundation depths and at three water table positions are presented in this section, along with a discussion of the findings. The numerical simulated results of five footing models of size 1.5x1.5m, 2x1.5m, 2.5X1.5m, 2x2m, and 2.5x2m at three different foundation depths (1.5m, 2m, and 3m below ground level) with three water table positions (GL, 1.5m, and 3m). The analysis and discussion of each footing model were taken separately.

3.1. Model size 1.5mx1.5m

The Model 1.5x1.5m was simulated 9 times with different water table positions and foundation embedment. The table 3 presented settlement results of 1.5x1.5m footing at foundation depth of 1.5m below GL and the water table position on GL, and foundation embedded 1.5m below ground level. The whole results of the models was presented in table 3.

Table 3. Settlement result for 1.5x1.5m footing.

Model size	Settlement (m)	D_F = 1.5m			D_F = 2m			D_F = 3m		
		W_T = 0m	W_T = 1.5m	W_T = 3m	W_T = 0m	W_T = 1.5m	W_T = 3m	W_T = 0m	W_T = 1.5m	W_T = 3m
1.5 x 1.5m	|U|	0.04664 m	0.04552 m	0.04506 m	0.03985 m	0.03728 m	0.03752 m	0.03472 m	0.03255 m	0.03209m
	Uz	Min =7.563x1$0^{-3}$m Max = -0.04575 m	Max =7.541X10^{-3}m Min = -0.04435 m	Max = 7.415X1$0^{-3}$m Min = -0.04392 m	Min = 6.956x1$0^{-3}$m Max = -0.03983 m	Min = 6.869x1$0^{-3}$m Max = -0.03724 m	Min = 6.793x10^{-3}m Max = -0.03748 m	Min = 5.731x1$0^{-3}$m Max = -0.03462 m	Min = 5.702x1$0^{-3}$m Max = -0.03243 m	Min = 5.374x10^{-3}m Max = -0.03194m

As shown in the result of table 3, the total settlement result of models with a depth of foundation 1.5m below GL, the greater displacement found is 46.64mm which is when the WT on GL. When WT lied on GL, the maximum total settlement of the depth of the foundation 2m below GL was 39.85mm. The total settlement of foundation depth of 3m is 34.72mm. The results show that the total settlement is always greater than the water table on GL in all three cases, and that the total settlement decreases sequentially with foundation depth. The settlement results from the various embedment show that increasing the depth of the foundation reduces the amount of settlement.

3.2. Model size 2mx1.5m

The Model 2x1.5m was modelled 9 times, with varying water table locations and foundation embedment, to look the depth of foundation, and water table impact on settlement. The results were presented in table 4.

Table 4. Settlement result for 2x1.5m footing.

Model size	Settlement(m)	D_F = 1.5m			D_F = 2m			D_F = 3m		
		W_T = 0m	W_T = 1.5m	W_T = 3m	W_T = 0m	W_T = 1.5m	W_T = 3m	W_T = 0m	W_T = 1.5m	W_T = 3m
2x1.5m	|U|	0.04061	0.03987 m	0.03959 m	0.02821 m	0.02654	0.02715 m	0.02840	0.02701	0.02676
	UZ	Min =7.417x 10^{-3}m	Max =7.446X 10^{-3}m	Max = 7.285X 10^{-3}m	Min =6.586x 10^{-3}m	Min =6.272x 10^{-3}m	Min =6.190x 10^{-3}m	Min =5.498x 10^{-3}m	Min =5.323x 10^{-3}m	Min =5.007 x10^{-3}m
		Max = -0.04006 m	Min = -0.03901 m	Min = -0.03873 m	Max = -0.02819 m	Max = -0.02647 m	Max = -0.02709 m	Max = -0.02837 m	Max = -0.02695 m	Max = -0.02666 m

The largest displacement discovered is 40.61mm when the WT is laid on GL, as indicated in table 4, the overall settlement result of models with a foundation depth of 1.5m below GL. When WT is laid on GL, the maximum total settlement of the depth of the foundation 2m below GL is 28.21mm. The total settlement at a depth of 3 meters is 28.40 millimetres. The total settlement is always greater when the water table is at GL in all three situations, and it also indicates that the total settlement decreases with the foundation depth.

3.3. Model size 2.5mx1.5m

To explore the depth of foundation and the effect of water table depth on settlement, the model was run 9 times, every instance adjusting the water table location and foundation embedment. Table 5 displayed the settlement outcomes.

Table 5. Settlement result for 2.5x1.5m footing.

Model size	Settlement (m)	D_F = 1.5m			D_F = 2m			D_F = 3m		
		W_T = 0m	W_T = 1.5m	W_T = 3m	W_T = 0m	W_T = 1.5m	W_T = 3m	W_T = 0m	W_T = 1.5m	W_T = 3m
2.5x1.5m	|U|	0.03448	0.03374 m	0.03361m	0.02820 m	0.02662 m	0.02714 m	0.02416 m	0.02310 m	0.02297
	Uz	Min =7.200x 10^{-3}m	Min =7.210X 10^{-3}m	Min = 7.118X 10^{-3}m	Min =6.592x 10^{-3}m	Min =6.340x 10^{-3}m	Min =6.327x 10^{-3}m	Min =5.233x 10^{-3}m	Min =4.871x 10^{-3}m	Min =4.606x 10^{-3}m
		Max = -0.03274 m	Max = -0.03193 m	Max = -0.03182m	Max = -0.02817 m	Max = -0.02657 m	Max = -0.02709 m	Max = -0.02411 m	Max = -0.02299 m	Max = -0.02282 m

The total settlement result of models with a depth of foundation of 1.5m below GL, as shown in table 5. The greatest displacement discovered is 33.74 mm, which is when the WT lies at 1.5m below GL. When WT is laid on GL, the maximum total settlement of the depth of the foundation at 2m below GL is 28.20mm. The total settlement for a 3m foundation depth is 24.16mm which is acceptable and safe structure. The results always demonstrate that the total settlement is greater when the water table on GL is higher in all three situations, and it also shows that the total settlement decreases sequentially with foundation depth.

3.4. Model size 2mx2m

To verify that the water table, and foundation depths had an influence on the settlement, the model was constructed 9 times, using various water table locations and foundation embedment, the settlement results presented in table 6.

Table 6. Settlement result for 2x2m footing.

Model size	Settlement (m)	DF = 1.5m			DF = 2m			DF = 3m		
		WT = 0m	WT = 1.5m	WT = 3m	WT = 0m	WT = 1.5m	WT = 3m	WT = 0m	WT = 1.5m	WT = 3m
2x2m	|U|	0.03356	0.03319 m	0.03302 m	0.02823m	0.02654	0.02715	0.02350	0.02246	0.02229
	UZ	Min =7.195x 10-3m Max = - 0.03242 m	Max =7.132X 10-3m Min = - 0.03178 m	Max = 7.103X1 0-3m Min = - 0.03156 m	Min =6.573x1 0-3m Max = - 0.02822m	Min =6.272x1 0-3m Max = - 0.02647 m	Min =6.190x1 0-3m Max = - 0.02709 m	Min =5.166x 10-3m Max = - 0.02342 m	Min =4.895 x10-3m Max = - 0.02231 m	Min =4.529 x10-3m Max = - 0.02208 m

As seen in table 6, the overall settlement result of models with a foundation depth of 1.5m below the GL is 33.56mm. That's when the WT is on the GL. When the WT lies on the GL, the maximum total depth of the foundation is 28.23mm below the GL. The overall depth of the base settlement is 23.50mm. When the water table in all three situations indicates GL, the total settlement is always greater, and it illustrates that the total settlement decreases with the foundation depth.

3.5. Model size 2.5mx2m

Table 7 demonstrates the 9 model findings with different foundation depths and water table position.

Table 7. Settlement result for 2.5x2m footing.

Model size	Settlement (m)	D_F = 1.5m			D_F = 2m			D_F = 3m		
		W_T = 0m	W_T = 1.5m	W_T = 3m	W_T = 0m	W_T = 1.5m	W_T = 3m	W_T = 0m	W_T = 1.5m	W_T = 3m
2.5x2m	|U|	0.02876	0.02828 m	0.02823 m	0.02434 m	0.02308	0.02366 m	0.02011 m	0.01920 m	0.01916
	Uz	Min =6.989x 10^{-3}m Max = - 0.02752 m	Max =6.930X 10^{-3}m Min = - 0.02700 m	Max = 6.867X 10^{-3}m Min = - 0.02697 m	Min =6.367x 10^{-3}m Max = - 0.02431 m	Min =5.894x 10^{-3}m Max = - 0.02278 m	Min =5.912x 10^{-3}m Max = - 0.02332 m	Min =4.788x 10^{-3}m Max = - 0.02002 m	Min =4.469x 10^{-3}m Max = - 0.01902 m	Min =4.175x 10^{-3}m Max = - 0.01891 m

The entire settlement results in models with a depth of foundation of 1.5m below GL, as shown in table 7. When the WT is on the GL (0m), the displacement is 28.76 mm. The maximum total settlement of the depth of the foundation at 2m below GL is 24.34mm. The overall settlement for a 3m foundation depth is 20.11mm. The results always demonstrate that the total settlement in case of depth of foundation was 1.5 is greater than the allowable settlement of 25mm. However, in the cases of 2m and 3m depth of foundation, the settlement results found were always less than 25mm. Therefore, when the water table on GL is higher in all three situations, it also shows that the total settlement decreases sequentially with the foundation depth.

3.6. Effect of footing size on settlement

Figure 2 gives typical area-settlement relationships for footings of different widths on the surface of a clay soil property. It can be seen that the of the size footings increase with the decrease in the settlement of the footings. However, the relationship indicated that the area of footing is significant for reduction of settlement.

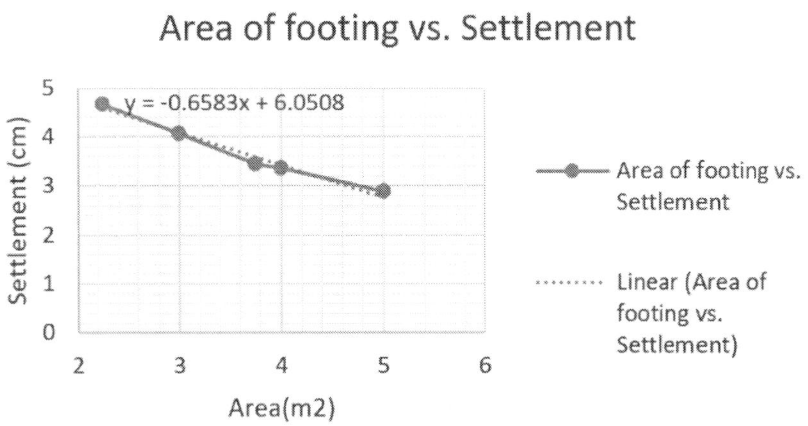

Figure 2. Area of foundation-Settlement curve.

3.7. Effect of depth of embedment on settlement

The depth of the footing is a key factor in determining how much the soil will settlement. In this study, the effect of increasing the embedment depth on total settlement under a concentrated load was investigated, and the embedment depth of the foundation was increased from 1.5 to 3 metres, as shown in figures 3. It can be shown that the maximum total settlement reduces as the embedment depth increases for all of the footing sizes under consideration in this study. These findings are in agreement with those of [9], and [6].who found that embedding results in a significant decrease in settlement. Consequently, based on the simulation results, it appears that the foundation embedment has a positive effect on the foundation's ability to reduce settlement for all of the foundation sizes considered in this study.

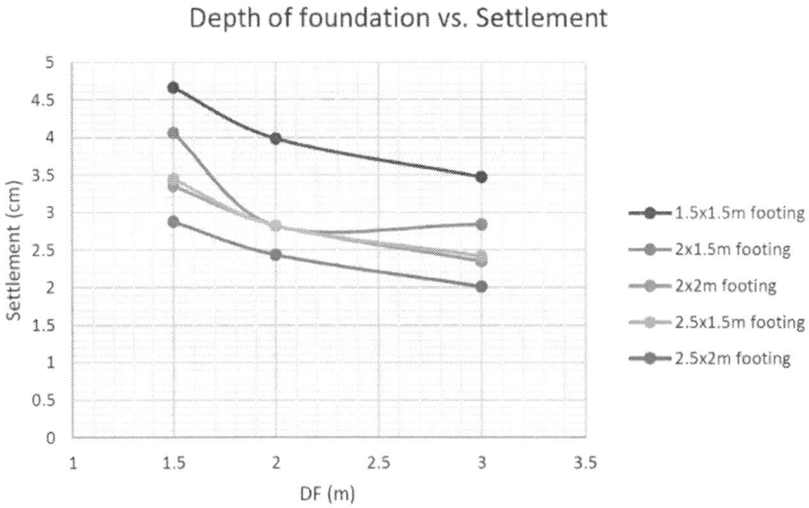

Figure 3. Settlement-depth of foundation.

3.8. Effect of water table on settlement

The water table influence on settlement was determined by measuring settlements at various water table depths. In figure 4, it can be shown that the water table has a very small effect on settlement, and as a result, it does not have a positive significant impact on settlement for the types of footing investigated in this study.

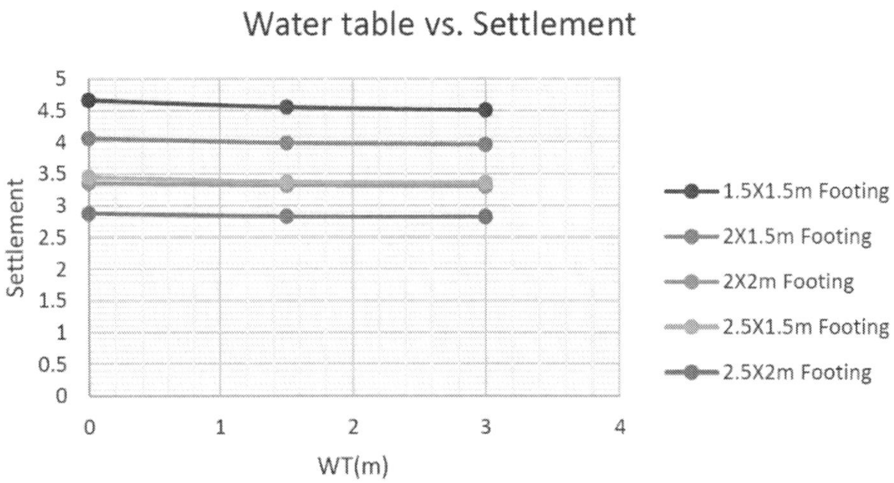

Figure 4. Settlement-water table.

3.9. The optimum footing

The five models were simulated 45 times in this study. Each footing was assessed nine times, three foundation depths and three water table positions were investigated, and settlement results were compared; smaller models of 1.5x1.5 and 2x1.5m failed in all cases when compared to the allowable settlement specified by [10]. Despite the fact that footings of 2.5x1.5m and 2x2m could withstand the structure's load to a depth of 3m. While the 2.5x2m footing, like all other footings, failed partially at 1.5m depth but could support the load at 2m and 3m depth, the maximum settlement at 2m depth was 24.34mm when the water table was at GL. The maximum settlement was determined to be 19.16mm for a 3m foundation depth. However, the settlement result at the 2m and 3m foundation embedment depths was acceptable in accordance with [11], which specifies a maximum settlement of 25mm. Thus, among the simulated models, the optimal footing was a 2.5x2m footing at 2m and a 3m foundation depth. The results presented in this paper is in accordance to the theories related to foundation engineering, where larger area of footing provide better stability. However, this study has laid a clear indication on the range of parameters that can be easily predicted with numerical analysis.

4. Conclusion

The Determination of the settlement of a pad foundation resting on soft clay, a series of numerical analyses was performed. Various settlements were found at varying water table depths, and each instance then received a varied foundation depth. In order to estimate the foundation settlement, a finite element analysis was performed using the PLAXIS 3D Connect. By using the Mohr Coulomb model, the analyses were performed. The following findings were obtained as a result of the research conducted:

1. Due to less settlement than the other footings, the 2.5x2m footing was determined to be the optimal footing among the five evaluated footing types. The depth of 3m proved to be the most successful.
2. The depth of the water table have a small impact on settlement.
3. The different size of footing gives different settlement results. It should also be noted that footing area increase, settlement decreases and vice versa.
4. The depth of embedment have noticeable influence on settlement.
5. The simulation takes less time when all essential data is available, it might be beneficial for gaining an understanding of the soil-structure interaction behaviour.
6. The study's findings can be used to the design of lightweight single-story structures.

5. References

[1] Yusof Z 2013 *Numerical Simulation of Bearing Capacity for Rigid Footing on Cohesive Soil* Masters Thesis (Skudai: Universiti Teknologi Malaysia)
[2] Prasanth S and Sankar N 2015 Prediction of Settlement of Shallow footings on Granular soils using Genetic Algorithm *Asian J. Eng. Technol.* **3** 376–383
[3] Hassan Allawi A M 2020 *Settlements Analysis of Pvd Installation in Soft Soil Using Plaxis 3D* Master Thesis (Skudai: Universiti Teknologi Malaysia)
[4] Kimmerling R 2002 *Shallow Foundations* Technical Report GEC No. 6 (Seattle Washington)
[5] Du P Liu X and Zhang Y 2017 Discussion of the Method to Determine the Ultimate Bearing Capacity of Soil Foundation *IOP Conf. Ser. Earth Environ* **100** 012007
[6] Salahudeen A B, Ijimdiya T S, Eberemu A O and Osinubi K J 2019 Assessment of Foundation Settlement Prediction using Analytical and Numerical Techniques *Niger. Res. J. Eng. Environ. Sci* **4** 152–165
[7] Salahudeen A B 2017 Investigation of Shallow Foundation Soil Bearing Capacity and Settlement Characteristics of Minna City Centre Development Site Using Plaxis 2D Software and Empirical Formulations *Niger. J. Technol* **36** 663-670
[8] Meghzili S A, Madun A Tajudin S A AYusof M F and Ismail M A M 2017 Numerical Evaluation on the Different Shapes of Gravelly Sand Columns to Increase the Loading Capacity of Soft Clay *MATEC Web Conf* **103**
[9] Belachew H 2017 *Evaluating Bearing Capacity of Layered Soils Using Finite element analysis software Evaluating Bearing Capacity of Layered Soils Using Finite Element Analysis Software* Master Thesis (Ethiopia: Ababa University)
[10] Belgaonkar P P 2016 Parametric Study on Settlement of Isolated Footings *Civ. Environ.* **8** 71–77
[11] Terzaghi K, Peck R B and Mesri G 1996 *Soil Mechanics in Engineering Practice* (New York: Wiley)

Acknowledgments

The authors would like to acknowledge Universiti Tun Hussein Onn Malaysia (UTHM) and the Ministry of Higher Education (MOHE) for the financial support for this study under Fundamental Research Grant Scheme (FRGS) Vot No. FRGS/1/2017/TK10/UTHM/02/12.

AUTHOR INDEX